High-Pressure Effects
in Molecular Biophysics
and Enzymology

High-Pressure Effects in Molecular Biophysics and Enzymology

Edited by
John L. Markley

Dexter B. Northrop

Catherine A. Royer

New York   Oxford
OXFORD UNIVERSITY PRESS
1996

Oxford University Press

Oxford   New York
Athens   Auckland   Bangkok   Bogota   Bombay
Buenos Aires   Calcutta   Cape Town   Dar es Salaam
Delhi   Florence   Hong Kong   Istanbul   Karachi
Kuala Lumpur   Madras   Madrid   Melbourne
Mexico City   Nairobi   Paris   Singapore
Taipei   Tokyo   Toronto

and associated companies in
Berlin   Ibadan

Copyright © 1996 by Oxford University Press, Inc.

Published by Oxford University Press, Inc.,
198 Madison Avenue, New York, New York 10016

Oxford is a registered trademark of Oxford University Press

*Chapters of this volume are based on presentations from
the Twenty-Third Steinbock Symposium, held May 15–19, 1994,
at the University of Wisconsin–Madison, U.S.A.*

Library of Congress Cataloging-in-Publication Data
High-pressure effects in molecular biophysics and enzymology / [edited]
   by John L. Markley, Dexter B. Northrop, and Catherine A. Royer.
      p.   cm.
   Includes bibliographical references and index.
   ISBN 0-19-509722-X
   1. High pressure biochemistry.   I. Markley, John L.
II. Northrop, Dexter B.   III. Royer, Catherine A.
QP517.H453H54   1996
574.19'25—dc20   96-22541

Printing (last digit): 9 8 7 6 5 4 3 2 1

Printed in the United States of America
on acid-free paper

Preface

To date, pressure perturbation has played second fiddle to temperature perturbation in the field of biophysical chemistry. Pressure has been perceived as the "other" thermodynamic variable. Nonetheless, pockets of interest in what could be gained by application of high pressure to biological molecules and systems have emerged in Japan, Europe, the former Soviet Union, and the United States. The earliest high-pressure studies on biological molecules date to the 1940s. Over the years, the database on pressure effects has built up slowly. Those working in the pressure perturbation field typically have had to justify the relevance of information and insights obtained from this approach to a much greater extent than those emplying variable temperature. One wonders at this attitude, given that these two variables are equivalent in the Gibbs function. Moreover, pressure studies yield a fundamental physical parameter of a reaction, namely its volume change. When applied to rates of reactions, pressure yields activation volumes, making it the singular probe of kinetics to return quantitative physical information about transition-state chemistry.

Perhaps the lack of familiarity with pressure as compared with temperature comes from our direct experience of hot and cold, but at only one atmosphere of pressure. Only scuba divers experience pressurization to a significant extent; few divers are biophysical chemists, and even divers confront changes of only a few atmospheres. Whatever the reason, pressure perturbation techniques have not been considered by most biophysical chemists. Students are introduced to the theoretical principles and equations of thermodynamics, which include ΔP, but are taught little by way of practical applications in chemistry, and only rarely anything about the specific effects of pressure on biological molecules. The result is that many biophysical chemists are not aware that high pressure will cause acidic groups to ionize, proteins to unfold, protein ligands to bind and others to dissociate. Unlike temperature perturbations, pressure perturbations can be of widely differing magnitudes and, more dramatically, of different signs for individual interactions within a multicomponent system, allowing specific assignments of the physical basis for certain biological activities. A particular distinction that sets pressure apart from other perturbants used in biological systems, such as salt, acid, urea, or organic solvents, is that pressure does not introduce new components to the system; it merely changes the equilibria among preexisting components. Because changes in equilibria can be many orders of magnitude, pressure can bring minor species out of the background noise and into the foreground for direct observation and study.

The Steenbock Symposium on High Pressure Effects in Molecular Biophysics and Enzymology, which brought together many of the leaders in the area of pressure perturbation of biological systems and led to this volume, documents the enormous advances that have been made in the past decade. The chapters that follow present an overview of current research on pressure perturbations on enzyme kinetics, protein folding and structure, lipid bilayer structure and organization, lipid-protein interactions, DNA structure, and protein-ligand interactions. The reader will surely marvel—as we do—at the successful combination of high-pressure perturbation with physical chemical techniques as diverse as Fourier transform infrared spectroscopy, multidimensional nuclear magnetic resonance, small-angle neutron scattering, fluorescence, and calorimetry. Given the extent to which the technical aspects of high-pressure technology have advanced, we predict that workers in this field will make contributions of an ever more fundamental nature to our understanding of bimolecular function. This will be coupled to practical applications of pressure in virology, bacteriology, and food sciences. It is our conviction that these technical advances, in combination with the recent availability of commercial equipment, and ever-increasing contributions of pressure studies to our fundamental understanding of biological processes will ultimately drive high-pressure techniques to the forefront of biophysics and biochemistry. This milestone volume is designed to acquaint investigators outside the field with the possibilities of research at high pressures. We hope this book will spur readers to apply these approaches to their own particular research interests, perhaps in new and unforeseen ways.

Madison, Wisconsin   J. L. M.
August 1995   C. A. R.
                                                                                      D. B. N.

Contents

Contributors

Lance E. Ballard
School of Chemical Sciences
University of Illinois–Urbana–
Champaign
Urbana, Illinois 61801

Claude Balny
INSERM Unité 128
Route de Mende
BP 5051
34 033 Montpellier Cedex 1, France

Bao-Shiang Lee
Department of Chemistry
School of Chemical Sciences
University of
Illinois–Urbana–Champaign
Urbana, Illinois 61801

Thomas Brauns
Department of Physical Chemistry
University of Dortmund
Otto-Hahn- Straße 6
D-44227 Dortmund, Germany

Parkson Lee Gau Chong
Department of Biochemistry
Temple University

School of Medicine
3420 N. Broad Street
Philadelphia, PA 19140

Robert M Clegg
Department of Molecular Biology
Max Planck Institute for Biophysical
Chemistry
Am Fassberg 11
D-37077 Göttingen, Germany

Claus Czeslik
Department of Physical Chemistry
University of Dortmund
Otto-Hahn Straße 6
D-44227 Dortmund, Germany

Andrea T. Da Poian
Departamento de Bioquimica Medica
Instituto de Ciencias Biomedicas
Universidade Federal do Rio de Janeiro
Cidade Universitaria Ilha do Fundao

Eric Deprez
Institut de Biologie Physico-Chimique
INSERM, France

Carmelo di Primo
Institut de Biologie Physico-Chimique
INSERM, France

Pierre Douzou
Institut de Biologie Physico-Chimique
INSERM, France

H. G. Drickamer
Department of Chemical Engineering
University of Illinois–Urbana–
Champaign
114 Roger Adams Lab,
600 S. Mathews Avenue
Urbana, Illinois 61801

Maurice R. Eftink
Department of Chemistry
University of Mississippi
University, Mississippi 38677

Jörg Erbes
Department of Physical Chemistry
University of Dortmund
Otto-Hahn Straße 6
D-44227 Dortmund, Germany

Debora Foguel
Departamento de Bioquimica Medica
Instituto de Ciencias Biomedicas
Universidade Federal do Rio de Janeiro
Cidade Universitaria Ilha do Fundao
21941 Rio de Janeiro, Brazil

Koen Goossens
Department of Chemistry
Katholiek Universiteit Leuven
Laboratory of Chemical and
Biological Dynamics
Celestijnenlaan 200 D
B-3001 Leuven, Belgium

Karel Heremans
Department of Chemistry
Katholiek Universiteit Leuven
Laboratory of Chemical and
Biological Dynamics
Celestijnenlaan 200 D
B-3001 Leuven, Belgium

Gaston Hui Bon Hoa
Institut de Biologie Physico-Chimique
INSERM Unité 310
13 Rue Pierre et Marie Curie
75005 Paris, France

Jason L. Johnson
Department of Biochemistry and
Biophysics
Texas A&M University
420 Bio/Bio Building
College Station, Texas 77843-2128

Ana Jonas
Department of Biochemistry
College of Medicine
University of Illinois–Urbana–
Champaign
Urbana, Illinois 61801

Jiri Jonas
School of Chemical Sciences and the
Beckman Institute for Advanced
Science and Technology
University of Illinois–Urbana–
Champaign
Urbana, Illinois 61801

Anne Landwehr
Department of Physical Chemistry
University of Dortmund
Otto-Hahn-Straße 6
D-44227 Dortmund, Germany

Horst Ludwig
Institut für Pharmazeutische
Technologie und Biopharmazie
Gruppe Physikalische Chemie
Universität Heidelberg
Im Neuenheimer Feld 346
D-69120 Heidelberg, Germany

Robert B. Macgregor Jr.
Faculty of Pharmacy
University of Toronto
Toronto, Ontario, Canada

John L. Markley
Department of Biochemistry
University of Wisconsin–Madison
420 Henry Mall
Madison, Wisconsin 53706–1569

Mark A. McLean
Beckman Institute for Advanced
Science and Technology
University of Illinois–Urbana–
Champaign
Urbana, Illinois 61801

Isao Morishima
Division of Molecular
Engineering
Graduate School of Engineering
Kyoto University
Kyoto 606, Japan

Reza Najaf-Zadeh
Faculty of Pharmacy
University of Toronto
Toronto, Ontario, Canada

Dexter B. Northrop
Department of Pharmacy
University of Wisconsin–Madison
2328 Chamberlin Hall
Madison, Wisconsin 53706

Xiangdong Peng
Department of Chemistry
School of Chemical Sciences
University of Illinois–Urbana–
Champaign
Urbana, Illinois 61801

Kenneth E. Prehoda
Department of Biochemistry
University of Wisconsin–Madison
420 Henry Hall
Madison, Wisconsin 53706–1569

Glen D. Ramsey
Aviv Associates
Lakewood, New Jersey 08701

Gregory D. Reinhart
Department of Biochemistry and
Biophysics
Texas A&M University
420 Bio/Bio Building
College Station,
Texas 77843-2128

Oliver Reis
Department of Physical Chemistry
University of Dortmund
Otto-Hahn- Straße 6
D-44227 Dortmund, Germany

Clifford R. Robinson
Massachusetts Institute for Technology
Cambridge, Massachusetts

Catherine A. Royer
Department of Pharmacy
University of Wisconsin–Madison
4330 Chamberlin Hall
Madison, Wisconsin 53706

Suzanne F. Scarlata
Department of Physiology and
Biophysics
Health Sciences Center
State University of New York–
Stony Brook
Stony Brook, New York,
11794–8661

Stephanie Schwer
Department of Chemistry
School of Chemical Sciences
University of Illinois–Urbana–
Champaign
Urbana, Illinois 61801

Wilhelm Scigalla
Institut für Pharmazeutische
Technologie und Biopharmazie
Gruppe Physikalische Chemie
Universität Heidelberg
D-69120 Heidelberg, Germany

Ana Sepulveda de Rezende
Department of Molecular Biology
Max Planck Institute for Biophysical
Chemistry
Am Fassberg 11
D-37077 Göttingen, Germany

Jerson L. Silva
Departamento Bioquimica Medica
Instituto Ciencias Biomedicas
Universidade Federal do
Rio de Janeiro
Cidade Universitaria Ilha do Fundao
21941 Rio de Janeiro, Brazil

Stephen G. Sligar
Department of Biochemistry and the
Beckman Institute for Advanced
Science and Technology
University of Illinois–Urbana–
Champaign
Urbana, Illinois 61801

Lásló Smeller
Institute of Biophysics
Semmelweis University of
Medicine
P.O. Box 263
H1444 Budapest, Hungary

Bernd Søjka
Institut für Pharmazeutische
Technologie und Biopharmazie
Gruppe Physikalische Chemie
Universität Heidelberg
Im Neuenheimer Feld 346
D-69120 Heidelberg, Germany

Naohira Takeda
Department of Chemistry
Faculty of Science and
Engineering
Ritsumeikan University
1916 Noji-cho
Kusatsu, Shiga 525, Japan

Yoshihiro Taniguchi
Department of Chemistry
Faculty of Science and
Engineering
Ritsumeikan University
1916 Noji-cho
Kusatsu, Shiga 525, Japan

Ana Theobald
Department of Molecular Biology
Max Planck Institute for Biophysical
Chemistry
Am Fassberg 11
D-37077 Göttingen, Germany

Raj Tomas
Department of Pharmacy
University of Wisconsin–Madison
4330 Chamberlin Hall
Madison, Wisconsin 53706

Gediminas J. A Vidugiris
Department of Pharmacy
University of Wisconsin–Madison
4330 Chamberlin Hall
Madison, Wisconsin 53706

Mauro Villas-Boas
Department of Molecular Biology
Max Planck Institute for Biophysical
Chemistry
Am Gassberg 11
D-37077 Göttingen, Germany

Gregorio Weber
Department of Biochemistry
University of Illinois–Urbana–
Champaign
Urbana, Illinois 61801

Roland Winter
Department of Physical Chemistry

University of Dortmund
Otto-Hahn-Straße 6
D-44227 Dortmund, Germany

Patrick T. T. Wong
Department of Biochemistry
Faculty of Medicine
Smyth Road Campus
Ottawa, Ontario, Canada

Annelies Zechel
Department of Molecular Biology
Max Planck Institute for Biophysical
Chemistry
Am Gassberg 11
D-37077 Göttingen, Germany

Jing Zhang
ADM Company
Lakeview Technical Center
Decatur, Illinois

High-Pressure Effects
in Molecular Biophysics
and Enzymology

1

Resolution of the Ambiguity of van't Hoff Plots by the Effect of Pressure on the Equilibrium

GREGORIO WEBER

The change in the Gibbs free energy function, ΔG, of chemical reaction is determined by the difference between the heats respectively released to and absorbed from the environment, and separation of the enthalpy and entropy changes that these changes represent cannot be achieved without specific hypotheses as to their relations. The determination of the enthalpy of reaction by the plot of $\Delta G/T$ against $1/T$ (van't Hoff plot) implicitly assumes that the enthalpy ΔH and entropy ΔS are temperature independent, and this assumption leads to very large errors when this is not the case and $\Delta H \ll T\Delta S$. It is therefore inapplicable to the reactions of molecules, such as proteins, that have thermally activated local motions. The concepts offered previously by the author to relate the entropy and enthalpy changes in protein associations are reviewed briefly and applied to account for the temperature dependence of ΔH and ΔS. It is shown that two different values of the enthalpy computed in that manner correspond to each value of the apparent van't Hoff enthalpy, but that the choice between the two is easily made by reference to the volume change on reaction. The enthalpies of association of subunit pairs of seven oligomers are all found to be positive and much more uniformly related to the size of the intersubunit surface than those previously assigned by use of the classical van't Hoff plot.

GIBBS FREE ENERGY AND THE BALANCE OF HEAT AND WORK

The change in the Gibbs free energy function, dG, of chemical reaction is defined as the difference between the change in total heat content, $d(TS)$, of the system of reagents and surroundings and the external work performed by the system at constant pressure, pdV, by the relationship

$$dG = -d(TS) + pdV \tag{1}$$

In any process, the change in heat content comprises that intrinsic to the system, $d(TS)_i$, and that of the passive environment, $d(TS)_x$, with the latter arising from effects starting originally in the system of reagents. Thus, following Planck (1932), we can write Eq. 1 as

$$dG = -d(TS)_x - d(TS)_i + pdV \qquad (2)$$

Equation 2 expresses a simple conservation relation between heat and work that is valid in all circumstances, not only at equilibrium. In the application of Eq. 2 to chemical reactions, Gibbs (1876) assumed that $d(TS)_x$ resulted from a change of opposite sign in the internal energy of the system, dE, and, as dH, the change in enthalpy at constant pressure, equals $dE + pdV$, the last equation becomes

$$dG = dH - d(TS)_i. \qquad (3)$$

Because $d(TS)_i = TdS_i + S_i dT$ and at constant temperature $dT = 0$, we have

$$dG = dH - TdS_i \qquad (4)$$

If Eq. 4 refers to the conversion of one mole of reactants into one mole of products under conditions in which the composition of the system is maintained constant, at the stable values characteristic of the chemical equilibrium, Eq. 4 takes the familiar form

$$\Delta G = \Delta H - T\Delta S_i \qquad (5)$$

where the deltas are standard molar changes in free energy, enthalpy, and entropy, respectively. I have detailed the conceptual derivation of the general relation of Eq. 4 and that applicable to equilibrium, Eq. 5, to indicate that ΔG, apart from the change in volume, equals the difference between two quantities of heat, those respectively released and absorbed by the reagents.

The difficulties in the separation of these two quantities are evident from the start. Calorimetry can measure only the difference between them and reproduces ΔG, with its proper sign, except for the contribution of $p\Delta V$, while ΔG, as derived from the proportions of the components at equilibrium, determines the magnitude of the difference between the heat absorbed and released by the system under specific equilibrium, but does not determine what predominates in the reaction: the release of heat into the surroundings (enthalpy-driven reactions) or the absorption of heat by the system from the surroundings (entropy-driven reactions). There is evidently no way to experimentally separate the changes in enthalpy and entropy in any chemical reaction carried out isothermally without some hypothesis about the relations that must exist between them.

RELATION OF THE ENERGIES OF THE EXCHANGED BONDS TO THE ENTHALPY AND ENTROPY CHANGES

In any isolated chemical system, there are only two sources of energy: one is thermal, or *caloric*, the product of the temperature T and a capacitive factor S, the entropy of the system; the other stems from the energies in the bonds that link the various particles of matter in the system. Of these, only the bond energies exchanged in the chemical reactions interest us in relation to Eq. 5. In the macromolecular reactions of the association of protein subunits to form an oligomer, or in the internal associations of a peptide chain upon folding, the bonds that undergo changes on reaction belong almost exclusively to three types, which from weaker to stronger are (1), those between the protein apolar structures that result from dispersion forces; (2), those between permanent dipoles and apolar structures, which are particularly relevant when there is an important change in the contacts of apolar structures and the permanent water dipoles, and (3), interactions between the permanent dipoles of water that are created or destroyed in the reactions generating the two other types. While all three types may be found in protein-water (P-W) and protein-protein (P-P) interactions, dispersion forces are expected to predominate in the latter (P-P) case, and permanent dipole-induced dipole interactions in the former (P-W) case. The energy of the bonds involved in W-W interaction energy are well known from the structure and properties of water, and if general rules for the relations of enthalpy and entropy are specified, one can derive, under suitable assumptions as to the fractions of the types of interactions involved in each case, the average bond energies involved in P-P and P-W interactions. I have developed these ideas extensively in other publications (Weber, 1993, 1995) and applied them to the analysis of the effects of pressure and temperature on the association of protein oligomers. The fundamental point concerns the relations of the energy and entropy with the bond energies that determine them. These can be understood as the simplest expression of a golden rule: the replacement of stronger bonds by weaker ones, by increasing the probability of the bonds being absent, increases the total heat capacity of the reagents by the amount TdS_i that is absorbed by them from the environment. The replacement of weaker bonds by stronger ones has the opposite effect, releasing an amount of heat TdS_x into the environment. These exchanges of heat owing to changes in the intrinsic entropy of the system add to those resulting from the strength of the bonds themselves and together determine the direction of spontaneous change in the system according to Eq. 3. One would expect that this reasonable golden rule ought to have been proposed previously by others, and more than once, but the only clear reference to it that I have seen is in a paper by Widom (1989). This rule is responsible for the often-noted compensation of enthalpy and entropy changes. I interpret the golden rule to imply that the entropy associated with the bonds derives exclusively from the existence of the multiple complexions generated by their temporary presence or absence. Its value is then given by the Boltzmann expression: $S = R \ln(Z)$, where Z is the number of complexions. One then needs only to relate the average probability P that a bond may be broken by the thermal energy to its contribution to the change in intrinsic entropy, dS_i. As I discuss elsewhere (Weber, 1995), the numerical value of S for a set of identical bonds given by the Boltzmann relation does not differ significantly from the much simpler form

$$S_i = R \ln 2[4P(1 - P)]^{1/2} \tag{6}$$

The fundamental assumption I make is that the entropy is determined by the *average probability P of a bond being absent at a given temperature.* Once this probability is given *for each of the reagents involved in the chemistry,* we can calculate without difficulty the changes in entropy and energy that take place during the reaction. We can qualitatively appreciate that if the bonds are of sufficient strength with respect to the thermal energy, P becomes close enough to zero to make the entropy change negligible, and the direction of the chemical reaction is then determined by the heat released when weaker bonds are replaced by stronger ones (Berthelot's rule). The simplest and must obvious probability rule of bond breakage states that if the average energy of the bond in thermal energy units is E/RT, its average probability of transient disappearance is $P = \exp(-E/RT)$, or $E/RT = -\ln(P)$. It follows that the total energy of the M equal bonds of a reactant or product exchanged in the course of the reaction is

$$E = ME(1 - P) \tag{7}$$

while their contribution to the exchangeable heat or caloric belonging to the reagent is given by

$$TS_i = MRT \ln 2[4P(1 - P)]^{1/2} \tag{8}$$

At atmospheric pressure, the product pdV may be neglected so that $H \approx E$ and the Gibbs free energy associated to the bond ensemble is

$$G = ME(1 - P) - MRT \ln 2[4P(1 - P)]^{1/2} \tag{9}$$

Equation 9 permits one to calculate the expected ΔG of Eq. 5, as well as the constitutive enthalpy and entropy changes, from the average bond energies attributed to the reactants and products. For the purposes of fit, it is often convenient to express G in Eq. 9 as a function of P alone, obtaining

$$G/RT = M\{[-\ln(P)(1 - P)] - \ln 2[4P(1 - P)]^{1/2}\} \tag{10}$$

While one can calculate the free energy difference, ΔG, at fixed temperature and pressure as that between reactants and products using Eq. 9 or 10 after assigning values to the Ps, or corresponding Es, we face an inverse problem: the assignment to products and reactants of values of P appropriate to generate the experimental free energy observed at the various temperatures or pressures. In the case of association of two identical subunits to form a protein dimer, we have the stoichiometric relation

$$2PW \rightarrow P - P + W - W \tag{11}$$

and the free energy change is then given by

$$\Delta G = M[G(P - P) + G(W - W) - 2G(P - W)] \tag{12}$$

in which we assume for simplicity that M is the same for all three reacting species. As the energy of the water–water bonds is known, fitting consists in varying the energies of the P—P and P—W bonds to satisfy the experimental free energy at constant temperature. Computation shows that many pairs of P—P and P—W bond energies can combine to give a single experimentally observed free energy of reaction (see Weber, 1993, Figure 5). Selecting from among these requires further information. In principle, calorimetry can inform us about the sign of the heat released or absorbed on reaction and therefore indicate whether the reaction is entropy or enthalpy driven, but it cannot go any further. The customary way of separating the enthalpy and entropy contributions to the free energy change on reaction is one I have also used in analyzing the data of the free energy of association of several oligomers (Weber, 1993). It applies the van't Hoff plot to the data of the standard free energy change at a series of temperatures. However, in view of the inherent impossibility of separating the sources of heat absorbed and evolved without additional hypotheses, which was discussed above, it becomes indispensable to examine in detail the general validity of van't Hoff plots for this purpose.

LIMITED RELIABILITY OF VAN'T HOFF PLOTS

If Eq. 5 is divided by T, one obtains

$$\Delta G/T = \Delta H/T - \Delta S_i \tag{13}$$

and if ΔH and ΔS are independent of temperature, then

$$d(\Delta G)/d(1/T) = \Delta H \tag{14}$$

Eq. 14 is the expression used in constructing the van't Hoff plot. In the more general case, ΔH and ΔS are to be considered as temperature dependent, and then

$$d(\Delta G)/d(1/T) = \Delta H + \Gamma(T) \tag{15}$$

where

$$\Gamma(T) = (1/T)d(\Delta H)/d(1/T) - d\Delta S_i/d(1/T) \tag{16}$$

Eviden ly Eq. 14 does not apply except in the limiting case in which

$$\Gamma(T) \ll \Delta H \tag{17}$$

Changes in heat content other than by bond exchanges involve the caloric $d(TS_i) = S_i dT + TdS_i$. The first term corresponds to the increase in entropy, by an amount equal to its product with the increase in temperature. It represents the increase in caloric owing to equipartition of the increased thermal energy among the degrees of freedom of the reagents, collectively responsible for the value of S_i at temperature T. The second term, TdS_i, requires the presence of *thermally activated motions* and depends on the absorption of heat from the surroundings owing to the appearance of new modes of motion, which are brought about by the increased probability of bond breakage. The presence in proteins of thermally activated motions has been demonstrated by many independent methods, so that one can deny outright the validity of van't Hoff plots according to Eq. 14 to analyze the macromolecular reactions of proteins—that is, the folding of peptide chains and their subsequent association into oligomers. These cases stand in sharp contrast to the classical use of van't Hoff plots in reactions involving the exchange of covalent bonds, of much greater energies, and therefore negligible values of entropy over the entire temperature range used in the determination of ΔH (e.g., Lewis & Randall, 1923). It is not surprising that in the late 1920s the variation of ΔS_i and ΔH with temperature was ignored because it was only after that time that the origin of bonds that are unstable at room temperatures was made clear by the 1924 work of Debye on polar molecules (1929) and the 1932 work of London (1936) on apolar molecules. Insufficient examination of the problem has meant that Eq. 14 has been employed without comment up to the present.

It follows from Eq. 15 that the largest errors in the enthalpy obtained by means of van't Hoff plots ought to occur when ΔH approaches zero and, in general, whenever the absolute values of $T\Delta S_i$ and ΔH are comparable. This is the case qualitatively in the entropy-driven association of subunits: in these cases, the bond energies of reactants and products must balance each other to the greatest extent to permit the comparatively small entropy change to impose the direction of chemical change. A quantitative decision about the validity of the approximation (Eq. 14) necessitates consideration of the effects of temperature on the changes in E, or P, as well as the general relations of these with G shown in Eqs. 7 and 8.

DEPENDENCE OF THE BOND ENERGIES ON PRESSURE AND TEMPERATURE

Temperature and pressure result in changes in the Gibbs (1876) function of each set of reagent bonds by, respectively, expansion and contraction of the bonds. The bond energies can be related to the volumes by the Born expression (Lennard Jones, 1931; Fowler, 1936; Weber, 1993). If the energy minimum of the bond E_o occurs at atmospheric pressure and temperature T_o, then

$$E(v) = E(v_o)\{[(v_o/v)^s - (s/t)(v_o/v)^t]/(1 - s/t)\} \qquad (18)$$

Increases above atmospheric pressure, as well as increases or decreases in temperature from the temperature T_o, at which $v = v_o$, result in loss of bond energy. The effects of pressure and temperature on the equilibrium reflect the differential effects that follow the changes in G of reactants and products, and these depend on the expansion of the bonds with increase in temperature or decrease in pressure, or on contraction with increase in pressure or decrease in temperature. A prediction of the change in ΔG as the temperature increases must then take into account all of the types of bonds according to their characteristic values of s and t and the expansivity or compressibility assigned to them. A t exponent must be assigned to bonds arising from dispersion forces ($t = 2$) and permanent dipoles ($t = 1$), which occur in liquid hexane and water, respectively. I have shown elsewhere (Weber, 1993) how the corresponding s exponents may be obtained

Table 1.1. Parameters used in computing the temperature and pressure dependence of bond energies[a]

Solvent	$10^4 \alpha$	$10^2 \beta$	s	t
Water	0.3	4.3	8	1
Hexane	1.4	9.5	10	2

[a] The parameters are α, bond expansivity; β, bond compressibility; s and t, exponents that enter into the expression (Eq. 18) that describes the dependence of bond energies on the volume.

from the compressibilities and expansivities of water and hexane reported by Bridgman (1931).

The expansivities α and compressibilities β of the bonds of hexane and water, which we take as representative of the expansivities and compressibilities of the two liquids, together with the s and t, exponents are shown in Table 1.1.

In principle, the expansivities, compressibilities, and s and t exponents for each reagent must be derivable from the bond energies alone, and scale according to this energy. To introduce such scaling, I shall take the values for hexane as representative of bond energies due to dispersion forces of 1 RT unit at T_o, and those of water as representative of bonds with energies of 12 RT at the same temperature. Intermediate values of α, β, s, and t can then be assigned by linear interpolation to bonds of intermediate energies. When the quantities α, β, s, and t are scaled as described, computation shows that a very good linear relation exists between the relative change in probability of bond breakage with temperature $(1/P)\,dPdT$ and the energy E of the bond in the interval of 0 and 20 °C (Figure 1.1).

With these relations, we can determine not only the P–P and P–W energies

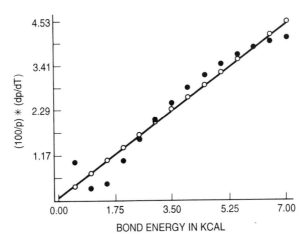

Figure 1.1. ($\bigcirc$) = Change in the relative probability of bond breakage with temperature, against the energy of the bond at temperature T_o for bonds of 0.5 to 7 kcal mol^{-1} by Eq. 15 with α, β, s, and t scaled as described in the text. ($\bullet$) = Coefficient of variation of the averages computed between 0 and 20 °C.

compatible with the experimental free energy of association at T_o but also the change in free energy with temperature for each reagent and the change in reaction-free energy in a given temperature interval. In this way, we eliminate those values of $P-P$ and $P-W$ bond energies that are incompatible with the experimental data. Because the $P-P$ and $P-W$ bond energies determine the entropy and enthalpy of the reaction by the rules expressed in Eqs. 10 and 11 together with Eq. 17, we can make a direct comparison of the enthalpy values derived by this means with those predicted by Eqs. 14 and 15.

Neither ΔH nor ΔS_i is constant with temperature, but they do not vary much more than the free energy, and to compare the enthalpies derived by Eqs. 14 and 15, I have taken as representative their average value in the temperature range of 0 to 20 °C.

Figure 1.2 shows a plot of the van't Hoff enthalpy (Eq. 14) against the enthalpy according to Eqs. 15 and 16 for the values of ΔH compatible with a free energy of -12.6 kcal mol at 0 °C, the free energy of association of the monomers of Rubisco (Erijman et al., 1993). The figure shows that ΔH according to Eq. 15 takes two different values for most values of ΔH according to Eq. 14. This can be understood by reference to Eq. 16. As ΔH increases from negative to positive values, $T\Delta S_i$ increases by a compensating amount and Γ increases rapidly; for sufficiently large positive values of the enthalpy, however, the computed van't Hoff enthalpy decreases again. These changes are responsible for the parabolic nature

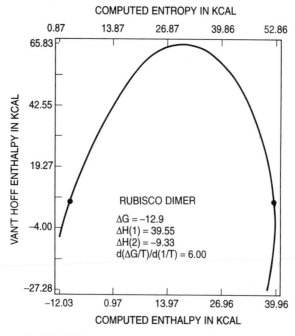

Figure 1.2. Plot of $d(\Delta G/T)/d(1/T)$ against the average enthalpy computed as described in text for the interval 0 to 20 °C, for a dimer with an intersubunit area of 1660 Å^2 and a free energy of association of -12.9 kcal mol^{-1}. The dark circles are the intersections for the experimental value of 6.0 kcal, derived from Eq. 12.

of the relation between the van't Hoff enthalpy and the true enthalpy of reaction. From the same plot, it follows that the ratio of van't Hoff enthalpy (Eq. 14) to computed enthalpy (Eq. 15), tends to -1 (heat released to the surroundings predominates) for those reactions in which the enthalpy change greatly predominates) for those reactions in which the enthalpy change greatly predominates in absolute value over the entropy change. The ratio tends to $+1$ (heat absorbed from the surroundings predominates) for entropy-driven reactions with sufficiently large positive enthalpy changes. Extraordinarily large values of this ratio occur in entropy-driven reactions in which the enthalpy change has a small absolute value, and in these cases even the sign of the van't Hoff plot may contradict that of the relatively small computed enthalpy.

The enthalpies of association of seven oligomers:—four dimers, one trimer, and two tetramers—were derived by Weber (1993) employing the van't Hoff plots for this purpose, and these results require reexamination in light of the concepts that we have just discussed. The fitting of the original experimental data to obtain the enthalpy values that follow from Eq. 15 rather than Eq. 14 involves three successive steps: (1) determination of the set of values of P—P and P—W bond energies that, together with W—$W = 7 \, \text{kcal mol}^{-1}$, can yield a free energy that differs from the experimental free energy at T_o by $\pm 0.2 \, \text{kcal mol}^{-1}$ or less; a set of corresponding enthalpies and entropies is thereby determined; (2) determination of the change of free energy with temperature expected from these bond energies, and the values of α, β, s, and t associated with them (Figure 1.1); and (3) selection of the cases that accord to the experimental van't Hoff plot (Eq. 14) and therefore of the enthalpy values with the corrections of Eq. 15 (Figure 1.2).

The enthalpy values according to Eq. 15 have been derived from the experimental plots of $d\Delta G/d(1/T)$ against $1/T$ and the procedure just described, and the results for the seven oligomers studied are reported in Table 1.2.

Two widely different values of the enthalpy, one positive and the other negative, are found for each of the seven protein oligomers studied (Table 1.2). Figure 1.2 shows these two values for the dimer of *Rhodobacter* Rubisco (Erijman et al., 1993), and entirely similar plots are obtained for all the cases shown in Table 1.2.

Table 1.2. Calculated enthalpies of subunit associations for various oligomeric proteins

Protein	No. of bonds	Enthalpies		
		v. Hoff	$\Delta H(+)$	$\Delta H(-)$
Yeast hexokinase	141	7.0	21.0	−12.6
E. coli β_2 tryptophan synthase	141	17.7	32.0	−5.4
Rhodobacter rubisco	166	6.0	39.6	−9.3
Glycogen phosphorylase A dimer	239	4.0	61.5	−7.0
Allophycocyanine trimer (sp)	118	14.0	28.5	−3.3
Glyceraldehyde phosphate dehydrogenase (sp)	120	−3.7	29.7	−9.5
Glycogen phosphorylase A tetramer (sp)	239	8.2	67.8	−5.1

[a] The quantities in the table are for interactions between subunit pairs (sp) of the oligomer, assumed all equivalent. The number of bonds is assigned from the volume of the subunits.

DISTINGUISHING BETWEEN ENTROPY-DRIVEN AND ENTHALPY-DRIVEN REACTIONS BY THE EFFECTS OF PRESSURE

As each value of the van't Hoff enthalpy corresponds to two very different values of the computed enthalpy, an unequivocal choice between them is required. For this purpose we need an independent measure of the stability of the aggregate at the different temperatures, which must be carried out under isothermal conditions. This measure of stability is readily available by the shift in the pressure of mid-dissociation with temperature, and the results obtained in a variety of aggregates are quite definitive in this respect (Silva & Weber, 1993): in the interval of 0 to 40 °C, an increase in temperature systematically increases the pressure necessary to reach a given degree of dissociation. Therefore, such reactions must all be entropy driven and, as shown in the data in Table 1.2, they must all have $\Delta H > 0$. The physical reasons for the correlation between increase in volume and entropy-driven character on association and its reciprocal decrease in volume and enthalpy-driven character on association are to be found in the physical origin of the principle of Le Chatelier. An increase in volume on association results from the replacement of shorter, and therefore stronger, bonds in the reactants by longer, and therefore weaker, bonds in the products, and the reaction that favors the weaker bonds can only occur if it is driven by entropy. The correlation between volume change and entropy- or enthalpy-driven character of the reaction is not limited to proteins: a number of complexes of small aromatic molecules are known in which stability is decreased by an increase in temperature, and they uniformly associate with a decrease in volume (Heremans, 1982; Weber & Drickamer, 1983).

With reference to the results on protein oligomers (Table 1.2), it appears that the enthalpy values selected according to Eq. 15 and the change in volume on association are far more uniform in character than those computed previously with the help of the van't Hoff plot (Weber, 1993). That analysis indicated that the enthalpy of association of glyceraldehyde phosphate dehydrogenase, unlike those of the other proteins studied, was negative. In the present analysis, this protein, as well the rest, have positive enthalpies of association with an average of 25.2 ± 4.7 kcal per 1000 Å^2 of intersubunit surface. In contrast, the van't Hoff enthalpies previously reported had an average of 6.8 ± 5.8 kcal mol^{-1}. In view of the uniformity in amino acid distribution and the similarities in the free energies of association of the various proteins, the much improved regularity of the values of the enthalpies derived by means of Eq. 15 constitutes strong evidence for the worth of the initial assumptions involved in the calculations.

REFERENCES

Bridgman, P. W. (1931). *The Physics of High Pressure*. New York, Dover, pp. 127–130.
Debye, P. (1929). *Polar Molecules*. New York, Dover.
Erijman, L., Lorimer, G. H., & Weber, G. (1993). Reversible dissociation and conformational stability of dimeric ribulose phosphate carboxylase. *Biochemistry* **32**, 5187–5195.
Fowler, R. H. (1936). Interatomic forces. In *Statistical Mechanics*. Cambridge, Cambridge University Press, chap. 10.

Gibbs, J. W. (1876). On the equilibrium of heterogeneous substances. In *The Scientific Papers of J. Willard Gibbs*. I. *Thermodynamics*. Woodbridge, OxBow Press, p. 85.

Heremans, K. (1982). High pressure effects on proteins and other biomolecules. *Annu. Rev. Biophys. Bioeng.* **11**, 1–21.

Lennard Jones, J. E. (1931). Cohesion. *Proc. Phys. Soc. London* **43**, 461–482.

Lewis, G. N., & Randall, M. (1923). *Thermodynamics*. New York, McGraw-Hill, pp. 298–301.

London, F. (1936). The general theory of molecular forces. *Trans. Faraday Soc.* **33**, 8–16.

Planck, M. (1932). *Theory of Heat* (Eng. trans. L. Brose). London, Macmillan, pp. 74–83.

Silva, J. L., & Weber, G. (1993). Pressure stability of proteins. *Annu. Rev. Phys. Chem.* **44**, 89–113.

Weber, G. (1993). Thermodynamics of the association and the pressure dissociation of oligomeric proteins, *J. Phys. Chem.* **97**, 7108–7115.

Weber, G. (1995). Van't Hoff revisited: The enthalpy of association of protein subunits. *J. Phys. Chem.* **99**, 1052–1059.

Weber, G., & Drickamer, H. G. (1983). The effects of pressure on proteins and other biomolecules. *Qt. Rev. Biophys.* **16**, 89–112.

Widom, B. (1989). Two ideas from Gibbs: The entropy inequality and the dividing surface. In *Proceedings of the Gibbs Symposium*, ed. D.G. Caldi & D.G. Mostow. New York, AMS, AIP, pp. 73–87.

2

Pressure-Tuning Spectroscopy: A Tool for Investigating Molecular Interactions

H. G. DRICKAMER

Pressure-tuning spectroscopy is a powerful tool for investigating molecular interactions. These interactions may involve organic or inorganic materials in liquid, polymeric, or crystalline media. In this article we confine our attention to organic molecules, largely in dilute solution in polymers or liquids. We demonstrate the use of high-pressure luminescence to study the effect of the environment on $\pi^* \to \pi$, $\pi^* \to n$ and charge-transfer excitations, as well as the interaction between singlet and triplet states. In addition, we provide tests of the energy gap law for nonradiative dissipation of excitation, the role of viscosity in luminescent efficiency, and the internal consistency of various means of predicting and correlating energy transfer.

Over the past 40 years, it has been amply demonstrated that high pressure is a powerful tool for studying electronic phenomena in condensed phases. The basic concept is as follows. The optical, electrical, magnetic, and chemical properties—collectively the electronic properties—of condensed phases depend on the interactions of the outer electrons on the atoms, molecules, or ions that make up the phase. Different kinds of electronic orbitals have different spatial characteristics—different radial extent, different shape (orbital angular momentum), and different diffuseness; therefore, pressure perturbs the energies associated with these orbitals in different degrees. This relative perturbation we call "pressure tuning," and the measurement and explanation of the tuning is "pressure-tuning spectroscopy." Pressure-tuning spectroscopy of the vibrational and rotational excitations of atoms in molecular and in crystal lattices is also an active and important field, but in this article we are concerned mainly with electronic phenomena.

We further limit this discussion primarily to organic molecules in solid polymers or liquid solutions, as these have the greatest relevance to biologically active systems. A variety of probes are used for studying electronic phenomena under high pressure, but the emphasis here is on luminescence.

The presentation consists of a series of examples of various types of excitations on interactions where high pressure has been an effective tool. Only references directly relevant to each example are included. Two general references to pressure studies of molecular luminescence have been published (Drickamer, 1982, 1990). Here I focus on results. For experimental techniques and the structures of the various molecules studied, the reader is referred to the original articles.

$\pi^* \to \pi$ EMISSIONS

Van der Waals Interactions

A ubiquitous aspect of molecular interactions involves the so-called van der Waals forces. The attractive part of the van der Waals interactions energy has the form

$$E \approx -\frac{\alpha_1 \alpha_2}{r^6} \tag{1}$$

where α_1 and α_2 are the polarizabilities of the interacting entities and r is the distance between them. The quantity r^3 varies with the density. Liquids compress ~ 30–35% in 10 kbar, whereas polymers and organic crystals compress $\sim 40\%$ in 100 kbar so that these interactions are strongly pressure dependent. The polarizability can be expressed in terms of the refractive index in the form

$$\alpha \approx \frac{n^2 - 1}{n^2 + 2} \tag{2}$$

where the refractive index can be calculated as a function of density (and, hence, pressure) from the well-known Lorenz-Lorentz relationship. The overall increase in the van der Waals interaction is of the order of 1.8 for water and 2.2 for organic liquids in 10 kbar. For polymers it is about 2.7–3.2 in 100 kbar. The importance of the polarizability is frequently underestimated by chemists. Even when one is dealing with polar molecules or media, the polarizability may provide the dominant interaction. The effect of polarizability on the emission energy for a $\pi^* \to \pi$ emission is shown in Figure 2.1 for three diphenyl polyenes (Brey et al., 1979). In a perfluorocarbon with $n = 1.27$ at one atmosphere, the peak is near $30,000 \text{ cm}^{-1}$. The data extrapolate smothly to methylcyclohexane, which overlaps toluene at high pressure. Toluene overlaps polymethylmethacrylate (PMMA), which in turn overlaps polystyrene (PS) at high pressure. At 100 kbar, polystyrene has $n = 1.9$, and the peak is at $\sim 25,500 \text{ cm}^{-1}$.

It is worth elaborating briefly on the red shift of the emission with pressure. A molecule is certainly larger when an electron is excited. However, when the excited and ground states have similar character, the excited state is more polarizable because the excited electron is more loosely bound. Thus, the volume of the system decreases near the excited molecule, and the π^* state is stabilized

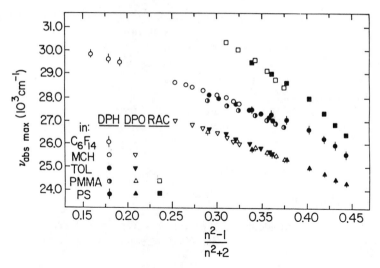

Figure 2.1. Energy of fluorescence emission versus $(n^2 - 1)/(n^2 + 2)$ (a measure of polarizability) for three diphenyl polyenes in a series of liquid and polymeric media. (Reprinted with permission from L. A. Bray, G. B. Schuster, and H. G. Drickamer, 1979, *J. Chem. Phys.* **71**, 2765–2772. Copyright 1979 American Physical Society.)

by pressure relative to the ground state. Where the two states differ in character, as for $\pi^* \to \pi$ or charge-transfer emissions, the situation is more complex. The difference in polarizability may be much smaller or may have to compete more intensively with other intermolecular interactions.

Azulene Derivatives

Certain derivatives of azulene can emit from two different excited states at the same time. The higher energy $S_2 \to S_0$ emission is a typical $\pi^* \to \pi$ emission, while the lower energy emission has charge-transfer character; the polarizabilities of S_1 and S_0 do not differ significantly. In this study (Mitchell et al., 1977a), in addition to the effects of polarizability we see the application of the energy gap law, which relates the rate of thermal dissipation of energy directly from the excited state to the ground state (internal conversion) to the difference in energy between these states. This then determines the intensity of emission in cases where intersystem crossing to a triplet state is not an important factor. This is the strong coupling case of Englmann and Jortner (1970). We follow the peak shift and change of luminescence efficiency in two media: PMMA, which is moderately polarizable, and PS, which is significantly more polarizable. In Figure 2.2 we see that the $S_2 \to S_0$ emission in PMMA shifts to lower energy by ~ 1500 cm^{-1} in 140 kbar, and the intensity drops by a factor of 50. For the $S_1 \to S_0$ emission (Figure 2.3), the energy increases by ~ 1350 cm^{-1}, and the intensity increases by a factor of 40. In PS, the $S_2 \to S_0$ emission shown in Figure 2.4 shifts red by over 3000 cm^{-1}, and the intensity drops by over 100. For the $S_1 \to S_0$ emission in PS, as shown in Figure 2.5, the initial shift is blue, but beyond ~ 70 kbar, it

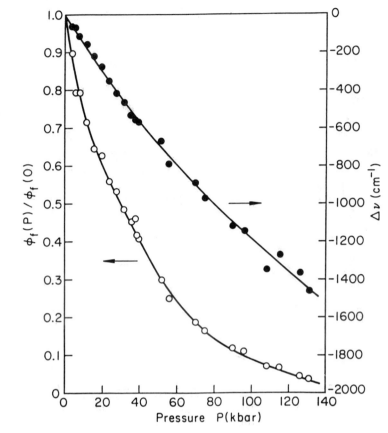

Figure 2.2. $S_2 \rightarrow S_0$ emission energy and efficiency versus pressure for an azulene derivative in PMMA. (Reprinted with permission from D. J. Mitchell, H. G. Drickamer, and G. B. Schuster, 1977, *J. Am. Chem. Soc.*, **99**, 7490–7495. Copyright 1977 American Chemical Society.

reverses sign. The intensity change precisely follows the peak shift. By 100 kbar, the compression is such that r^{-6} has more than doubled—hence the increased importance of the van der Waals forces at high pressure.

Singlets versus Triplets

There have been a number of studies of emission from triplet and singlet states of the same symmetry (e.g., Shaw & Nicol, 1976). In general for $\pi^* \rightarrow \pi$ systems, the triplet emission shifts red by a factor of 4–5 less than the corresponding singlet emission. This is understandable in that the necessity for the two electrons with parallel spins to stay apart adds a boundary condition which reduces the result of this relative shift as a tendency to quench the fluorescence because of increased intersystem crossing as the $S_1 - T_1$ energy difference decreases. An interesting case involves a series of 9-carbonyl derivatives of anthracene. At one atmosphere, they exhibit no fluorescence, even at low temperature. By 3 kbar at 295 K, measurable

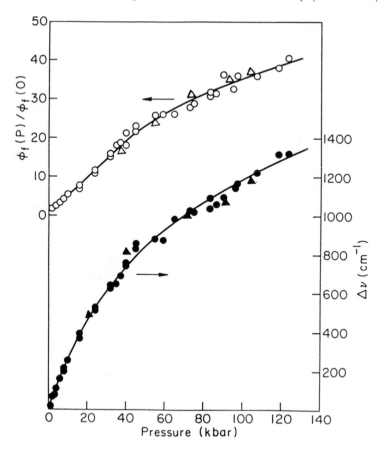

Figure 2.3. $S_1 \to S_0$ emission energy and efficiency versus pressure for an azulene derivative in PMMA. (Reprinted with permission from D. J. Mitchell, H. G. Drickamer, and G. B. Schuster, 1977, *J. Am. Chem. Soc.*, **99**, 7490–7495. Copyright 1977 American Chemical Society.)

fluorescence is seen, and this increases by a factor of several hundred by 150 kbar, as shown for example in Figure 2.6 (Mitchell et al., 1977b). The explanation for this phenomenon is that at one atmosphere the T_2 state lies only slightly below S_1. With increasing pressure, S_1 becomes lower in energy than T_2, and the molecule fluoresces, limited only by the energy gap law and the relatively ineffective intersystem crossing to T_1, which lies well below S_1 in energy.

$\pi^* \to n$ EMISSIONS

The characteristics of $\pi^* \to n$ emissions are more variable than those of the $\pi^* \to \pi$ case. The oscillator strength, and thus the radiative rate and the intensity, is, in general, lower for $\pi^* \to n$ emissions. The difference in interaction with the environment between n and π^* is frequently less, so the pressure shift of the $\pi^* \to n$ emission is usually much less than for $\pi^* \to \pi$ and may even have the opposite

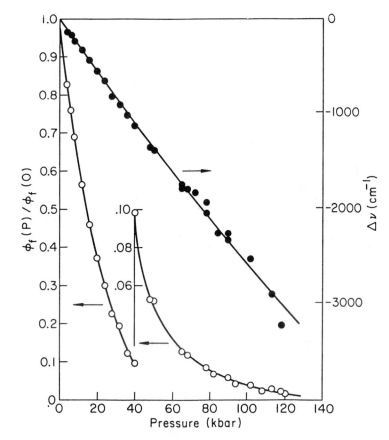

Figure 2.4. $S_2 \rightarrow S_0$ emission energy and efficiency versus pressure for an azulene derivative in PS. (Reprinted with permission from D. J. Mitchell, H. G. Drickamer, and G. B. Schuster, 1977, *J. Am. Chem. Soc.*, **99**, 7490–7495. Copyright 1977 American Chemical Society.)

sign. A study of fluorenone in different environments illustrates this clearly (Mitchell et al., 1977c). Figure 2.7 is a schematic diagram of the effect of the environment on different types of transitions. It is not implied that there is no perturbation of $\pi^* \rightarrow n$ or $T_1 \rightarrow S_0$ emission, but only that the effect is small compared with the large environmental effects on typical $S_1 \rightarrow S_0$ transitions. In region I, the $S(n\pi^*)$ transition dominates. In regions II and III, the $S(\pi\pi^*)$ is observed. In region II, there should be an increase in emission intensity, but in region III this competes with loss of intensity through intersystem crossing.

The pressure-induced changes for fluorescent quantum yield for fluorenone in four polymeric media, PS, PMMA, PIB (polyisobutylene), and PMP (poly-4-methyl-1-pentene), as well as from the crystalline solid, are shown in Figure 2.8. PIB and PMP have low polarizability, so the emission lies initially in the middle of region I. With increasing pressure and increasing van der Waals interaction, the environment is described by a movement to the right in Figure 2.7, through region II and into region III, where intersystem crossing to a triplet state reduces emission efficiency. PMMA starts in region I near the I-II boundary and moves

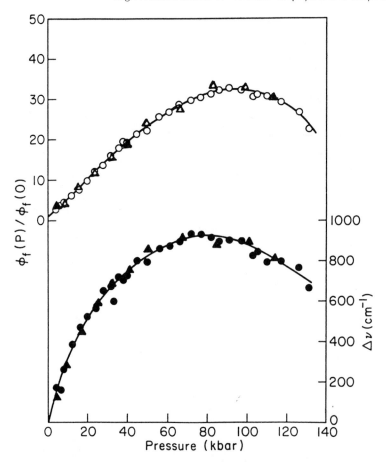

Figure 2.5. $S_1 \rightarrow S_0$ emission energy and efficiency versus pressure for an azulene derivative in PS. (Reprinted with permission from D. J. Mitchell, H. G. Drickamer, and G. B. Schuster, 1977, *J. Am. Chem. Soc.*, **99**, 7490–7495. Copyright 1977 American Chemical Society.)

into region III at a lower pressure than PIB or PMP. PS starts nearer region II and passes into region III at a little lower pressure than PMMA. The crystalline solid, with the strongest interaction, is already in region III at one atmosphere and shows only a continuous decrease in luminescent efficiency.

CHARGE-TRANSFER EMISSIONS

Molecules with excitations involving substantial intramolecular transfer of charge during excitation and emission form an important class of materials frequently used as ligand probes of protein conformation. From an extensive study of such molecules (Rollinson & Drickamer, 1980) we select PRODAN, a molecule used in biochemical fluorescence, for presentation here. A thorough discussion would involve a more extensive presentation of quantum yields, lifetimes, and so on, than is practical here. The peak shifts in a series of liquid solvents are shown in

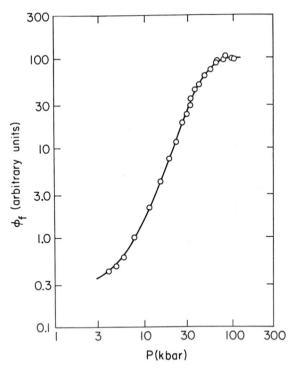

Figure 2.6. Pressure dependence of the relative fluorescence yield for 9-benzoylanthracene in PS. (Reprinted with permission from D. J. Mitchell, G. B. Schuster, and H. G. Drickamer, 1977, *J. Am. Chem. Soc.,* **99,** 1145–1148. Copyright 1977 American Chemical Society.)

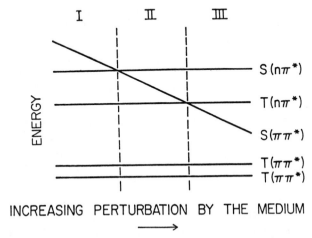

Figure 2.7. Schematic representation of the change in energy of $S(\pi\pi^*)$ relative to energies associated with other excitations as a function of degree of perturbation by the medium. (Reprinted with permission from D. J. Mitchell, G. B. Schuster, and H. G. Drickamer, 1977, *J. Chem. Phys.* **67,** 4832–4835. Copyright 1977 American Physical Society.)

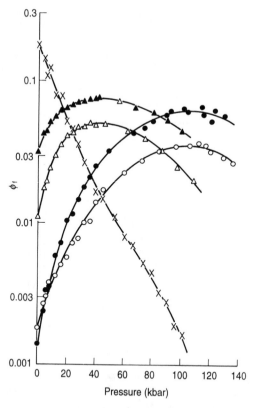

Figure 2.8. Pressure dependence of the fluorescence quantum yield for fluorenone in the crystalline form (×) and dissolved in PMMA (▲) PS (△), PMP (●) and PIB (○). (Reprinted with permission from D. J. Mitchell, G. B. Schuster, and H. G. Drickamer, 1977, *J. Chem. Phys.* **67**, 4832–4835. Copyright 1977 American Physical Society.)(

Figure 2.9. In methylcyclohexane (MCH) and toluene, the peaks lie at relatively high energy, and shift to lower energy with pressure. In methanol, the peak lies 4000–5000 cm^{-1} lower in energy than in the hydrocarbons, but also shifts red with pressure. In isobutanol, the shift is initially to lower energy, but above 5–6 kbar it reverses sign. In glycerol, the shift is to higher energies at all pressures. These observations, together with quantum yield and lifetime data presented in the original article, indicate the following conclusions. PRODAN can emit either from a locally excited (LE) state or a charge-transfer (CT) state, which may involve a change in molecular geometry. Increasing solvent polarity, as in the alcohols and glycerol, stabilizes the charge-transfer state. At one atmosphere in the hydrocarbons, the emission is entirely from the LE state, while from the alcohols it is the CT state, which is observed because the polar CT state is stabilized by the OH of the alcohols. The initial viscosities of the alcohols and glycerol are MeOH = 0.5 cp, iBuOH = 2.9 cp, gly = 800 cp (cp = centipoise). In 10 kbars, the factor of increase in viscosity is for MeOH, ~9; for iBuOH, ~100; for gly, ~250. In glycerol, a double exponential decay indicates some mixture of emissions at all

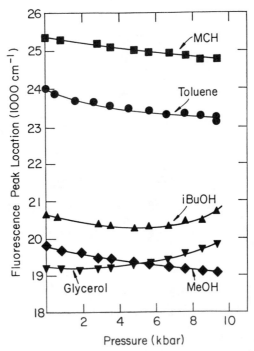

Figure 2.9. Effect of pressure on the fluorescence peak location of PRODAN in various liquid solvents. (Reprinted with permission from A. M. Rollinson and H. G. Drickamer, 1980, *J. Chem. Phys.*, **73**, 5981–5995. Copyright 1980 American Physical Society.)

pressures, while above ~ 4 kbar the same situation occurs as for iBuOH. Evidently, high viscosity inhibits the formation of the CT state, which is reasonable if, indeed, some change in geometry is involved. In Figure 2.10 we exhibit the pressure shift for PRODAN in PMMA, in PS, and in the crystal. Evidently, the rigidity of the medium inhibits formation of the CT state at all pressures. The large red shift observed in the crystal is consistent with the fact that the molecules orient themselves in the crystal to maximize attractive interactions so there is a much stronger polarizability effect.

The study of locally excited versus charge-transfer emissions, as well as changes in excited state geometry (TICT molecules), is a very active field in which high-pressure studies are playing an increasing role.

Metalloporphyrins

There has been an extensive study of the luminescence of a number of metalloporphyrins dissolved in the polymers PMMA and PS at a concentration of 10^{-2} mol/mol of monomer (Politis & Drickamer, 1981). Here we limit our discussion to one facet of their behavior. In particular, we emphasize the correlation between the relative shift in peak location in the two polymers and the relative rate of thermal dissipation of energy from the excited state for the two

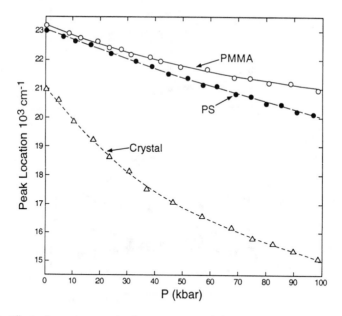

Figure 2.10. Effect of pressure on the fluorescence peak location of PRODAN in solid PMMA, PS, and the crystalline solid.

polymers. We do this for the fluorescent nonradiative rate k_2 in Figure 2.11, and for the phosphorescent nonradiative rate k_3 in Figure 2.12. PMMA has a refractive index $n = 1.43$ at one atmosphere, while for PS $n = 1.58$, so there is a significant difference in polarizability which controls the peak shift which, in turn, determines the nonradiative rates. This correlation holds both for fluorescence and phosphorescence, even though these two phenomena individually demonstrate very different degrees of interactions with the environment. The results in this section bring together and recapitulate features discussed in several of the previous sections.

Viscosity and Luminescence Efficiency

Many nonrigid molecules, which can assume more than one configuration in the ground or excited state, show a marked increase in luminescent efficiency as the viscosity of the solution increases. Förster and Hoffmann (1971) developed a model for the viscosity effect which predicted that, over a considerable range of viscosity (η), the dependence should be of the form

$$\phi = C\eta^{2/3} \tag{3}$$

The usual way of varying viscosity is by varying the temperature or composition of the solution. It is, very difficult however, to separate viscosity effects

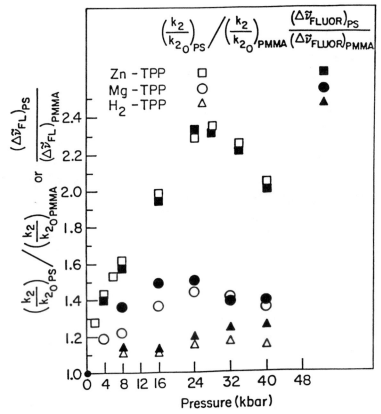

Figure 2.11. $\left[\dfrac{k_2(p)}{k_2(o)}(PS)\right]\Big/\left[\dfrac{k_2(p)}{k_2(o)}(PMMA)\right]$ and $[\Delta V_{FL}(PS)]/[\Delta V_{FL}(PMMA)]$ versus pressure for Zn-TPP, Mg-TPP, and H$_2$-TPP. (Reprinted with permission from T. G. Politis and H. G. Drickamer, 1981, *J. Chem. Phys.* **74**, 263–292. Copyright 1981 American Physical Society.)

from the other effects of temperature or composition; furthermore, it is difficult to cover a sufficient range of viscosity in this manner.

Two dyes that are useful for such an investigation are crystal violet and auramine-O. We measured their emission efficiencies in terms of relative intensity in methanol, isopropanol, isobutanol, and glycerol over a pressure range of 11 kbar (Brey et al., 1977). From Bridgman's measurement of the pressure effects on viscosities (Bridgman, 1926), we covered a range of viscosity from $\sim 5 \times 10^{-3}$ poise to $\sim 10^3$ poise, with considerable overlap between the values of η in the high-pressure range of one solvent and the low-pressure range of another.

The result for crystal violet (CV) is shown in Figure 2.13. The solid line represents $\eta^{2/3}$. Over a range of at least 3.5 orders of magnitude, the agreement is excellent. The deviation at the highest viscosities is expected. Förster and Hoffmann predict a smaller slope at very low viscosities. The initial slope is actually a little larger than $\eta^{2/3}$. The data for auramine-O fall exactly on top the CV data

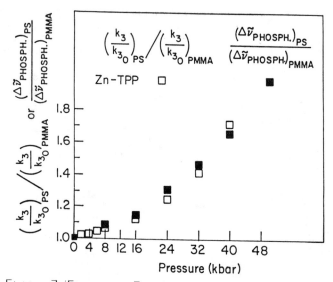

Figure 2.12. $\left[\dfrac{k_3(p)}{k_3(o)}(PS)\right]\bigg/\left[\dfrac{k_3(p)}{k_3(o)}(PMMA)\right]$ and $[\Delta V_{PHOS}(PS)]/[\Delta V_{PHOS}(PMMA)]$ versus pressure for Zn-TPP. (Reprinted with permission from T. G. Politis and H. G. Drickamer, 1981, *J. Chem. Phys.* **74,** 263–292. Copyright 1981 American Physical Society.)

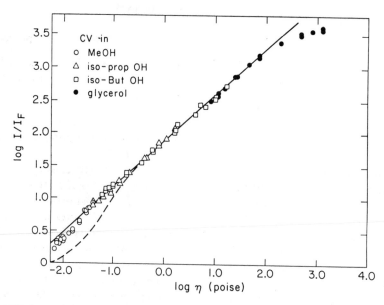

Figure 2.13. Log relative emission intensity versus log viscosity [poise] for crystal violet. (Reprinted with permission from L. A. Bray, G. B. Schuster, and H. G. Drickamer, 1977, *J. Chem. Phys.* **67,** 2648–2650. Copyright 1977 American Physical Society.)

at viscosities above ~ 0.2 poise. The significant deviation at lower viscosities is not explained at present.

This study demonstrates the advantage of changing bulk properties with pressure at constant temperature as it relates to an electronic property and to test the validity of a model.

ENERGY TRANSFER

Energy transfer has for many years been a process of great theoretical and applied importance, with applications from fluorescent lighting and laser development to photosynthesis and biopolymer properties. The basic theoretical treatment was by Förster (1959); also see references to earlier work in this article. This analysis provides three methods of predicting the transfer efficiency between donor and acceptor, two based on measurements in the time domain, and one on steady-state data. In addition, at least one empirical approach, not related in any clear way to the Förster model, has recently been applied (Wiczk et al., 1990). There have been two previous pressure tests of the theory. One involved an inorganic crystal over a pressure range of 18 kbar (Bieg & Drickamer, 1977). The second used pyrene-perylene energy transfer in PMMA and involved time-dependent measurements only (Johnson & Offen, 1972). In both cases, the pressure dependence was monotonic. Here we report a recent study in which the nonmonotonic behavior of the transfer efficiency with pressure and the use of both time domain and steady-state measurements provide an opportunity to compare all four methods of prediction (Lang & Drickamer, 1993). Experimental data are presented for transfer from the donor Coumarine-138 (C-138) to the acceptor Rhodamine B (RhB) in dilute solution in a matrix of polyacrylic acid (PAA) over a pressure range of 80 kbar.

The development of the Förster approach is discussed in detail in the above reference. We present only a few essential equations here. The time dependence of the decay of the donor can be expressed in the form

$$I(t) = I_o \exp[-t/t_D - 2\gamma(t/t_D)^{1/2}] \tag{4}$$

when τ_D is the decay of the isolated donor. One can extract γ from this measurement and calculate the efficiency in the most direct way from

$$E = \pi^{1/2}\gamma[\exp \gamma^2][1 - \text{erf}(\gamma)] \tag{5}$$

where

$$\text{erf}(\gamma) = \frac{2}{\sqrt{\pi}} \int_0^\gamma \exp(-x^2)dx$$

Using this approach, we extract an efficiency of about 70–71% at 1 atm, which increases to 84–85% near 40–50 kbar, and then drops to 79–80% at 80 kbar. We use this as our fiducial result and compare the predictions of the other methods with it. A second expression for the efficiency given by Förster is

$$E = \frac{R_o^6}{R_o^6 + r^6} \tag{6}$$

where R_0 is the so-called Förster distance—the distance between donor and acceptor at which there is a 50% chance of D–A radiationless energy transfer and r is the actual D–A distance. We assume that $r = \bar{r}$—that is, that the molecules are randomly distributed. Then

$$\left[\frac{\bar{r}}{R_o}\right]_o^6 = \frac{1}{E_o} - 1 \tag{7}$$

For any assumed value of $E = E(0)$, we can extract $(\bar{r}/R_o)^6$. The change of $\bar{r}^6$ with pressure can be obtained from the compressibility

$$\left[\frac{\bar{r}(p)}{\bar{r}(o)}\right]^6 = \left[\frac{V(p)}{V(o)}\right]^2 \tag{8}$$

We can extract R_o as a function of pressure most straightforwardly from the time-dependent measurements

$$\gamma = \frac{2}{3}\pi^{3/2}n_A R_o^3 \tag{9}$$

where n_A is the number density of acceptors, a number which increases proportionally to the compression. R_o^6 can also be determined from steady-state emission data

$$R_o^6 = \alpha \frac{\Phi_D}{n^4} \frac{1}{\bar{v}^4} \int f_D(v)E_A(v)dv \tag{10}$$

where Φ_D = quantum efficiency of donor, n = refractive index of medium, $\int = f_D(v)$ $E_A(v)dv = OL$ = overlap integral, $\bar{v}$ = mean energy of overlap area, and α is a combination of constant coefficients.

Since Φ_D, n, OL, and $\bar{v}$ all vary with pressure, this provides a very severe test of the data and the theory. The relevant parameters are presented as a function of pressure in Figure 2.14 along with the change in r^{-6}. In Figure 2.15, we compare the values of R_o obtained from time-dependent data (solid line) and steady-state data (dashed line). While these are small quantitative differences, the agreement is remarkably good. The small initial dip is due to the large initial change in n^4 with pressure. In the region ~ 7 to 45 kbar, the large increase in overlap dominates.

The drop at high pressure is a function primarily of a small decrease in overlap and the change in n^4. The fourth (unrelated) method of expressing the efficiency mentioned above has the form

$$E = 1 - \frac{I_{DA}}{I_D} \tag{11}$$

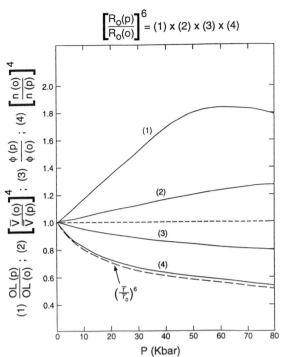

Figure 2.14. Pressure dependence of the relevant parameters for Eq. 10 and of $(\bar{r} \neq \bar{r}_0^6)^6$. (Reprinted with permission from J. M. Lang and H. G. Drickamer, 1993, *J. Phys. Chem.* **97,** 5058–5064. Copyright 1993 American Chemical Society.)

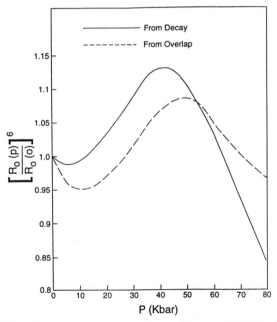

Figure 2.15. $[R_0(o)]/[R_0(p)]$ versus pressure as obtained from Eq. 9 (solid line) and from Eq. (10) (dashed line). (Reprinted with permission from J. M. Lang and H. G. Drickamer, 1993, *J. Phys. Chem.* **97,** 5058–5064. Copyright 1993 American Chemical Society.)

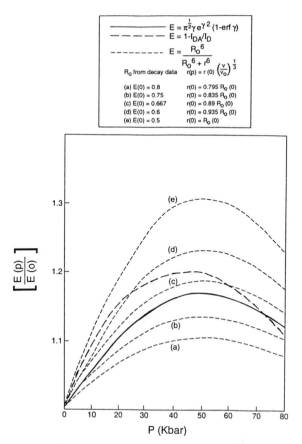

Figure 2.16. Comparison of four methods of predicting energy transfer as a function of pressure. The box above the figure identifies the various lines. (Reprinted with permission from J. M. Lang and H. G. Drickamer, 1993, *J. Phys. Chem.* **97**, 5058–5064. Copyright 1993 American Chemical Society.)

where I_{DA} = intensity of donor emission with acceptor present; I_D = intensity of donor emission with no acceptor present (Wiczk et al., 1990).

The four methods of prediction are compared in Figure 2.16. The solid curve represents the change in the fiducial value of efficiency extracted from Eq. 4. The lightly dashed lines represent the calculation from Eq. 6. Using R_0 extracted either from time-dependent or steady-state data, the small difference shown in Figure 2.15 has essentially no effect. As indicated above, to apply Eq. 6 it was necessary to assume an initial value for $E(0)$. The different dashed lines are associated with different values of $E(0)$. The fiducial curve lies between the curves for $E(0) = 0.67$ and $E(0) = 0.75$, but a little nearer the latter. The $E(0)$ from Eq. 4 was 0.70–0.71, so the agreement is excellent. The heavy dashed line results from the application of Eq. 11. There are clear quantitative differences, but in a crude way the qualitative trends are preserved; however, the initial value (~ 0.65) is significantly different. Especially given the nonmonotonic behavior of E and R_0 with pressure, these results provide a severe and satisfactory test of the Förster model.

CONCLUSION

These vignettes constitute only a limited subset of the wide variety of examples of the use of high pressure to elucidate electronic phenomena, and even these are presented only in outline, eliding a number of details. I hope, however, that they convey an accurate and stimulating picture of the power and versatility of pressure-tuning spectroscopy.

ACKNOWLEDGMENTS: The work of a large group of students over a considerable period plus the collaboration of my colleague Professor G. B. Schuster made these contributions possible. It is a pleasure to acknowledge the continuing support of the Materials Science Division of the Department of Energy under Contract DEFG02-91ER45439.

REFERENCES

Bieg, K. W., & Drickamer, H. G. (1977). A high pressure test of the theory of energy transfer by sensitized luminescence. *J. Chem. Phys.* **66**, 1437–1442.

Brey, L. A., Schuster, G. B., & Drickamer, H. G. (1977). High pressure studies of the effect of viscosity on fluorescence efficiency in crystal violet and auramine O. *J. Chem. Phys.* **67**, 2648–2650.

Brey, L. A., Schuster, G. B., & Drickamer, H. G. (1979). High pressure studies of radiative and non-radiative processes in diphenyl hexatriene, diphenyloctatetraene and retinyl acetate. *J. Chem. Phys.* **71**, 2765–2772.

Bridgman, P. W. (1926). The effect of pressure on the viscosity of 43 pure liquids. *Proc. Am. Acad. Arts Sci.* **61**, 5791.

Drickamer, H. G. (1982). High pressure studies of molecular luminescence. *Annu. Rev. Phys. Chem.* **33**, 25–47.

Drickamer, H. G. (1990). High pressure luminescence spectroscopy in polymers. In *Photochemistry and Photophysics*, (ed. J. F. Rabek. *Boca Raton, CRC Press*, Fla., chap. 1, 137–175.

Englmann, R., & Jortner, J. (1970). The energy gap law for radiationless transitions in laye molecules. *Mol. Phys.* **18**, 145–164.

Förster, T. (1959). Transfer mechanisms of electronic excitation. *Faraday Disc.—Chem. Soc.* **27**, 7–17.

Förster, T., & Hoffmann, G. (1971). Die Viskositätabhängigkeit der Fluoreszenzquanten-ausbeuten einiger Farbstoffsysteme. *Zeit. Physik. Chemie NE* **75**, 63–76.

Johnson, P. C., & Offen, H. W. (1972). Pressure dependence of energy transfer from pyrene to perylene. *J. Chem. Phys.* **57**, 1473–1475.

Lang, J. M., & Drickamer, H. G. (1993). High pressure study of energy transfer between Coumarin 138 and Rhodamine B in a solid polymeric matrix, *J. Phys. Chem.* **97**, 5058–5064.

Mitchell, D. J., Drickamer, H. G., & Schuster, G. B. (1977a). The effect of pressure on the fluorescence of 9-carbonyl substituted anthracenes. *J. Am. Chem. Soc.* **99**, 7490–7495.

Mitchell, D. J., Schuster, G. B., & Drickamer, H. G. (1977b). Energy dependence of the radiationless deactivation rate for the azulene system probed by the effect of external pressure. *J. Am. Chem. Soc.* **99**, 1145–1148.

Mitchell, D. J., Schuster, G. B., & Drickamer, H. G. (1977c). High pressure studies of fluorenone emission in plastic media. *J. Chem. Phys.* **67**, 4832–4835.

Politis, T. G., & Drickamer, H. G. (1981). High pressure luminescence studies of metalloporphyrins in polymeric media. *J. Chem. Phys.* **74**, 263–272.

Rollinson, A. M., & Drickamer, H. G. (1980). High pressure study of luminescence from intermolecular CT compounds. *J. Chem. Phys.* **73**, 5981–5995.

Shaw, R. W., & Nicol, M. (1976). Phosphorescence and triplet-triplet absorption spectra of anthracene-$_{d10}$ in PMMA under high pressure. *Chem. Phys. Lett.* **39**, 108–112.

Wiczk, W., Eis, P. S., Fishman, M. N., Johnson, M. L., & Lakowicz, J. R. (1990). Global analysis of distance distribution data for donor-acceptor pairs with different Förster distance. *SPIE* **1204**, 645–657.

3

Use of Partial Molar Volumes of Model Compounds in the Interpretation of High-Pressure Effects on Proteins

KENNETH E. PREHODA and JOHN L. MARKLEY

The transfer of liquid hydrocarbons into water is accompanied by a large decrease in volume at 25 °C and atmospheric pressure, with typical values for ΔV_{tr}° of -2.0 ml mol methylene^{-1}. Considering the large amount of apolar surface that is exposed when a globular protein unfolds, the hydrocarbon transfer results imply that the change in volume accompanying the unfolding process (ΔV_{obs}°) should be highly negative under these conditions. However, experimental data on the pressure denaturation of proteins typically yield relatively small values of ΔV_{obs}° at atmospheric pressure and 25 °C. We analyze this apparent inconsistency in terms of a simple thermodynamic dissection of the partial molar volume. This approach allows the volume effects that result from solute-solvent interactions to be determined from experimental partial molar volumes. The use of absolute quantities (partial molar volumes) circumvents assumptions associated with the use of results from transfer experiments. An important finding is that hydration of apolar species is less dense than bulk water. This discovery leads to the conclusion that the contribution to ΔV_{obs}° for protein unfolding from the hydration of apolar surfaces is highly positive, contrary to predictions based on transfer data. Further, hydration of polar surfaces makes a positive contribution to ΔV_{obs}°. The large, positive term from the differential hydration of the folded and unfolded states is compensated by the difference in free volume of the protein in the two states. This finding provides a new framework for interpreting pressure effects on macromolecules.

The full characterization of a macromolecular system requires knowledge of the effect of pressure on the system. The thermodynamic information obtained from using pressure as a perturbation is a volume change for the particular reaction being studied. The observed volume change, ΔV°_{obs}, for protein unfolding may provide insight into the mechanisms that determine the three-dimensional structure of the folded state. Pressure denaturation experiments have been demonstrated for a number of proteins, including ribonuclease A (Gill & Glogovsky, 1965; Brandts et al., 1970), chymotrypsinogen (Hawley, 1971), metmyoglobin (Zipp & Kauzmann, 1973), and, more recently, lysozyme (Samarasinghe et al., 1992) and staphylococcal nuclease (Royer et al., 1993). From these studies, ΔV°_{obs} is generally between -50 and -100 ml mol^{-1} at room temperature and atmospheric pressure (Figure 3.1). These changes are relatively small considering that the partial molar volumes of these proteins are greater than 10,000 ml mol^{-1}. They represent volume changes on the order of 0.5%. The magnitude and sign of ΔV°_{obs} indicate that the partial molar volumes of the folded and unfolded states are similar, with that of the unfolded state being slightly smaller.

While values of ΔV°_{obs} have been determined for a number of proteins, the physical basis of this quantity is still unclear. Without a detailed understanding of the underlying physical processes that contribute to ΔV°_{obs}, the usefulness of pressure as a perturbing variable is limited. The goal of this study is to provide a consistent framework for interpreting pressure effects on macromolecules that relies on thermodynamic information from model systems.

Since protein unfolding is accompanied by a large increase in solvent-exposed, nonpolar surface area, the transfer of liquid hydrocarbons into water has been invoked as a model for this process. The pressure dependence of the solubility of liquid hydrocarbons in water provides information in the form of a transfer volume, ΔV°_{tr}. Results from a number of hydrocarbons indicate that ΔV°_{tr} is

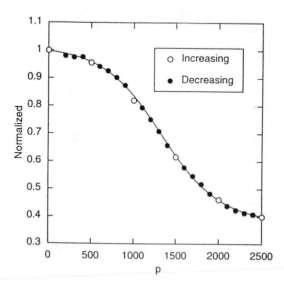

Figure 3.1. High-pressure denaturation of staphylococcal nuclease. The intrinsic fluorescence of the single tryptophan of nuclease is plotted as a function of increasing and decreasing pressures. The data were collected at 25 °C and H 5.5. Nonlinear least squares analysis of the unfolding transition yields a volume change of -60 ml mol^{-1}. The transition if highly reversible and shows little hysteresis.

approximately $-2.0 \, \text{ml mol}^{-1} \, \text{methylene}^{-1}$ at atmospheric pressure and 25 °C. Thus, the transfer of a pure liquid hydrocarbon into water is accompanied by a decrease in volume. The traditional interpretation of ΔV_{tr}° describes it as a collapse of the water around the hydrophobic solute. This implies that hydration of apolar solutes is more dense than bulk water and that the exposure of apolar surfaces leads to a decrease in volume (Heremans, 1982; Weber & Drickamer, 1983; Dill, 1990).

On extrapolating this model to protein unfolding, it would be expected that hydration of nonpolar surfaces should make a significant negative contribution to ΔV_{obs}°. For example, from the difference in nonpolar surface area for the folded and unfolded states of staphylococcal nuclease (Spolar et al., 1992) and an average value for ΔV_{tr}° for a number of apolar solutes (Sawamura et al., 1989), one calculates the contribution from differential hydration of nonpolar surfaces alone as approximately $-2400 \, \text{ml mol}^{-1}$. This is significantly more negative than the observed volume change of $-60 \, \text{ml mol}^{-1}$. What other contributions to ΔV_{obs}° can be expected?

Another factor to be considered in the unfolding volume change is the packing of the protein atoms themselves. Previous workers (Weber & Drickamer, 1983) have noted that packing imperfections present in the folded state of a protein (that is, regions of free volume) will not contribute to the excluded volume of the unfolded state. In other words, the excluded volume decreases on going from the folded to the unfolded state. This effect is also expected to make a significant negative contribution to ΔV_{obs}°.

At this point, we are faced with two large, negative contributions to ΔV_{obs}° and with no obvious positive factors that could compensate to yield the observed volume change. This prompts us to reexamine the validity of the liquid hydrocarbon transfer model. Although the transfer process is used to gain information about hydration of solutes, ΔV_{tr}° is not just a function of solvation but depends on the physical characteristics of the pure reference state (that is, pure liquid hydrocarbon, in this case).

One of the advantages of the pressure-volume regime is the ability to determine absolute partial molar volumes. Whereas it is very difficult, if not impossible, to obtain absolute partial molar enthalpies, entropies, or free energies of solutes, this valuable piece of information can be obtained for volume. It should be possible to circumvent problems associated with transfer quantities (which are relative values) by using the partial molar volumes of model compounds.

METHODOLOGY

First, it is necessary to have a model for the partial molar volume, since the partial molar volume contains two pieces of information. The goal for the model is to separate the contributions to the total volume into those from the solute itself and from the solute-solvent interaction. It is the solute-solvent interaction that we are interested in. With this in mind, the total volume of solution can be written as

$$V = n_1 \bar{V}_1^* + n_2 \bar{V}_2^* + V_{int} \tag{1}$$

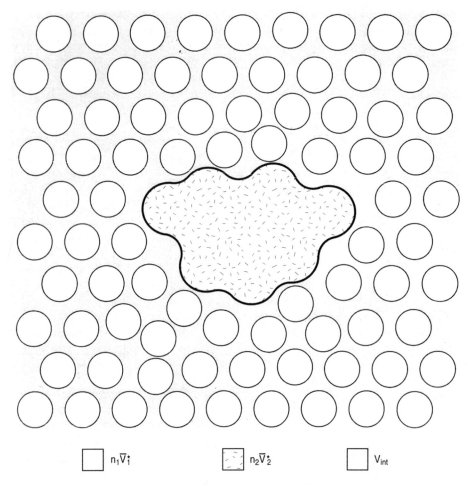

$n_1\bar{V}_1^*$ $n_2\bar{V}_2^*$ V_{int}

Figure 3.2. Definition of terms in the partial molar volume model. In this model, the total volume is composed of three terms: the volume excluded by solvent atoms ($n_1\bar{V}_1^*$), the volume excluded by solute atoms ($n_2\bar{V}_2^*$), and the void volume interstitial to the solvent ($\bar{V}_{int}$).

where n_1 and n_2 are the moles and $\bar{V}_1^*$ and $\bar{V}_2^*$ are the excluded volumes of solvent and solute, respectively (Figure 3.2). V_{int} is the amount of volume interstitial to the solvent atoms. Equation 1 describes the total volume of the solution in terms of the volume displaced by the atoms (first two terms) and the voids interstitial to those atoms (third term). The partial molar volume of the solute is then

$$\bar{V}_2 \equiv \left(\frac{\partial V}{\partial n_2}\right)_{n_1, T, P} = \left(\frac{\partial V_{int}}{\partial n_2}\right) + \bar{V}_2^* \qquad (2)$$

The derivative of the interstitial volume with respect to n_2 is the partial molar interstitial volume, $\bar{V}_{int}$. This term is due to the solute-solvent interaction in dilute solution. Equation 2 describes the partial molar volume of the solute as the volume that is displaced by the solute itself ($\bar{V}_2^*$) and the change in interstitial volume due to interaction of the solute with solvent ($\bar{V}_{int}$). The partial molar volume is an experimental quantity, while $\bar{V}_2^*$ is determined computationally from the structure of the solute. Since we are mainly interested in the solute-solvent interaction, we can rewrite Eq. 2 as

$$\bar{V}_{int} = \bar{V}_2 - \bar{V}_2^* \tag{3}$$

It is expected that the amount of solvent that the solute interacts with will be proportional to the surface area of the solute (Edward et al., 1977). Figure 3.3 shows that the partial molar interstitial volume for a number of hydrocarbons in water is indeed a function of their surface area.

Equation 2 can be extended to proteins, since ΔV_{obs}° is the difference in partial molar volumes of the unfolded and folded states. On applying the model to ΔV_{obs}°, we obtain

$$\Delta V_{obs}^\circ \approx \Delta \bar{V}_2^* + \Delta \bar{V}_{int, np} + \Delta \bar{V}_{int, p}$$

where $\Delta \bar{V}_2^*$ is the difference in solvent excluded volumes between the unfolded and folded states, $\Delta \bar{V}_{int, np}$ and $\Delta \bar{V}_{int, p}$ are the contributions from hydration of nonpolar and polar surfaces, respectively. ΔV_{obs}° is approximate in this case, owing to the assumption of additivity between hydrating molecules (see the Results). Additionally, electrostriction of water molecules by charged groups on the protein is neglected. Electrostriction, the binding of water dipoles by charged species, is accompanied by a volume decrease (Morild, 1981). Although proteins can be highly charged molecules, they contain relatively few buried charges. Therefore, the amount of electrostricted water should be roughly the same in both the folded

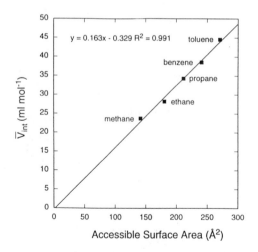

Figure 3.3. Dependence of partial molar interstitial volume on solvent accessible surface area. The slope of this plot gives the amount of interstitial void that is introduced in the solution per unit of nonpolar surface area. This can be used to approximate the contribution of hydration of nonpolar surfaces to the pressure denaturation of proteins.

and unfolded states. Thus, although the contribution of electrostriction to the partial molar volumes of each individual state may be significant, the net contribution to ΔV_{obs}° would not be.

To calculate the contribution to ΔV_{obs}° from apolar hydration, the average value of $\bar{V}_{int}/A$ for the hydrocarbons was multiplied by ΔA_{np}, the difference in nonpolar surface area between the unfolded and folded states. For the series of amides, the contribution from apolar surfaces was subtracted to calculate the average value of $\bar{V}_{int}/A$ for polar surfaces. Experimental partial molar volumes were taken from the literature (Masterton, 1954; Shahidi et al., 1977; Shahidi, 1981).

RESULTS

The volume parameters in Eq. 3 have been evaluated for a number of model compounds (Table 3.1). The most relevant term is the normalized partial molar interstitial volume, $(\bar{V}_{int}/A)$. Terms in this column can be interpreted by comparison to the value for pure water (Figure 3.4). Solutes with values of $\bar{V}_{int}/A$ greater than that for water have hydration that is less dense than bulk water. In contrast, solutes with values of $\bar{V}_{int}/A$ less than that for water have hydration with density greater than bulk water. The results of Table 3.1 clearly indicate that apolar solutes have values of $\bar{V}_{int}/A$ greater than that of bulk water and therefore have hydration densities less than bulk water.

How can this result be reconciled with interpretations of the transfer volume data which indicate that the hydration of nonpolar solutes is more dense than bulk water? The answer to this question can be seen by looking at the behavior of ΔV_{tr}° with pressure (Figure 3.5). Whereas ΔV_{tr}° is negative at atmospheric pressure, it gradually becomes less negative, reaching zero at approximately 1500 bar and turning positive thereafter. The observed negative transfer volumes at low pressures are due to the relatively lower density of hydrocarbons in the neat liquid than in aqueous solution. This emphasizes how difficult it is to draw conclusions about solvation from transfer experiments.

Table 3.1. Volume parameters for hydrocarbons and amides in water at 25 °C and atmospheric pressure

Solute	$\bar{V}_2$ (ml)	$\bar{V}_2^{*}$ (ml)	$\bar{V}_{int}$ (ml)	A_{np} (Å^2)	A_p (Å^2)	$(\bar{V}_{int}/A)$ (ml/Å^2)
Water	18.0	8.5	9.5	—	137	0.069
Methane	37.3[a]	13.6	23.7	142	—	0.167
Ethane	51.2[a]	23.0	28.2	181	—	0.156
Propane	67.1[a]	32.8	34.3	212	—	0.162
Benzene	81.3[a]	42.7	38.6	241	—	0.160
Toluene	97.0[b]	52.4	44.6	271	—	0.164
Formamide	39.2[c]	28.6	10.6	38	108	0.042
Acetamide	56.1[c]	39.3	16.8	70	100	0.055

[a] Taken from Masterton (1954).
[b] Taken from Shahidi (1981).
[c] Taken from Shahidi et al. (1977).

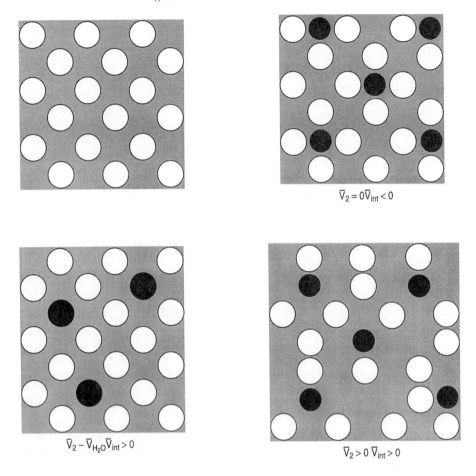

$$\bar{V}_2 = 0\, \bar{V}_{int} < 0$$

$$\bar{V}_2 - \bar{V}_{H_2O}\, \bar{V}_{int} > 0$$

$$\bar{V}_2 > 0\ \bar{V}_{int} > 0$$

Figure 3.4. Description of different hypothetical solutes in terms of the partial molar volume model: (**A**) Solvent before addition of solute. (**B**) A solute that has a partial molar volume of zero. Therefore, the partial molar interstitial volume is negative, and the solvent is more dense than bulk. (**C**) A solute that is the same size and has the same interactions as water itself. We define this as a reference state whose hydration density is equal to bulk water. (**D**) A solute that has a partial molar interstitial volume greater than that for water itself. The solvent therefore has a hydration density less than bulk.

To apply the information from model systems, a number of assumptions must be made. First, interactions between nonpolar and polar hydration are neglected. It is expected that cooperativity between different types of hydration will also contribute to the overall volume change for unfolding (Ben-Naim, 1993). Since proteins have very inhomogeneous surfaces in this respect, the magnitude of this term is not clear; however, since calculations of enthalpy, and free energy of hydration have been reasonably successful, even though they have neglected additivity between hydration effects, (Makhatadze & Privalov, 1993; Privalov & Makhatadze, 1993), these factors are expected to be small.

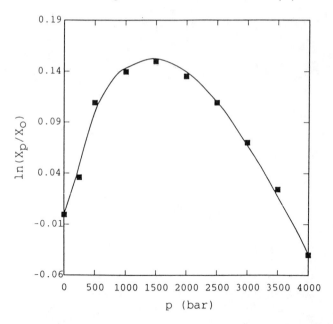

Figure 3.5. Pressure dependence of the solubility of ethylbenzene in water. Adapted from Sawamura et al (1989). The mole fraction solubility of ethylbenzene in water at 25 °C is plotted as a function of pressure. The slope is proportional to the transfer volume. At atmospheric pressure, the volume change (slope of the curve) is negative. Near 1500 bar, the volume change becomes zero and turns positive.

Additionally, structural models are required for the folded and unfolded states. The folded state can be reasonably modeled in solution by NMR spectroscopy or in crystals by diffraction studies. The unfolded state presents something of a problem, however. The structure of the pressure-unfolded state is not well defined, with some authors going so far as to say that this state should be regarded as only partially unfolded (Balny et al., 1989). Results from high-pressure NMR experiments suggest that the extent of pressure-induced unfolding of staphylococcal nuclease resemble those of temperature or guanidine unfolding (Figure 3.6). Indeed, the lack of dispersion in the one-dimensional ^{1}H NMR spectrum of nuclease at denaturing pressures indicates that the protein is highly solvated under these conditions. Therefore, we have used an extended chain to model the unfolded form of the protein. Although the pressure-unfolded state is most likely not a true extended state, for the purposes of calculating excluded volumes and surface areas, an extended state and a somewhat more collapsed state are indistinguishable.

The application of the results from the model compounds to proteins is shown in Table 3.2. The difference in excluded volume of the folded and unfolded states, $\Delta \bar{V}_2^*$, is highly negative. Because hydration of nonpolar surfaces introduces void into the solution (that is, $\bar{V}_{int,np}$ is positive), the effect from differential hydration of nonpolar surfaces in the folded and unfolded states is positive. Similarly, the contribution to the observed volume change from hydration of polar surfaces ($\bar{V}_{int,p}$) is also positive.

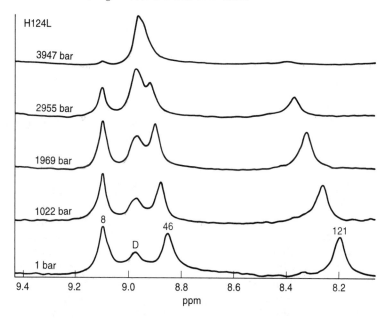

Figure 3.6. High-pressure NMR denaturation of nuclease H124L. The histidine region of the one-dimensional ^{1}H-NMR spectrum of staphylococcal nuclease is shown as a function of increasing pressure. Conditions were 3 mM protein, pH 5.5, and 310 K. High-pressure denaturation resembles that of guanidine- or temperature-induced denaturation, in that signals from a given residue type become coincident and sharper as the protein unfolds. The numbers above the bottom spectrum label the NMR signals from the H$^{\varepsilon 1}$ protons of histidines 8, 46, and 105 of the folded (native) protein; the letter D refers to the overlapped signals from these histidines in the unfolded (denatured) protein. Adapted from Royer et al. (1993).

The calculated volume change for unfolding is more negative than the observed volume change. The difference may arise from nonadditivity of hydration components, which were neglected in this analysis. Since the error scales with the molecular weight of the proteins, nonadditivity may be the more likely source. Electrostriction, which was also neglected, would cause less agreement between experimental and calculated quantities, if it were included; however, electrostriction is expected to be a small component relative to the excluded volume or the hydration components calculated here.

Table 3.2. Calculated and experimental volume parameters for proteins

Protein	$10^{-3}\,M$	$10^{-3}\,\Delta\bar{V}_2^*$ (ml)	$10^{-3}\,\Delta\bar{V}_{\text{int, np}}$ (ml)	$10^{-3}\,\Delta\bar{V}_{\text{int, p}}$ (ml)	$\Delta V^\circ_{\text{calc}}$ (ml mol)	$\Delta V^\circ_{\text{obs}}$ (ml mol)
Ribonuclease A	13.7	−1.39	0.95	0.217	−225	−35
Lysozyme	14.3	−1.79	1.13	0.423	−237	−20
Staphylococcal nuclease	16.8	−1.85	1.28	0.284	−286	−60
Chymotrypsyinogen	25.7	−3.52	2.37	0.705	−445	−30

DISCUSSION

Pressure provides an additional piece of information in the study of protein chemistry in the form of the unfolding volume change. This thermodynamic quantity governs the behavior of proteins under pressure. Indeed, it is because the volume change is negative that proteins unfold with pressure at all. Therefore, if the physical basis of the volume change can be determined, the behavior of proteins under pressure can be investigated in a more enlightened manner. To this end, we have used model systems to elucidate the physical basis of volume effects in proteins.

The transfer of hydrocarbons into the aqueous phase has been used extensively as a model for protein unfolding (Spolar et al., 1992; Makhatadze & Privalov, 1993; Privalov & Makhatadze, 1993). Depending on the reference state used in these calculations (that is, solid, liquid, or gaseous hydrocarbon), the resulting enthalpies and entropies can vary widely (this is not so for heat capacity, since pure hydrocarbons have a relatively low heat capacity regardless of the reference state). The transfer quantities have also been used in an attempt to understand pressure effects on proteins (Weber & Drickamer, 1983).

Because the transfer quantities depend on the reference state used, it is apparent that conclusions regarding solvation are problematic. Nevertheless, the negative volume change upon transfer of liquid hydrocarbons in the aqueous phase has been readily interpreted as meaning hydration of apolar solutes is more dense than bulk water (Weber & Drickamer, 1983; Dill, 1990). This leads to the expectation that solvation of nonpolar surfaces will be a significant negative contribution to the unfolding volume.

The current results, derived from experimental absolute partial molar volumes of model compounds, show that the opposite is true: hydration of apolar solutes is less dense than bulk water. If we ask why the volume change is so small relative to the volumes of the individual states, these results tell us that the answer lies in the compensation of excluded volume and solvation effects. The change in volume resulting from the protein atoms themselves (that is, the excluded volume change) is relatively large and negative; however, the observed volume change is quite small because the positive volume changes that result from differential hydration of the folded and unfolded states partially compensate for the excluded volume change. These results are completely general and should apply to pressure effects on proteins, as well as to any other macromolecular process such as protein-protein interactions or ligand binding.

ACKNOWLEDGMENTS: Supported by grant GM35976 from the National Institutes of Health. K.E.P. was supported in part by a traineeship from the NIH Molecular Biophysics Training Grant (GM08293).

REFERENCES

Balny, C., Masson, P., & Travers, F. (1989). Some recent aspects of the use of high pressure for protein investigations in solution. *High Press. Res.* **2**, 1.

Ben-Naim, A. (1993). Solvation thermodynamics of biopolymers. In *Water and Biological Macromolecules*. Boca Raton, Fla., ed. E. Westhof, CRC Press, pp. 431–459.

Brandts, J. F., Oliveira, R. J., & Westort, C. (1970). Thermodynamics of protein denaturation: effect of pressure on the denaturation of ribonuclease A. *Biochemistry* **9**, 1038.

Dill, K. (1990). Dominant forces in protein folding. *Biochemistry* **29**, 2357.

Edward, J. T., Farrel, P. G., & Shahidi, F. (1977). Partial molar volumes of organic compounds in water: Part 1. Ethers, ketones, esters and alcohols. *J. Chem. Soc., Faraday Trans.* **173**, 705.

Gill, S. J., & Glogovsky, R. L. (1965). Pressure denaturation of ribonuclease A. *J. Phys. Chem.* **69**, 1515.

Hawley, S. A. (1971). Reversible pressure-temperature denaturation of chymotrypsinogen. *Biochemistry* **10**, 2436.

Heremans, K. (1982). High pressure effects on proteins and other biomolecules. *Annu. Rev. Biophys. Bioeng.* **11**, 1.

Makhatadze, G. I., & Privalov, P. L. (1993). Contribution of hydration to protein folding thermodynamics: I. The enthalpy of hydration. *J. Mol. Biol.* **232**, 639.

Masterton, W. L. (1954). Partial molal volumes of hydrocarbons in water solution. *J. Chem. Phys.* **22**, 1830.

Morild, E. (1981). The theory of pressure effects on enzymes. *Adv. Prot. Chem.* **34**, 93.

Privalov, P. L., & Makhatadze, G. I. (1993). Contribution of hydration to protein folding thermodynamics: II. The entropy and Gibbs energy of hydration. *J. Mol. Biol.* **232**, 660.

Royer, C. A., Hinck, A. P., Loh, S. N., Prehoda, K. E., Peng, X., Jonas, J., & Markley, J. L. (1993). Effects of amino acid substitutions on the pressure denaturation of staphylococcal nuclease as monitored by fluorescence and NMR spectroscopy. *Biochemistry* **32**, 5222.

Samarasinghe, S. D., Campbell, D. M., Jonas, A., & Jonas, J. (1992). High-resolution NMR study of the pressure-induced unfolding of lysozyme. *Biochemistry* **31**, 7773.

Sawamura, S., Kitamura, K., & Taniguchi, Y. (1989). Effect of pressure on the solubilities of benzene and alkylbenzenes in water. *J. Phys. Chem.* **93**, 4931.

Shahidi, F. (1981). Partial molar volumes of organic compounds in water: Part 8. Benzene derivatives. *J. Chem. Soc., Faraday Trans.* **77**, 1511.

Shahidi, F., Farrel, P. G., & Edward, J. T. (1977). Partial molar volumes of organic compounds in water: Part 2. Amines and amides. *J. Chem. Soc., Faraday Trans.* **73**, 715.

Spolar, R. S., Livingstone, J. R., & Record, M. T. (1992). Use of liquid hydrocarbon and amide transfer data to estimate contributions to thermodynamic functions of protein folding from the removal of nonpolar and polar surface from water. *Biochemistry* **31**, 3947.

Weber, G., & Drickamer, H. G. (1983). The effect of high pressure upon proteins and other biomolecules. *Q. Rev. Biophys.* **16**, 89.

Zipp, A., & Kauzmann, W. (1973). Pressure denaturation of metmyoglobin. *Biochemistry* **12**, 4217.

4

Pressure-Tuning Spectroscopy of Proteins: Fourier Transform Infrared Studies in the Diamond Anvil Cell

KAREL HEREMANS, KOEN GOOSSENS, and LÁSZLÓ SMELLER

The effect of hydrostatic pressure on the secondary structure of proteins can be followed by Fourier transform infrared (FTIR) spectroscopy in the diamond anvil cell. Pressure-induced changes in the amide I′ region of the deconvolved spectrum are used to follow the features of the secondary structure up to 20 kbar. The changes in the side chains such as tyrosine also can be followed. A self-deconvolution and fitting procedure is presented that allows the determination of both pressure-induced and temperature-induced changes in the secondary structure of proteins. The method takes into account the elastic, as well as the possible conformational, effects on the spectral bands of the protein. Applications are presented on pressure-induced changes in several proteins. Attention is also given to the influence of inert cosolvents. The fundamental principles of the phase diagram of proteins are presented to clarify their importance for understanding the behavior of proteins under pressure at different temperatures. Our results show that the infrared technique explores unique aspects of the behavior of proteins under these extreme conditions.

The study of the effects of pressure has received considerable attention in recent years (Balny et al., 1992; Silva & Weber, 1993). In general, low pressures induce reversible changes such as the dissociation of protein-protein complexes, the binding of ligands, and conformational changes. Pressures higher than about 5 kbar induce denaturation, which in most cases is irreversible. However, reports on a few proteins indicate that such high pressures may also cause reversible changes. One such protein is horse serum albumin (Chen & Heremans, 1990). A molecular interpretation of these phenomenon is based on the fact that pressure mainly affects the volume of a system, thus damping the molecular fluctuations. Temperature effects are known to affect both the kinetic energy and the volume of the system.

Early in this century, it was shown that one can cook an egg by subjecting it to high pressure (Bridgman, 1914). The appearance of the pressure-induced coagulum of egg white is quite different from the coagulum induced by temperature. Bridgman also observed that the effect of temperature "seems to be such that the ease of (the pressure-induced) coagulation increases at low temperatures, contrary to what one might expect" (511). It is now clear that these observations are the consequence of the unique behavior of proteins (Suzuki, 1960; Hawley, 1971; Zipp & Kauzmann, 1973). The phase diagram for the conditions under which the native and the denatured conformation occur reflects the importance of the role of the change in heat capacity and compressibility between the native and the denatured state. These quantities reflect the change in energy and volume fluctuations, respectively, of the protein in the solvent (Cooper, 1976).

In this article we illustrate some of the consequences of the phase diagram for the denaturation of proteins. We first look into the difference between the temperature and pressure denaturation of proteins. Then we ask about the correlation between the denaturation temperature and pressure for homologous proteins. Finally, we explore the effect of organic cosolvents on the denaturation pressure and correlate this with the effect of the solvent on the thermal denaturation (Timasheff, 1993). The data allow a deeper insight into the thermodynamics of protein denaturation and the contribution of the protein and the solvent. A unified approach toward understanding protein-protein interactions, protein folding, and the role of solvent in these processes is now possible.

The emphasis here is on the use of Fourier transform infrared (FTIR) spectroscopy to study pressure- and temperature-induced changes in proteins. Infrared spectroscopy allows the analysis of aspects of these processes that are different and complementary to those of other techniques such as fluorescence and NMR.

Two responses may be expected from a study of the effect of pressure on the vibrational frequencies: an elastic and a conformational effect (Noguti & Go, 1989; Frauenfelder et al., 1990). The elastic effect causes a shift in the spectral band of the secondary structures due to the compressibility of the system. In the absence of any conformational effects, it is expected to be reversible. The conformational effect is the consequence of changes in the conformation which may be either reversible or irreversible, depending on the conditions and the protein. In addition to frequency shifts, which allow the determination of denaturation pressure and temperature, analysis of the amide I' band allows the determination of the contribution of the secondary structure components of the proteins. This provides a global view of the protein under pressure which is equivalent to the results obtained with the circular dichroism technique. For technical reasons, however, the latter technique is not easily accessible for high-pressure studies.

METHODOLOGY

In this section, we first present our methodology for studying pressure-induced changes in the diamond anvil cell with Fourier transform infrared spectroscopy. We then discuss how the experimental data are analyzed to extract information on the contribution of the elastic and the conformational components of the spectrum. We present results showing the differences between the temperature and pressure denaturation of lipoxygenase, along with the correlation between the denaturation temperature and pressure of a novel neuropolypeptide H3 isolated from different tissues in different species. Finally, we present some results on the effect of organic cosolvents on the denaturation temperatures and pressures of chymotrypsinogen, ribonuclease A, and β-lactoglobulin.

High-Pressure Infrared Spectroscopy with the Diamond Anvil Cell

The combined use of the diamond anvil cell (DAC) and vibrational spectroscopy such as Fourier transform infrared absorption and Raman scattering proves to have distinct advantages for the study of pressure-induced changes in bio-molecules, as well as in living tissues (Wong et al., 1993). Raman scattering has the advantage that experiments can be performed in water and that virtually no sample preparation is needed. The main disadvantage is the possible inter-ference of fluorescent impurities in protein samples. Although this interference can be overcome in some cases by photobleaching, in many cases infrared spectroscopy is the only solution: colored samples may be studied conveniently under resonance conditions, and the Raman scattering is enhanced considerably because it is coupled with electronic transitions. The advantages of infrared absorption are also clear: there is no interference from fluorescence, and there is no contribution from colored centers in proteins, such as in heme proteins. A minor disadvantage is the need to work in D_2O because of the strong absorption of water in the protein absorption region. Sample preparation is needed to find the convenient range of absorption. The advantages of infrared and Raman scattering are, at least in principle, combined in the Fourier transform Raman technique, but at present there are no reported pressure studied on biomolecules with this technique.

Proteins are dissolved in D_2O or any desired buffer solution and mounted in a stainless steel gasket of a diamond anvil cell. We found the minicell from Diacell Products (Leicester, U.K.), which has a rated maximum of about 50 kbar, quite convenient. Pressure is calculated from the ruby fluoresence obtained on a Spex Raman spectrometer and predetermined ruby fluorescence pressure calibra-tion curve. The ruby technique has the advantage that it allows easier inspection of the sample under the microscope. Infrared spectra were obtained with a Bruker IFS66 FTIR spectrometer equipped with a broadband MCT detector. The infrared light was focused on the sample by a NaCl lens (Wong, 1991). A total of 350 interferograms were coadded after registration at a resolution of $2\,cm^{-1}$. The time to take one data point at each pressure is about 15 minutes.

Analysis of the Infrared Spectra: Amide I′ Band and Side-Chain Bands

The secondary structure of a protein may be determined from the analysis of the amide I′ bandshape of the infrared or the Raman spectrum using two different approaches. First, one may compare the spectrum with a database of the amide bands of several proteins with known secondary structure from x-ray data. This approach, sometimes called factor analysis, was pioneered by Williams (1986) for

Raman spectroscopy. Second, one may use curve fitting after Fourier self-deconvolution. This approach was introduced by Susi and Byler (1986) for infrared spectroscopy. Both methods have been discussed critically by Surewicz et al. (1993).

For the analysis of the pressure-induced changes according to the factor analysis approach, one needs a data set of reference proteins as a function of pressure in order to take the elastic component into account. This is not an easy task. One possible complication is pressure-induced H/D (hydrogen/deuterium) exchange. This will not be treated here. An estimate may be made from an analysis of the shift and the intensity changes of the amide II' band (Wong, 1991). The only practical approach is curve fitting after self-deconvolution. The advantage of this procedure is that shifts in the band frequencies are allowed, which are independent of any assumptions about the assignments of the subcomponents. In many instances, however, the assignments are not unequivocal (Surewicz et al., 1993).

We have therefore developed a method to determine the secondary structure of proteins and peptides as a function of pressure (Smeller et al., 1995). Although the method of self-deconvolution combined with band fitting has been applied before by Susi and Byler (1986), we apply it for the first time to pressure-induced effects in proteins and peptides. Our method gives new information on the effect of pressure and temperature on proteins from an analysis of the frequency shifts, as well as the changes in the area of the subcomponents of the amide I' band.

The first step in the analysis of the spectra is the solvent substraction. Experiments in the diamond anvil cell show that solvent substraction has very little effect on the subsequent analysis. Solvent substraction leads to larger effects on temperature experiments carried out in a conventional infrared cell, probably because of the longer pathlength and the lower protein concentration used in conventional cells. Fourier self-deconvolution, a mathematical technique of band narrowing, was performed with the Bruker software. The deconvolution and noise reduction factor to be used depend on the quality of the spectral data. A Blackman-Harris apodization function was used. In general, we use the deconvolution factor that gives the smallest errors in the determination of the individual components of the amide I' band and that gives the smallest residual for the overall fitting procedure. The fractional composition in secondary structure was calculated by fitting Gaussian curves to the deconvolved spectrum with a program developed in our laboratory that fits all the parameters simultaneously, in contrast to the method of Susi and Byler (1986), in which the parameters are fitted consecutively. The initial values for our fitting were obtained from the second derivative of the deconvolved spectrum. Our approach allows a detailed analysis of the pressure-induced changes in the secondary structure on the molecular level.

In addition to analysis of the amide I' band, infrared spectroscopy allows the analysis of some amino acid side chains. We have found that in practice this is restricted to the analysis of tyrosines. A quite different situation is encountered in Raman spectroscopy.

Temperature- and Pressure-Induced Denaturation in Lipoxygenase

Soybean lipoxygenase is a protein of about 95 kDa. It is a nonheme iron protein and a key enzyme in the biosynthesis of leukotrienes, which play an important

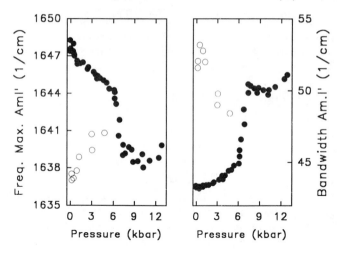

Figure 4.1. Pressure effect on the frequency maximum (*left*) and the bandwidth at half height (*right*) in lipoxygenase. ● = pressure increase; ○ = pressure decrease.

role as bioregulators in inflammation and allergy. The primary structure of soybean lipoxygenase is known, and because the crystallization and preliminary x-ray analysis has been reported recently (Boyington et al., 1990), a three-dimensional structure can be expected in the near future. The enzyme was kindly provided by Dr. J. Frank (Delf University of Technology, The Netherlands). The lyophilized protein was dissolved in Tris-DCl buffer pD7.6 at a concentration of about 30 mg ml^{-1}.

The pressure dependence of the frequency maximum of the amide I' band of lipoxygenase is shown in Figure 4.1. The 1648 cm^{-1} band at 1 bar is indicative of the large amount of α-helix present. The deconvolved spectrum reveals the presence of α-helix and β-structure in good agreement with circular dichroism data (Spaapen et al., 1979). One main transition is observed at 6 kbar when the pressure is increased. The general trend is that the frequency maximum of the band decreases with increasing pressure, indicating a strengthening of the hydrogen bonds. The fact that the effect of pressure on lipoxygenase is irreversible is clearly seen from the pressure dependence of the frequency, as well as the width at half height of the amide I' band. The transitions observed in the frequency dependence correlate with those observed in the change in the bandwidth. The changes are more clearly visible from observations on the bandwidth. We have observed this effect in other proteins. Sometimes the denaturation process seems to be reversible from the frequency data, whereas it is clearly irreversible from the bandwidth data. The changes of the conformation of the protein may also be followed by observing characteristic frequencies of the side chains of the polypeptide. In the case of lipoxygenase, it is possible to follow the tyrosines at 1515 cm^{-1} (data not shown). The transition pressure observed by the changes in the amide I' band and the tyrosine side chain band is, within experimental error, the same.

From Figure 4.2 it is clear that the pressure-denatured protein is quite

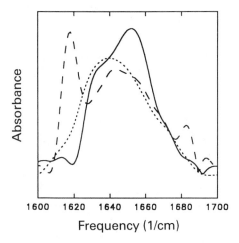

Figure 4.2. Temperature and pressure effect on the deconvolved spectrum of lipoxygenase. Solid line (—) = spectrum at ambient temperature and pressure; dotted line (...) = spectrum at 7.4 kbar (ambient temperature); broken line (---) = spectrum at 75 °C (ambient pressure).

different from the temperature-denatured protein. The deconvolved spectrum of the pressure-denatured lipoxygenase at 7.4 kbar is broad and does not show any features. The deconvolved spectrum at 75°C shows the appearance of two additional bands at low and high frequency of the main band. This indicates the occurrence of intermolecular hydrogen bonds, suggesting a strong interaction between the denatured molecules (Ismail et al., 1992).

Correlation between Denaturation Temperature and Pressure in Homologous Neuropolypeptides H3

In this section we compare the effect of pressure and temperature on the secondary structure of two very similar novel proteins. The novel proteins were isolated initially from the human brain (hence the term neuropolypeptide H3) and then from ox liver (Bollengier et al., 1993). The samples were kindly provided by Dr. F. Bollengier (Free University of Brussels). Figure 4.3 shows the pressure- and temperature-induced changes in the amide I′ region of the experimental spectrum. Under normal conditions of temperature and pressure, the amide I′ band of the two proteins is almost identical. However, the human protein denatures at a lower pressure than the ox protein. This is also observed in the tyrosine band (data not shown). In contrast, the denaturation temperature of the human protein is higher than that of the ox protein. Moreover, the pressure denaturation of both proteins seems to be reversible while the temperature denaturation is irreversible. These results show that the infrared technique explores fairly well the small differences in behaviour of nearly related proteins under extreme conditions, demonstrating the sensitivity of FTIR to detect pressure-induced and temperature-induced phenomena in proteins.

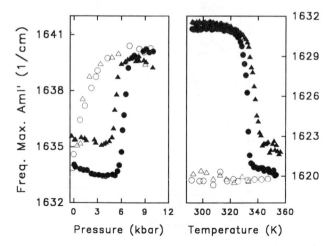

Figure 4.3. Pressure (*left*) and temperature (*right*) denaturation of neuropeptide H3 in ox liver (○) and human (△) brains. Closed symbols: increasing pressure and temperature; open symbols: decreasing pressure and temperature.

Effect of Organic Cosolvents on the Pressure-Induced Denaturation

The effect of various organic cosolvents on the temperature of denaturation has been investigated for chymotrypsinogen and β-lactoglobulin (sample provided by Dr. J. C. Cheftel, University of Montpellier II) in the presence of polyethylene glycol (PEG) (Lee & Lee, 1987) and for ribonuclease A in the presence of glycerol (Gekko & Timasheff, 1981). The presence of PEG lowers the denaturation temperature of chymotrypsinogen and β-lactoglobulin, whereas the presence of glycerol increases the denaturation temperature of ribonuclease.

We measured the effect of 30% (w/v) of PEG 1000 on the denaturation temperature and pressure of chymotrypsinogen and β-lactoglobulin with infrared spectroscopy. Protein concentrations were about 100 mg ml^{-1} and pD 3.0. For chymotrypsinogen, a small effect on the denaturation temperature could be observed ($\Delta T = -1.5\,°C$) in contrast to the observation of Lee and Lee (1987). For β-lactoglobulin, $\Delta T = -7.4\,°C$ was observed. Under similar conditions, the effect of PEG 1000 on the pressure denaturation was found to be small for chymotrypsinogen but marked for β-lactoglobulin ($\Delta p = -0.3$ kbar). From these experiments we see that a reduction of the temperature of denaturation is coupled with a decrease in the denaturation pressure.

Similar experiments were performed on ribonuclease A in the presence of 40% (v/v) deuterated glycerol at pD 3.0. We found an increase in the denaturation temperature ($\Delta = +3.6\,°C$) and an increase in denaturation pressure ($\Delta p = +0.8$ kbar). Under similar solvent conditions, Gekko and Timasheff (1981) found an increase of about 7 °C for the denaturation temperature when measured with ultraviolet absorption; however, a much lower protein concentration was used in their experiment (1 mg ml^{-1}).

During our studies on the solvent effects on ribonuclease we observed a pretransition around 5 kbar which is visible from the frequency decrease in the

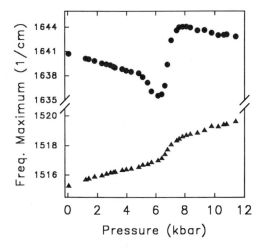

Figure 4.4. Pressure effect on the frequency maximum of the amide I' (●) and tyrosine (▲) bands of ribonuclease A at room temperature.

spectrum, as shown in Figure 4.4. It occurs before the main transition, which takes place at 7 kbar and which shows a frequency increase. From Figure 4.4 it may also be seen that the transition in the tyrosine bands around 1515 cm^{-1} (at 1 bar) coincides with the main transition and not with the pretransition.

DISCUSSION

We have attempted to explore the consequences of the phase diagram for the denaturation of proteins. With infrared spectroscopy, we have looked into differences between temperature- and pressure-denatured proteins, at the effect of small changes in one protein on its denaturation by pressure and temperature to establish possible correlations, and at the effect of organic cosolvents on the pressure and temperature stability of proteins.

Thermodynamics of Denaturation

The pressure-temperature phase diagram for proteins has been determined in a few cases: the kinetics of inactivation of ovalbumin and hemoglobin (Suzuki, 1960), the denaturation of chymotrypsinogen (Hawley, 1971) and ribonuclease (Brandts et al., 1970; Hawley, 1971), and the pressure denaturation of myoglobin (Zipp & Kauzmann, 1973), a study which includes the effect of pH. A schematic representation of the diagram that results from these studies is given in Figure 4.5. It can be seen that the denaturation temperature of proteins increases with increasing pressure up to about 1 kbar. When the pressure is increased further, the temperature of denaturation decreases. At sufficient pressure, the protein denatures below room temperature. The denaturation of a protein may be analyzed starting from the pressure and temperature dependence of the free energy:

$$d(\Delta G) = -\Delta S dT + \Delta V dP \tag{1}$$

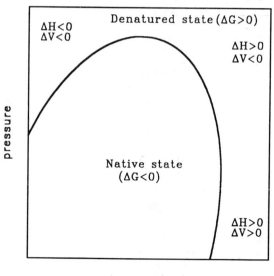

temperature

Figure 4.5. Typical temperature-pressure phase diagram of proteins. At low pressure, pressure stabilizes the protein against heat denaturation. At room temperature, pressure-induced denaturation takes place at higher pressures than at low temperature. After Suzuki (1960) and Hawley (1971).

Integration of this equation gives the dependence of the free energy on temperature, pressure, isothermal expansion ($\Delta\alpha$), isothermal compressibility ($\Delta\beta$), and heat capacity at constant pressure (ΔCp) (Hawley, 1971):

$$\Delta G = \Delta G^\circ + \Delta V^\circ(P - P^\circ) - \Delta S^\circ(T - T^\circ)$$
$$+ (\Delta\beta/2)(P - P^\circ)^2 + (\Delta Cp/2T^\circ)(T - T^\circ)^2$$
$$+ \Delta\alpha(P - P^\circ)(T - T^\circ)$$

This equation can be expanded to include the solvent composition (Illian et al. 1988), and this phenomenon of reentrant transitions, which results in elliptic phase diagrams, is also observed in liquid crystals (Klug & Whalley, 1979). There is also evidence that this phenomenon occurs in polysaccharides such as starch (Thevelein et al., 1981; Muhr & Blanshard, 1982; Muhr et al., 1982) and in lipids (Winter & Pilgrim, 1989) as well as in proteins. It has also been observed in the destruction of bacteriophages (Gross & Ludwig, 1992) and bacteria (Sonoike et al., 1992).

One possible explanation for the reentrant phase behavior of proteins comes from a model that has been proposed to explain the melting-curve maxima observed at high pressure in the case of metals, intermetallic compounds, and inorganic salts. It is assumed that two species in the liquid state exist in an equilibrium that may be shifted by the action of temperature and pressure (Rapoport, 1967). By transferring this model to protein denaturation we see that the denatured state of the protein exists in two or more conformational states, the distribution of which can be changed by temperature and pressure. This idea is consistent with the observed increase in heat capacity and compressibility upon denauration, which implies increased energy and volume fluctuations in the denatured state (Kharakoz & Sarvazyan, 1993).

Pressures versus Temperature of Denaturation

It is not surprising to observe a difference in the denatured state of a protein by the action of temperature and pressure. What is interesting is that a number of instances the pressure denaturation seems to be reversible while the temperature denaturation is not. In the case of lipoxygenase, both the temperature and the pressure denaturation are irreversible. An aspect of considerable interest is that the temperature denaturation induces a precipitation of the protein while the pressure denatured protein does not. Because of the small volume of the diamond anvil cell, it is not possible to decide whether the protein has formed any transparent gel.

The denaturation of the neuropeptide H3 presents a different picture. Here the temperature-induced denaturation is irreversible while the pressure denaturation seems to be reversible. Clearly, there is considerable hysteresis on the reversibility for the pressure denaturation, but when the possible influence of the pressure-induced H/D exchange is taken into account, the pressure denaturation, even at the high concentration used in the infrared experiments, seems reversible. We (Chen & Heremans, 1990) as well as others (Taniguchi & Takeda, 1992) have observed similar phenomena in other proteins.

The case of ribonuclease presents some interesting aspects of the pressure-induced denaturation in a protein. As can be seen from Figure 4.4, there is small change in the frequency of the main band around 5 kbar. The main transition is seen at about 7 kbar. Only the second main transition can be observed from the tyrosine band.

It should also be noted that a number of high-pressure absorption and fluorescence studies have been performed which show that pressure-induced denaturation is reversible (Weber & Drickamer, 1983). Studies with Raman (Heremans & Wong, 1985) and infrared spectroscopy (Wong & Heremans, 1988) have shown that, at high concentrations used in these techniques, the denaturation is irreversible and is accompanied by the formation of a gel or a precipitate. These experimental conditions may be considered a disadvantage. On the other hand, it should be kept in mind that NMR and calorimetric studies also need high concentrations. We consider this an advantage in the sense that infrared explores different aspects of the denaturation process, those not easily accessible by the other techniques. In any case, in the few instances in which the high- and low-concentration techniques have been applied to the same protein, almost identical denaturation pressures have been found. This clearly indicates that the techniques explore different aspects of the same phenomenon.

Studying the effect of temperature on pressure-induced denaturation is also of considerable interest. Temperature may have substantial effects on protein-protein interactions (Payens & Heremans, 1969; Silva & Weber, 1993). The analogy between pressure-induced protein-protein dissocation and protein unfolding has recently been stressed by Silva and Weber (1993). Such experiments could also give information on the effect of pressure on the cold denaturation that has been observed in a number of proteins. An advanatge of such studies is the fact that water remains liquid under pressure down to about $-20\,°C$ at 2 kbar.

Correlation between the Pressure and Temperature of Denaturation

Numerous studies have been reported in the literature on the effect of mutations of amino acids in proteins and enzymes on the denaturation temperature. A correlative question is the effect of these mutations on the pressure denaturation. We plan to explore this issue in more detail. The first results which we present on the neuro-peptide H3 are very interesting in this respect and should stimulate further work in this domain. We observe that the protein with the highest denaturation tempera-ture shows the lowest denaturation pressure. It is clear that a detailed analysis of natural mutants is a task even more difficult than the analysis of laboratory-made mutants. Pressure effects on the denaturation of mutants of staphylococcal nuclease have been reported recently by Royer et al. (1993), and they show equally interesting possibilities for correlation studies with temperature denaturation.

Solvent Effects on Pressure Denaturation

In the previous section we noted that the position of the ellipse of the phase diagram depends on the protein. This point has been noted by earlier investigators (Suzuki, 1960; Hawley, 1971). The position of the ellipse may also be shifted by changing the solvent conditions. Zipp and Kauzmann (1973) have noted that the center of the ellipse for metmyoglobin moves in a complex way with pH. At the extreme of the spectrum, recent reports indicate that dry proteins are found to be extremely resistant to temperature (Nicolini et al., 1993) and pressure denaturation (Murphy, 1978). We have observed in our laboratory that the pressure effect on dry proteins may easily be studied in the diamond anvil cell (unpublished observations). Polyols, such as sugars and glycerol, protect proteins against temperature denaturation, and we find similar protection against pressure de-naturation for ribonuclease. The pressure-induced state of this protein before the onset of the denaturation is also shifted to higher pressures in the presence of glycerol. This observation is in agreement with the results of several studies in which the protein-protein interactions are stabilized by glycerol (Silva & Weber, 1993). The effect of glycerol and other polyols on protein reactions has been analyzed in terms of the osmotic pressure of the solution (Kornblatt et al., 1993a; Kornblatt et al., 1993b; Robinson & Sligar, 1994). In some instances, osmotic pressure works against hydrostatic pressure, and this has been interpreted in terms of hydration/dehydration phenomena. Our observations on the effect of glycerol and high pressure suggest similar mechanisms.

Organic cosolvents such as PEG destabilize proteins against temperature denaturation (Timasheff, 1993). Steric exclusion is believed to be the principal mechanism of action of these compounds. We do not have a satisfactory explanation for the absence of any clear effect of PEG on chymotrypsinogen in contrast to the effects of the temperature of denaturation observed by Lee and Lee (1987). The absence of any effect on the temperature denaturation may also explain the absence of an effect on the denaturation by pressure. For β-lactoglobulin, the effect is clearer. In both cases, the temperature of denaturation is much lower than reported in the literature.

In contrast to the amount of data available on the effect of solvent composition on the denaturation and other protein processes, much less in known about the effect of cosolvents on other biopolymers. We know of one instance in which the effect of pressure has been studied on the gelatinization of starch Thevelein et al., 1981). It was found that 2 M n-propanol decreases the temperature of gelatinization. A pressure of 1 kbar is sufficient to partially counteract this effect.

Molecular Details: Pressure Effect on the Secondary Structures

When proteins are compressed in solution, two effects may occur: compression without any distortion of the bond angles, and compression with changes in conformation. This first (elastic) effect may be expected to occur at low pressures (Noguti & Go, 1989; Frauenfelder et al., 1990). Depending on the conditions, the second (plastic) effect may be reversible or irreversible. In this section we explore both effects as expressed as possible pressure-induced shifts of the amide I' band components of proteins.

Proteins in Solution in the Elastic Regime

When molecules are compressed in the liquid state, one observes an increase in the frequency of the stretching vibrations with increasing pressure. This is true for C–H bonds (Moon & Drickamer, 1974; Zakin & Herschbach, 1988). The situation is different in hydrogen-bonded systems. Here one observes a pressure-induced decrease of the O—H and N—H stretching frequency (Moon & Drickamer, 1974). This can be interpreted as a bond elongation in terms of a simple hydrogen bond model. The frequency is correlated with the oxygen-oxygen distance. The stronger the frequency shift, the shorter the hydrogen bond becomes. In proteins, we are interested in hydrogen bonds between C=O and amide groups. Various model systems show that increasing pressure reduces the frequency of the C=O bond. This has been studied in a number of crystalline compounds: N-methylacetamide (Moon & Drickamer, 1974), DCOOD (Shimizu & Ikuta, 1985) CH$_3$COOD (Shimizu, 1986), and formamide (Shimizu et al., 1988). The pressure dependence of the C=O and O—D (N—H) stretching frequencies are collected in Table 4.1.

Table 4.1. Pressure effect on the stretching frequencies C=O and O—D (N—H) (cm^{-1} kbar^{-1}) of model compounds for the hydrogen-bonded amide I bond

Compound	C=O	O—D (N—H)
D—COOD[a]	−1.7	−3.1
CH$_3$—COOD[b]	−0.9	(−1.8)[e]
H—COND$_2$[c]	−1	−2
CH$_3$—CONHCH$_3$[d]	(−0.5)[e]	−0.9

[a] Shimizu & Ikuta (1985).
[b] Shimizu (1986).
[c] Shimizu et al., (1988).
[d] Moon & Drickamer (1974).
[e] Figures in parentheses are estimated.

Table 4.2. Pressure-induced frequency shifts ($cm^{-1} kbar^{-1}$) in the secondary structure components of proteins

Protein[a]	α-helix		β-sheet		Unordered	
	v	$\delta v/dp$	v	dv/dp	v	dv/dp
Myoglobin sw (α)	1650	−0.29	1629	−0.20		
Lipoxygenase (α = β)	1648	−0.62	1629	−0.10		
Chymotrypsinogen (β/α)	1652	−0.24	1638	−0.46		
Ribonuclease (β/α)	1652	+0.30	1638	−0.39		
β-lactoglobulin (β/α)	1649	−0.27	1635	−1.7		
Neuropeptide (β/U) ox			1629	−0.45	1646	+0.37
Neuropeptide (β/U) human			1628	−0.38	1646	+0.42

[a] Main structural components in parentheses.

Additional arguments in favor of the strengthening of the hydrogen bonds come from solution studies. We have observed that there is a substantial difference in the effect of pressure on the low-frequency Raman spectrum of liquid amides that form intermolecular hydrogen bonds compared to those that do not form hydrogen bonds (Smeller et al., 1993; Goossens et al., 1993).

On the basis of the model systems, we may thus expect that the elastic deformations observed in proteins via the C=O peptide vibration would be visible as a negative pressure shift. For the proteins we have studied and reported here, this statement seems to be valid, as may be judged from Table 4.2, except for the unordered component of the neuropeptides. We have evidence that the positive value for the α-helix in ribonuclease is indicative for a plastic effect. In the absence of any conformational change in a protein, the presence of weak and medium-strong hydrogen bonds would indicate a simple compression of the hydrogen bonding of the secondary structures. The differences that are found between the secondary structures and for the same secondary structure between proteins may reflect differences in the compressibility, H/D exchange, or conformational changes. For it may indicate the influence of local effects within the protein. This point needs further investigation.

Proteins in Solution in the Plastic Regime

In the plastic regime, we expect two possibilities: a change in conformation may be reversible or irreversible. We are aware of only one study of the effect of pressure on the reversible conformational change of a protein under the influence of pressure: the pressure-induced destabilization of the salt bridge in chymotrypsin-related protease, as studied by Raman spectroscopy (Heremans & Heremans, 1989). Unfortunately this study did not concentrate on the pressure-induced frequency changes since it used the method of factor analysis as first proposed by Williams (1986). As we show here, in principle this method is not applicable to pressure-induced conformational changes, and it should be worthwhile to repeat this study with infrared spectroscopy.

Absorption and fluorescence studies of pressure-induced denaturation in proteins have shown reversible changes in many cases (Hawley, 1971; Zipp & Kauzmann, 1973; Weber & Drickamer, 1983) whereas Raman spectroscopy studies for lysozyme (Heremans & Wong, 1985) and Fourier transform infrared studies for chymotrypsinogen (Wong & Heremans, 1988; Wong, 1991) show irreversible pressure effects. This may be due to the large concentrations used in the latter methods so that the intermolecular interactions become more pronounced. In the study of serum albumin, it was found that the protein reversibly aggregates at a pressure of 6 kbar (Suzuki et al., 1963). In a previous Raman spectroscopy study we found that, in contrast to the behavior of lysozyme and chymotrypsinogen, the pressure-induced changes in bovine serum albumin are reversible (Chen & Heremans, 1990).

On the basis of pressure-induced transformations in model polypeptides, it is clear that the different classes of secondary structure have different volumes in solution (Carrier et al., 1990). The advantage of the infrared technique is that it allows one to simultaneously follow both the transitions and the change in secondary structure. On the basis of the arguments presented in the previous section, we may expect a fairly uniform elastic effect on these structures, which is observed as a decrease of the amide I′ frequency with increasing pressure as shown in Table 4.1 for model compounds. Severe distortions in these structures will result in changes in either the slope or the sign of the dv/dp value. This is observed in the α-helix of ribonuclease and the unordered structure of the neuropeptides (see Table 4.2). The effect of these distortions on the frequency is unclear at the present time, but studies of the effect of pressure on model systems may prove to be fruitful in this respect. A detailed quantitative analysis of the pressure-induced changes that take place in proteins showing reversible changes is in progress in our laboratory. At present, only a qualitative analysis is possible.

An interesting result from our studies is the striking difference between the temperature- and pressure-induced phenomena as found in lipoxygenase and neuropeptide H3. The temperature denaturation is irreversible in both cases, but the extent of intermolecular interaction is quite different. The pressure denaturation of the neuropeptides is reversible, which is in sharp contrast to lipoxygenase.

The most important result from our study is the pressure-induced transformation that occurs in ribonuclease before the onset of the main transition (see Figure 4.4). We have observed this pretransition repeatedly. Since this transition is not seen in the tyrosine side chains and is accompanied by a change in the amide II′ band, which is absent in the main transition, we think this is good evidence for a pressure-induced intermediate state with the characteristics of a molten globule state. It should be noted that the pressure (5 kbar) at which the transformation occurs is much higher than the previously observed molten globule state which takes place below 2.5 kbar in the dissociation of the dimer of arc repressor (Silva et al., 1992). We also note that the induction of the molten globule state occurs at a higher pressure in the presence of glycerol. This is consistent with the idea that the partial unfolding of the protein is disfavored by the presence of the organic cosolvent (Timasheff, 1993). Recently, a temperature-induced premelting was also reported in ribonuclease T1 with infrared spectroscopy (Fabian et al., 1993).

CONCLUSIONS

While pressure-induced protein denaturation has been a topic of research for many years, it is clear that there is little agreement on the definition of the denatured state thus obtained. Care should be taken when making extrapolations from studies on model systems to proteins. Vibrational spectroscopy reveals that secondary structures such as α-helix and β-sheet may be discernible up to very high pressures, but it is also clear that in the denatured state these structures disappear (Wong & Heremans, 1988). The analysis of the amide I' band by deconvolution and band fitting may be used to study pressure-induced changes up to about 20 kbar. The observed denaturation pressures correspond to those observed with fluorescence techniques (Weber & Drickamer, 1983). The most remarkable observation is that for a few proteins the changes seem to be reversible; however, in most cases either a precipitate (for example, in lysozyme) or a gel-like structure (for instance, for chymotrypsinogen) is formed. Although infrared spectroscopy gives information mainly on the backbone of the protein, its information content is rich in terms of secondary structures. The method we present in this paper allows analysis not only of the elastic effect but also of the conformational (plastic) effect. Studies on a number of proteins that differ in size and composition of secondary structure are ongoing in our laboratory. The method is also applicable to membrane-bound proteins and peptides (Smeller et al., 1995).

Recently, the pressure-induced denaturation of lysozyme up to 5 kbar has been studied by Samarasinghe et al. (1992), and Royer et al. (1993), who looked at the pressure denaturation of several mutants of staphylococcal nuclease. NMR spectroscopy, up to 10 kbar, may provide a possible tool for obtaining more spatial information on amino acid side chains. Such studies may be technically feasible in the future. A comparison between experiments and theoretical models should also prove useful. A recent molecular dynamics study of a small peptide, bovine pancreatic trypsin inhibitor, in solution has been performed at 10 kbar (Kitchen et al., 1992). The most important result was the observation of large, pressure-induced changes in the structure of the hydration shell. Over the 100-ps duration of the simulation, there was no evidence for unfolding. This is confirmed by fluorescence experiments reported in the same paper. Our high-pressure FTIR studies on this protein do show that the protein undergoes changes that are only partially reversible (Goossens et al., 1996). Our results show that the infrared technique explores aspects of the behavior of proteins under extreme conditions that cannot be detected by other methods, and they demonstrate the sensitivity of FTIR in detecting minor, pressure-induced phenomena in proteins.

ACKNOWLEDGMENTS: We thank Dr. F. Bollengier, Dr. J. C. Cheftel, and Dr. J. Frank for protein samples. Our research is supported by the Research Fund of the Leuven University, the National Fund for Scientific Research (N.F.W.O), and the European Union (AIRI-CT92-0296).

REFERENCES

Balny, C., Hayashi, R., Heremans, K., & Masson, P. (eds.) (1992). *High Pressure and Biotechnology*, vol. 224. London, Colloque INSERM/John Libbey Eurotext.

Bollengier, F., Mahler, A., Andries, R., & Bourgain, R. (1993). Peptide mapping of mammalian brain protein H3 in subsets of tissues and ligand binding studies. *Arch. Int. Physiolog. Biochim. Biophys.* **101**, 63.

Boyington, J. C., Gafney, B. J., & Amzel, L. M. (1990). Crystallization and preliminary x-ray analysis of soybean lipoxygenase-1, a non-heme iron-contaning dioxygenase. *J. Biol. Chem.* **265**, 12771.

Brandts, J. F., Olivera, R. J., & Westort, C. (1970). Thermodynamics of protein denaturation: effect of pressure on the denaturation of ribonuclease A. *Biochemistry* **9**, 1038.

Bridgman, P. W. (1914). The coagulation of albumen by pressure. *J. Biol. Chem.* **19**, 511.

Carrier, D., Mantsch, H. H., & Wong, P. T. T. (1990). Pressure-induced reversible changes in secondary structure of poly(L-lysine): an FTIR spectroscopy study. *Biopolymers* **29**, 837.

Chen, G., & Heremans, K. (1990). Pressure induced changes in proteins studied in the diamond anvil cell with Raman spectroscopy. *High Press, Res.* **5**, 749.

Cooper, A. (1976). Thermodynamic fluctuations in protein molecules. *Proc. Natl. Acad. Sci. USA* **73**, 2740.

Fabian, H., Schultz, C., Naumann, D., Landt, O., Hahn, U., & Saenger, W. (1993). Secondary structure and temperature-induced unfolding and refolding of ribonuclease T1 in aqueous solution. *J. Mol. Biol.* **232**, 967.

Frauenfelder, H., Alberding, N. A., Ansari, A., Braunstein, D., Cowen, B. R., Hong, M. K., Iben, I. E. T., Johnson, J. B., Luck, S., Marden, M. C., Mourant, J. R., Ormos, P., Reinisch, L., Scholl, R., Schulte, A., Shyamsunder, E., Sorensen, L. B., Steinbach, P. J., Xie, A., Young, R. D., & Yue, K. T. (1990). Proteins and pressure. *J. Phys. Chem.* **94**, 1024.

Gekko, K., & Timasheff, S. N. (1981). Mechanism of protein stabilization by glycerol: preferential hydration in glycerol-mixtures. *Biochemistry* **20**, 4667.

Goossens, K., Smeller, L., & Heremans, K. (1993). Pressure-tuning spectroscopy of the low-frequency Raman spectrum of liquid amides. *J. Chem. Phys.* **99**, 5736.

Goossens, K., Smeller, L., Frank, J., & Heremans, K. (1996). Pressure-tuning spectroscopy of Bovine pancreatic trypsin inhibitor: an FTIR study. *Eur. J. Biochem.* **236**, 254–262.

Gross, P., & Ludwig, H. (1992). Pressure-temperature phase diagram for the stability of bacteriophage T4. In *High Pressure and Biotechnology*, ed. C. Balny et al., London, Colloque INSERM/John Libbey Eurotext, **224**, 151.

Hawley, S. A. (1971). Reversible pressure-temperature denaturation of chymotrypsinogen. *Biochemistry* **10**, 2436.

Heremans, L., & Heremans, K. (1989). Raman spectroscopic study of the changes in secondary structure of chymotrypsin: effect of pH and pressure on the salt bridge. *Biochem. Biophys. Acta* **999**, 192.

Heremans, K., & Wong, P. T. T. (1985). Pressure effects on the Raman spectrum of proteins: pressure-induced changes in the conformation of lysozyme in aqueous solutions. *Chem. Phys. Lett.* **118**, 101.

Illian, G., Kneppe, H., & Schneider, F. (1988). Phase diagrams with closed loop smectic-nematic phase boundaries. *Ber. Bunsenges. Phys. Chem.* **92**, 776.

Ismail, A. A., Mantsch, H. H., & Wong, P. T. T. (1992). Aggregation of chymotrypsinogen: portrait by infrared spectrosopy. *Biochem. Biophys. Acta* **1121**, 183.

Kharakoz, D. P., & Saravazyan, A. P. (1993). Hydrational and intrinsic compressibility of globular proteins. *Biopolymers* **33**, 11.

Kitchen, D. B., Reed, L. H., & Levy, R. M. (1992). Molecular dynamics simulation of solvated protein at high pressure. *Biochemistry* **31**, 10083.

Klug, D. D., & Whalley, E. (1979). Elliptic phase boundaries between smectic and nematic phases. *J. Chem. Phys.* **71**, 1874.

Kornblatt, J. A., Kornblatt, M. J., Hui Bon Hoa, G., & Mauk, A. G. (1993a). Responses of two protein-protein complexes to solvent stress: Does water play a role at the interfaces. *Biophys. J.* **65**, 1059.

Kornblatt, M. J., Kornblatt, J. A., & Hui Bon Hoa, G. (1993b). The role of water in the dissociation of enolase, a dimeric enzyme. *Arch. Biochem. Biophys.* **306**, 495.

Lee, J. C., & Lee, L. L. Y. (1987). Thermal stability of proteins in the presence of poly(ethylene glycole). *Biochemistry* **26**, 7813.

Moon, S. H., & Drickamer, H. G. (1974). Effect of pressure on hydrogen bonds in organic solids. *J. Chem. Phys.* **61**, 48.

Muhr, A. H., & Blanshard, J. M. V. (1982). Effect of hydrostatic pressure on starch gelatinisation. *Carbohydr. Polym.* **2**, 61.

Muhr, A. H., Wetton, R. E., & Blanshard, J. M. V. (1982). Effect of hydrostatic pressure on starch gelatinisation, as determined by DTA. *Carbohydr. Polym.* **2**, 91.

Murphy, R. B. (1978). Anomalous stability of insulin at very high pressure. *Experientia* **34**, 188.

Nicolini, C., Erokhin, V., Antolini, F., Catasti, P., & Facci, P. (1993). Thermal stability of protein secondary structure in Langmuir-Blodgett films. *Biochim. Biophys. Acta* **1158**, 273.

Noguti, T., & Go, N. (1989). Structural basis of hierarchial substrates of a protein: I. Introduction, *Proteins. Structure, Funct. Genet.* **5**, 97.

Payens, T. A. J., & Heremans, K. (1969). Effect of pressure on the temperature-dependent association of α-casein. *Biopolymers* **8**, 335.

Rapoport, E. (1967). Model for melting-curve maxima at high pressure. *J. Chem. Phys.* **46**, 2891.

Robinson, C. R., & Sligar, S. G. (1994). Hydrostatic pressure reverses osmotic effects on the specificity of *Eco*-RI-DNA interactions. *Biochemistry* **33**, 3787.

Royer, C. A., Hinck, A. P., Loh, S. N., Prehoda, K. E., Peng, X., Jonas, J., & Markley, J. L. (1993). Effect of amino acid substitutions on the pressure denaturation of staphylococcal nuclease as monitored by fluorescence and nuclear magnetic resonance spectroscopy. *Biochemistry* **32**, 5222.

Samarasinghe, S., Campbell, D. M., Jonas, A., & Jonas, J. (1992). High resolution NMR study of the pressure induced unfolding in lysozyme. *Biochemistry* **31**, 7773.

Shimizu, H. (1986). High-pressure Raman study of the hydrogen bonded crystalline formic acid. *Physica* **139 & 140B**, 479.

Shimizu, H., & Ikuta, F. (1985). Pressure-induced phase transition in crystalline DCOOD by Raman scattering in the diamond cell. *J. Phys. Soc. Jpn.* **54**, 2812.

Shimizu, H., Nagata, K., & Sasaki, S. (1988). High-pressure Raman study of the hydrogen-bonded crystalline formamide. *J. Chem. Phys.* **89**, 2743.

Silva, J. L., & Weber, G. (1993). Pressure stability of proteins. *Annu. Rev. Phys. Chem.* **44**, 89.

Silva, J. L., Silveira, C. F., Correia, A., & Pontes, L. (1992). Dissociation of a native dimer to a molten globule monomer: effects of pressure and dilution on the association equilibrium of Arc repressor. *J. Mol. Biol.* **223**, 545.

Smeller, L., Goossens, K., & Heremans, K. (1993). A theoretical model that predicts the effect of pressure and hydrogen-bonding on the low frequency Raman spectrum of liquid amides. *J. Mol. Struct.* **298**, 155.

Smeller, L., Goossens, K., & Heremans, K. (1995). The determination of the secondary structure of proteins at high pressure. *Vibrational Spectrosc.* **8**, 199.

Sonoike, K., Setoyama, T., Kuma, Y., & Kobayashi, S. (1992). The effect of pressure and temperature on the death rates of Lactobacillis casei and Escherichia coli, In *High Pressure and Biotechnology*, ed. C. Balny et al., London, Collogue INSERM/John Libbey Eurotext, **224**, 297.

Spaapen, L. J. M., Veldink, G. A., Liefkens, T. J., Vliegenthart, J. F. G., & Kay, C. M. (1979). Circular dichroism of lipoxygenase-1 from soybeans. *Biochim. Biophys. Acta* **574**, 301.

Surewicz, W. K., Mantsch, H. H., & Chapman, D. (1993). Determination of protein secondary structure by Fourier transform infrared spectroscopy: a critical assessment. *Biochemistry* **32**, 389.

Susi, H., & Byler, D. M. (1986). Resolution enhanced Fourier transform infrared spectroscopy of enzymes. *Methods Enzymol.* **130**, 290.

Suzuki, K. (1960). Studies on the kinetics of protein denaturation under high pressure. *Rev. Phys. Chem. Jpn.* **29**, 91.

Suzuki, K., Miyosawa, Y., & Suzuki, C. (1963). Protein denaturation by high pressure: measurements of turbidity of isolectric ovalbumin and horse serum albumin under high pressure. *Arch. Biochem. Biophys.* **101**, 225.

Taniguchi, Y., & Takeda, N. (1992). Pressure-induced secondary structure of proteins studied by FTIR spectroscopy. In *High Pressure and Biotechnology*, ed. C. Balny et al. London Colloque INSERM/John Libbey Eurotext, **224**, 115.

Thevelein, J. M., Van Assche, J. A., Heremans, K., & Gerlsma, S. Y. (1981). Gelatinisation temperature of starch as influenced by high pressure. *Carbohydr. Res.* **93**, 304.

Timasheff, S. N. (1993). The control of protein stability and association by weak interactions with water: How do solvents affect these processes. *Annu. Rev. Biophys. Biomol. Struct.* **22**, 67.

Weber, G., & Drickamer, H. G. (1983). The effect of pressure upon proteins and other biomolecules. *Q. Rev. Biophys.* **16**, 89.

Williams, R. W. (1986). Proteins secondary structure analysis using Raman amide I and II spectra. *Meth. Enzymol.* **130**, 311.

Winter, R., & Pilgrim, W.-C. (1989). A SANS study of high pressure phase transitions in model biomembranes. *Ber. Bunseng. Phys. Chem.* **93**, 708.

Wong, P. T. T. (1991). Pressure effect on hydrogen isotope exchange kinetics in chymotrypsinogen investigated by FT-IR spectroscopy. *Can. J. Chem.* **69**, 1699.

Wong, P. T. T., & Heremans, K. (1988). Pressure effect on protein secondary structure and deuterium exchange in chymotrypsinogen: a Fourier transform infrared spectroscopic study. *Biochim. Biophys. Acta* **956**, 1.

Wong, P. T. T., Lacelle, S., & Yazdi, H. M. (1993). Normal and malignant human colonic tissues investigated by pressure-tuning FT-IR spectroscopy. *Applied Spectr.* **47**, 1830.

Zakin, M. R., & Herschbach, D. R., (1988). Density dependence of attractive forces for hydrogen stretching vibrations of molecules in compressed liquids. *J. Chem. Phys.* **89**, 2380.

Zipp, A., & Kauzmann, W. (1973). Pressure denaturation of metmyoglobin. *Biochemistry* **12**, 4217.

5

Temperature- and Pressure-Induced Unfolding of a Mutant of Staphylococcal Nuclease A

MAURICE R. EFTINK and GLEN D. RAMSAY

Nuclease conA is a hybrid version of staphylococcal nuclease that contains a six amino acid β-turn substitute from concanavalin A. This hybrid protein has a much lower thermodynamic stability than does the wild-type protein. This enables the unfolding of the protein to be achieved easily by several types of perturbations. From temperature-, pressure-, and denaturant-induced unfolding studies, we have found the free energy change for unfolding, ΔG_{un}°, to be approximately 1.4 kcal mol^{-1} at pH 7, 0.1 M NaCl, and 20 °C, as compared to a thermodynamic stability of approximately 5.5–6 kcal mol^{-1} for wild-type nuclease A. Due to its reduced thermodynamic stability, nuclease conA also shows evidence of unfolding at low-temperature (cold denaturation), with a temperature of maximum stability of 13–15 °C. The thermal unfolding of nuclease conA is shown to be two-state by simultaneous measurement of fluorescence and CD changes as a function of temperature, using a modified AVIV CD instrument. Increased hydrostatic pressure unfolds nuclease conA in what appears to be a two-state manner, with an apparent ΔV_{un} of approximately -100 ml mol^{-1}. From studies of the pressure (p)-induced unfolding of this hybrid protein as a function of temperature (T), we can define the complete p-T free energy surface for the unfolding transition.

In auxiliary studies, we have characterized the fluorescence intensity decay and anisotropy decay of the single tryptophan residue (Trp-140) of nuclease conA in the native state and in the unfolded state induced by temperature, pressure, and denaturant. For each type of perturbation, there is a red shift in fluorescence, a lowering of the mean fluorescence lifetime, and a lowering of the rotational correlation time of the tryptophan residue to a value of ~ 1 ns (compared to 10–15 ns for the native state).

The thermodynamics of the unfolding of proteins has received renewed interest in recent years, owing to the availability of a rich variety of mutant proteins and to advances in our understanding of their structural features. Among the questions being asked are, What are the relative energetic contributions of the hydrophobic effect and other interaction forces? What is the basis for the high degree of cooperativity in unfolding? To what extent is the unfolding of subdomains coupled to the unfolding of other subdomains? Can we achieve a satisfactory and self-consistent interpretation of the thermodynamic parameters (that is, ΔH, ΔS, ΔC_p, and ΔV) for protein unfolding? Are these interpretations similar for different folding motifs?

The pressure-induced unfolding of monomeric globular proteins has been studied infrequently, for technical reasons. In principle, such studies can provide information about the free energy change (ΔG_{un}°) and volume change (ΔV_{un}°) for unfolding, and they may also enable determination of higher order parameters, the compressibility change ($\Delta \beta$), and the thermal expansivity ($\Delta \alpha$, the latter from combined temperature and pressure dependence studies). These are fundamental thermodynamic parameters needed for a complete description of the unfolding process.

A problem with studying the pressure-induced unfolding of monomeric proteins is that very high pressures usually are needed to promote the transition, or the transition must be poised by working at acidic pH or by adding denaturant. In this article we use a thermodynamically unstable mutant form of the nuclease A from Staphylococcus aureus. Because this mutant is quite unstable, we show that it is possible to induce its complete unfolding within an experimentally accessible pressure range at neutral, nonperturbing solution conditions. We have also performed unfolding studies, as a function of both pressure and temperature, to characterize the stability of the mutant protein along both of these perturbant axes. From these results we will generate a three-dimensional free energy surface (ΔG_{un} versus T and P) for the stability of the protein.

METHODOLOGY

Frequency domain fluorescence intensity and anisotropy decay measurements were made on an instrument described elsewhere (Eftink et al., 1991a). For pressure studies, the sample was placed in a high-pressure cell made after the design of Paladini and Weber (1981). The hydrostatic pressure in the cell was controlled using a High Pressure Products pump, which had been modified by the addition of a computer-controlled electric motor. This motor was linked to the data acquisition computer via an RS232 interface. A program to control the pressure pump was incorporated into the ISS acquisition software (with help from Dr. Brett Fedderson, ISS Inc., Champaign, IL). The system includes a digital pressure sensor. The acquisition routine allows for a user-specified waiting time after each pressure adjustment for pressure equilibration (usually 5 minutes). At each pressure, the instrument measures the phase angle and fluorescence intensity of the sample. The exciting laser beam is modulated at 50 MHz in all of our single-frequency studies.

The change in fluorescence intensity, F (or CD signal, phase angle), with temperature (or pressure) was analyzed using the following equation

$$F = \sum X_i(F_{oi} + s_i T) \tag{1}$$

where X_i is the mole fraction of state i, F_{oi} is the fluorescence intensity of state i at the standard condition (0 °C, 1 atm), and s_i is the dependence of the fluorescence intensity of state i on the variable T (or p). For a two-state transition, $N \rightleftharpoons U$, i will be equal to 2, and the two mole fractions will be referred to as X_N and X_U, for the native (N) and unfolded (U) states. The equilibrium constant and free-energy change for a two-state transition will depend on T and P according to the following relationships:

$$\Delta G_{un}(T, p) = \Delta H_{o,\,un}(1 - T/T_G) + \Delta V_{o,\,un}(p - 1)$$
$$- \Delta C_p[T_G - T + T \cdot \ln(T/T_G)]$$
$$+ \Delta\beta \cdot (p - 1)^2 + \Delta\alpha \cdot (p - 1) \cdot (T - T_G) \tag{2}$$

Here T_G is the high-temperature unfolding temperature (where $\Delta G_{un} = 0$ at 1 atm), $\Delta H_{o,\,un}^{\circ}$ is the enthalpy change for the unfolding transition at the reference condition (that is, at $T = T_G$ and $p = 1$ atm), $\Delta V_{o,\,un}$ is the volume change, $\Delta C_p(= \delta\Delta H_{un}/\delta T)$ is the heat capacity change, $\Delta\beta(= \Delta V_{un}/\delta p)$ is the compressibility change, and $\Delta\alpha(= \delta\Delta V_{un}/\delta T = -\delta\Delta S_{un}/\delta p)$ is the thermal expansivity change for the $N \rightleftharpoons U$ transition. At a constant pressure of 1 atm, Eq. 2 simplifies to the familiar $\Delta G_{un} = \Delta H_{o,\,un} + \Delta C_p \cdot (T - T_G + T \cdot \ln(T/T_G)$. Likewise, at a constant temperature, the equation becomes $\Delta G_{un} = \Delta G_{o,\,un} + \Delta V_{un} \cdot (p - 1)$, where $\Delta G_{o,\,un}$ is the free energy for unfolding at 1 atm and where ΔV_{un} may itself be pressure and temperature dependent through the values of $\Delta\alpha$ and $\Delta\beta$. Note that pressure is actually expressed in units of cal/ml for the calculations (1 cal/ml = 41 atm).

For pressure-dependence studies, we have found the frequency domain phase angle to be a more reproducible signal than the fluorescence intensity. Consequently, we have used the phase angle (with modulation at 50 MHz) as the observed signal. As discussed elsewhere, the phase angle does not directly track the population of states, for a two-state transition, since the more dominantly fluorescing species will disproportionately contribute to the phase angle (Eftink, 1994). We have developed a fitting routine to take into account the difference in the fluorescence intensity (that is, differences in average lifetime) for the native and unfolded states. When data (such as in Figure 5.2) are analyzed in terms of the more correct model, the fitted ΔV_{un} values are nearly the same and the fitted ΔG_{un}° values are slightly smaller than the values using the assumption of linear tracking. In the results presented here, we have not performed the more complete analysis due to lack of time.

Some unfolding studies were performed with an AVIV 62DS circular dichroism spectrophotometer that has been modified to permit the simultaneous measurement of CD and steady-state fluorescence signals for a sample as a function of temperature (Ramsay & Eftink, 1994a,b). These data have been globally analyzed to test the two-state model for thermal unfolding. Protein samples were obtained as described elsewhere (Eftink et al., 1991a). Solutions were generally prepared in 0.01 M tris buffer, which included 0.1 M NaCl.

RESULTS

In studies with the hybrid mutant protein, nuclease conA (which contains a six amino acid β-turn substitute from concanavalin A at residues 27–31), we have shown that this protein (and a related S28G hybrid mutant) is much less thermodynamically stable than the wild-type nuclease A (Eftink et al., 1991a). Whereas the wild type has a stability of 5.5–6 kcal mol^{-1} (at 20 °C, pH 7), nuclease conA has a stability of only 1–1.5 kcal mol^{-1}. We have also shown that the fluorescence intensity of this protein increases slightly with increasing temperature between 2 and 10 °C in a manner that is suggestive of the existence of cold unfolding (Eftink et al., 1991a). To adequately fit such data for nuclease conA, it was necessary to include a ΔC_p term (as, in Eq. 2). A value of $\Delta C_p = 2.3$ kcal (mol$^{-1} \cdot$°C^{-1}) was required, which leads to the calculation of a temperature of maximum stability of ~ 13 °C and cold and high temperature melting temperatures of ~ -8 °C and ~ 32 °C at pH 7.

In previous time-resolved fluorescence studies, we determined that the native and mutant proteins have an average fluorescence lifetime of approximately 5–5.5 ns and a rotational correlation time, ϕ_1 (the dominant, long ϕ value), of approximately 10–12 ns at 20 °C (Eftink et al. 1991b). Urea, guanidine-HCl, and thermal unfolding of the proteins result in a decrease in the average fluorescence lifetime and rotational correlation time. For example, the urea unfolded state of the mutant has an average fluorescence decay time of 2.3 ns and a dominant rotational correlation time of 0.75 ns at 20 °C, and the thermal unfolded state has an average fluorescence decay time of 1.5 ns and a dominant rotational correlation time of less than 1 ns at 40 °C (Eftink et al., 1991a, 1991b; Eftink & Wasylewski, 1992).

Use of Multidimensional Spectrophotometer to Test Two-State Unfolding Model

A concern when studying such unstable proteins is that the proteins may no longer be unfolding in a two-state manner; instead, a multistate unfolding process may occur, with the formation of partially unfolded intermediates. One method for verifying whether this may be the case is by use of our multidimensional spectrophotometer, in which we simultaneously monitor the thermal unfolding by circular dichroism (which views the entire protein) and fluorescence (which views only changes in the environment of Trp-140 of this protein). By globally analyzing such multidimensional data, we can determine with increased confidence whether the unfolding is two-state or more complex.

Shown in Figure 5.1A are simultaneously collected CD and fluorescence data for the thermal unfolding of nuclease conA at pH 7. The fluorescence drops with unfolding; the CD signals at 222 and 235 nm both decrease in absolute magnitude with unfolding. A global fit of the data was first performed using a two-state model

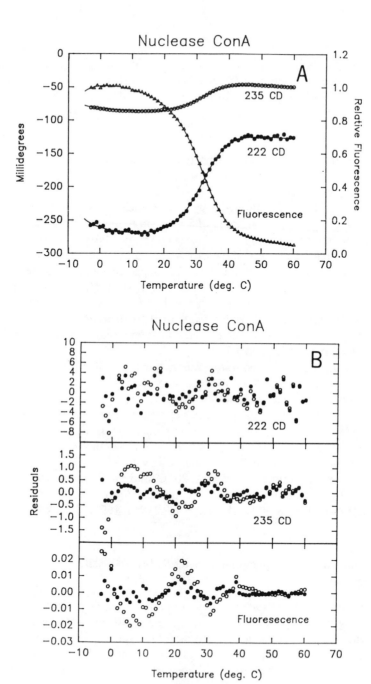

Figure 5.1. (**A**) Multidimensional spectroscopic study of the thermal unfolding of nuclease conA at pH 7. Simultaneous measurements were made of the fluorescence at 340 nm (ex at 290 nm) and the CD signals at 222 nm and 235 nm. A global fit is shown with the parameters given in Table 5.1. (**B**) Deviation plots for the above signals and fits. The open symbols are for a fit without a ΔC_p term and the closed symbols are a fit including at ΔC_p.

that does not contain a ΔC_p term. As can be seen in Figure 5.1B, the deviation patterns are nonrandom. The next most complicated model is to include a ΔC_p term. Shown as the solid lines in Figure 5.1A is such a fit with $\Delta C_p = 1.74 \pm .09 \text{ kcal} \cdot \text{mol}^{-1} \cdot \text{K}^{-1}$, $\Delta H^{\circ}_{o, \text{ un}} = 40.5 \pm 8 \text{ kcal mol}^{-1}$, and $T_G = 308.1 \pm 1.5 \text{ K}$. This fit describes the data quite well (notice the more random deviation patterns in Figure 5.1B), and thus we conclude that the thermal unfolding of nuclease conA remains a two-state process, with the caveat that a ΔC_p term be included.

Pressure-Induced Unfolding

Next, we studied the pressure-induced unfolding of nuclease conA. Shown in Figure 5.2 are fluorescence phase angles as a function of pressure at 1, 16, and 32 °C). The phase angle decreases with increasing pressure. The solid line through the data at 16 °C is a fit with $\Delta V_{\text{un}} = -107 \pm 6 \text{ ml/mol}$ and $\Delta G^{\circ}_{\text{un}} = 1.78 \pm 0.11 \text{ kcal mol}^{-1}$.

A couple of practical concerns when performing such pressure-induced unfolding reactions is whether the transition is reversible and whether equilibrium is achieved during the measurements. Reversibility is demonstrated by the observation of similar transitions with either increasing or decreasing pressure scans and by our ability to repeatedly make similar pressure scans with the same sample. While we have not performed a thorough study, equilibration appears to occur with a relaxation time of a few minutes near room temperature and approximately 10 to 15 minutes at 2 °C. We have included a pause of 5 to 20 minutes after each pressure step so that the total time required for the measurements ranged from 120 minutes to nearly 500 minutes (longer times used

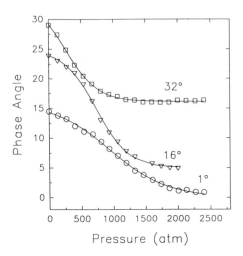

Figure 5.2. Phase angle data for the pressure-induced unfolding of nuclease conA at pH 7. Data at 1 °C (○), 16 °C (▽), and 32 °C (□). The phase angle data were obtained using a modulation frequency of 50 MHz. The raw phase angle data are offset for improved display. The fits are with Eq. 2 (where the $\Delta H^{\circ}_{o, \text{ un}}$ and $\Delta S^{\circ}_{o, \text{ un}}$ and ΔC_p terms are lumped together as a $\Delta G^{\circ}_{o, \text{ un}}$ and the $\Delta \alpha$ and $\Delta \beta$ are ignored to yield apparent values of ΔV_{un}).

at lower temperatures). While we are not certain that the curves are not skewed somewhat by this effect, our measurements at different scan rates do not show an obvious trend.

Pressure versus Temperature Unfolding Profiles

The pressure-induced unfolding was also studied as a function of temperature from -1 to $41\,°C$. Shown in Figures 5.2 and 5.3 are phase angle data and theoretical fits as a function of pressure for several temperatures. (The data in Figure 5.3 have been adjusted to have the same final baseline, which makes the individual curves hard to follow; the purpose of this figure is to illustrate the global analysis discussed next.) At the higher temperatures, the transition amplitude is reduced because a fraction of the protein is thermally unfolded at $p = 1$ atm. At the lower temperatures, the transition is less sharp (which is described by a lower magnitude for the $-\Delta V_{un}$). Figure 5.4 shows a plot of the resulting ΔG_{un}° and $p_{1/2}$ (pressure at which $K_{un} = 1$) determined from the pressure-induced studies at various temperatures. The ΔG_{un}° pattern shows a maximum between 5 and 20 °C and appears to drop at both lower and higher temperature, consistent with both cold and hot unfolding (that is, a positive value for the ΔC_p). Figures 5.5A and 5.5B show plots of the apparent ΔV_{un} versus the experimental temperature and versus the $p_{1/2}$. The ΔV_{un} appears to become more negative at higher temperature, indicating that there is a negative $\Delta \alpha$ $(= \delta \Delta V_{un}/\delta T)$. The ΔV_{un} appears to become more positive at higher pressure. Although there are significant error estimates for the ΔV_{un} values at high T and low $p_{1/2}$, the trends along both the temperature and pressure axes do not appear to be linear.

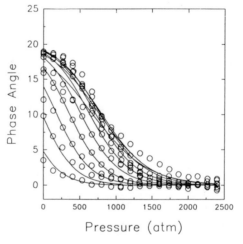

Figure 5.3. Temperature dependence of the pressure-induced unfolding of nuclease conA. The family of data curves are for temperatures of 1, 5, 8, 12, 16, 20, 24, 28, 32, 36, and 41 °C. The solid lines are for a global analysis of the 10 data sets with Eq. 2 with the parameters in Table 5.1. The global analysis assumed that there is a common change in phase angle $\Delta\theta$ for the unfolding transition and that this signal change is the same at each temperature.

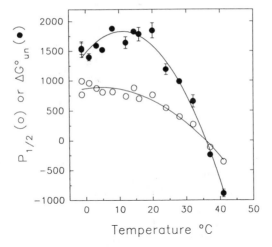

Figure 5.4. Plot of the apparent ΔG°_{un} (●) and $p_{1/2}$ (○) for the pressure-induced unfolding of nuclease conA as a function of temperature.

The pressure-induced unfolding data at the 11 temperatures were analyzed globally using Eq. 2 to obtain the lines shown in Figure 5.3. The fits were obtained with values of $\Delta C_p = 1.67$ kcal/(mol·K) and $\Delta\alpha = -1.42$ ml/(mol·K), along with $\Delta H^\circ_{un} = 38.3$ kcal mol^{-1}, $T_G = 308.9$ K, and $\Delta V^\circ_{un} = -107.0$ ml/mol. In this fit, the value of $\Delta\beta$ was fixed at zero. We also performed a global analysis in which we floated the value of $\Delta\beta$ and fixed $\Delta\alpha$ at zero, which gave a value of $\Delta\beta = 0.023$ ml^2/(cal·mol) and was not quite as good as when $\Delta\alpha$ was floated. Finally, we performed a global analysis in which both $\Delta\alpha$ and $\Delta\beta$ were floated; this fit did not lead to a lowering of the sum of squared residuals, and the corresponding value of ΔH°_{un} departed significantly from the value obtained in the independent study in Figure 5.1. Thus, we do not feel that the data in Figure 5.3 allow simultaneous determination of both $\Delta\alpha$ and $\Delta\beta$, and our fitted $\Delta\beta$ values are very near zero.

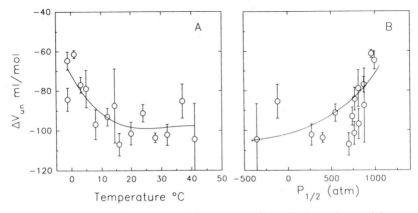

Figure 5.5. (**A**) Temperature dependence of the apparent ΔV_{un}. (**B**) Dependence of the apparent ΔV_{un} on the $p_{1/2}$.

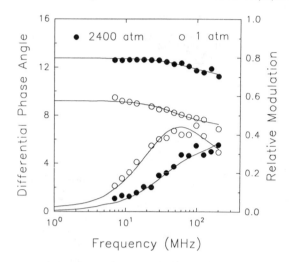

Figure 5.6. Differential polarized phase-modulation data (anisotropy decay) for nuclease conA at 1 atm (○) and 2400 atm (●).

Time-Resolved Fluorescence Studies of the Native and Pressure Unfolded State

To determine whether the pressure-induced transition actually leads to an unfolded state, we measured the fluorescence lifetime, anisotropy decay, and emission spectra for the protein at 1 atm and at 2400 atm. The fluorescence decays (data not shown) are non exponential at both low and high pressure; for example, at 24 °C, the intensity decay time has lifetime components of 5.9 ns and 0.7 ns at 1 atm and lifetime components of 3.8 ns and 0.44 ns at 2400 atm. Figure 5.6 shows anisotropy decay data at 16 °C for the sample at 1 atm and 2400 atm. At low pressure, the anisotropy decay is described as a biexponential with a long rotational correlation time, ϕ_1, of 12.5 ns and a short rotational correlation time, ϕ_2, of less than 1 ns. A fluorescence red shift is seen upon pressure unfolding; the emission maximum of Trp-140 is 335 nm at ambient pressure and 345 nm at 2400 atm.

DISCUSSION

The fact that nuclease conA is a relatively unstable protein makes it possible to observe the complete pressure-induced unfolding transition at neutral pH over a wide range of temperatures. In contrast, wild-type nuclease A does not unfold below 3000 atm at neutral pH. Hence, research at the high pressures required to unfold the wild-type protein is quite difficult. All indications are that hydrostatic pressure leads to the unfolding of this mutant of nuclease A in a two-state manner (Royer et al., 1993). Whereas the single tryptophan residue is largely immobilized in the native state, the high-pressure state has a mobile and solvent relaxed Trp-140. The fluorescence intensity and average fluorescence lifetime decrease with the application of pressure, providing signals that can be conveniently used to observe the transition.

The most obvious result of the pressure-induced unfolding studies with nuclease conA is that the ΔG°_{un} value reaches a maximum at approximately 1.5 kcal mol^{-1} between 5 and 20 °C. The shape of the ΔG°_{un} versus T data in Figure 5.4 indicates the presence of a positive heat capacity change, ΔC_p, for unfolding and the existence of both high temperature and low temperature unfolding (cold denaturation). The ΔC_p, $\Delta H^{\circ}_{o,un}$, and T_G needed to fit the pressure-unfolding data in Figure 5.3 are similar to the values determined independently in the temperature-induced unfolding studies in Figure 5.1 (see Table 5.1 for a comparison of the thermodynamic parameters obtained from the separate studies). The global fit of the pressure-unfolding data yields a value of -1.42 ml/(mol·K) for the difference in thermal expansivity, $\Delta\alpha$, between the unfolded and native states. The negative $\Delta\alpha$ for nuclease conA is of opposite sign to the previously reported value for the T- and p-induced unfolding of ribonuclease A and chymotrypsinogen (Brandts et al., 1970; Hawley, 1971). A negative $\Delta\alpha$ indicates that the thermal expansivity of the native state is larger than that for the unfolded state. An alternative fit of Eq. 2 to the data in Figure 5.3 (with $\Delta\alpha$ fixed at zero and $\Delta\beta$ floated) gives a near zero value for $\Delta\beta$, the difference in compressibility between the unfolded and native states. A negative $\Delta\beta$ of approximately 1 ml^2/(cal·mol) has been reported for the pressure-induced unfolding of ribonuclease and chymotrypsinogen (Brandts et al., 1970; Hawley, 1971). The molecular interpretation of such a negative $\Delta\beta$ is somewhat problematic. As pointed out by Kautzman (1987), studies of model systems lead one to expect a positive $\Delta\beta$ for a process involving the exposure of hydrophobic side chains. At this time, we believe that our data show the existence of a negative $\Delta\alpha$, but the value of $\Delta\beta$ appears to be so small that its sign is uncertain.

Finally, the fits of the temperature versus pressure-unfolding data enable us to construct the free energy surface shown in Figure 5.7, to describe the unfolding of this mutant protein along both perturbant axes. Figure 5.7A shows a mesh plot for the experimental values of ΔG_{un} as a function of p and T. Although it is a little difficult to visualize, the data are concave toward the T axis and slope downward at higher p. The latter slope (the ΔV_{un}) is smaller at low temperature. Figure 5.7B is a theoretical mesh plot using the thermodynamic parameters in Table 5.1.

Table 5.1. Thermodynamic parameters for the temperature- and pressure-induced/unfolding of nuclease conA.

	$\Delta G^{\circ}_{o,\,un}$	$\Delta H^{\circ}_{o,\,un}$	T_G	ΔC_p	ΔV°_{un}	$\Delta\beta$	$\Delta\alpha$
Spectroscopic studies[a]	1.33	40.5	30.8°	1.74			
Fluorescence studies[b]	1.29	38.3	35.9°	1.67	-1.07	—	-1.42

[a] Multidimensional spectroscopic studies at $p = 1$ atm.

[b] Fluorescence phase angle studies as a function of T and p.

ΔG°_{un} and ΔH°_{un} are in kcal mol^{-1}; ΔC_p is in kcal mol^{-1}·K^{-1}; ΔV°_{un} is in ml/mol; and $\Delta\alpha$ is in ml/mol·K. Values of $\Delta G^{\circ}_{o,\,un}$, $\Delta H^{\circ}_{o,\,un}$, and ΔV°_{un} are defined as the values at 273 K and 1 atm. T_G is the high melting temperature; the low-temperature transition temperature can be calculated using Eq. 2.

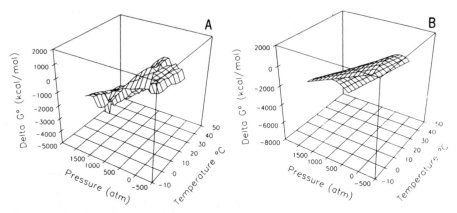

Figure 5.7. (**A**) Three-dimensional free-energy surface for the *p*- and *T*-induced unfolding of nuclease conA at pH 7 (experimental ΔG_{un} values). (**B**) Three-dimensional surface using the thermodynamic parameters in Table 5.1.

ACKNOWLEDGMENTS: This research was supported by NSF grant MCB 94-07167. We thank Drs. Robert Fox and Roger Kautz, Yale University, for supplying a sample of nuclease conA.

REFERENCES

Brandts, J. F., Oliveira, R. J., & Westort, C. (1970). Thermodynamics of protein denaturation. Effect of pressure on the denaturation of ribonuclease A. *Biochemistry* **9**, 1038.

Eftink, M. R. (1994). The use of fluorescence methods to monitor unfolding transitions in proteins. *Biophys. J.* **66**, 482.

Eftink, M. R., & Ramsay, G. D. (in press) Evidence for cold unfolding of an unstable mutant of staphylococcal nuclease. Unpublished manuscript.

Eftink, M. R., & Wasylewski, Z. (1992). Time-resolved fluorescence studies of the thermal and guanidine induced unfolding of nuclease A and its unstable mutants. *SPIE Symposium Proc.* **1640**, 579.

Eftink, M. R., Ghiron, C. A., Kautz, R. A., & Fox, R. O. (1991a). Fluorescence and conformational stability studies of *Staphylococcus* nuclease and its mutants, including the less stable nuclease-conA hybrids. *Biochemistry* **30**, 1193.

Eftink, M. R., Gryczynski, I., Wiczk, W., Laczko, G., & Lakowicz, J. R. (1991b). Effects of temperature on the fluorescence intensity and anisotropy decays of staphylococcal nuclease and the less stable nuclease-conA-S28G mutant. *Biochemistry* **30**, 8945.

Hawley, S. A. (1971). Reversible pressure-temperature denaturation of chymotrypsinogen. *Biochemistry* **10**, 2436.

Kauzmann, W. (1987). Thermodynamics of unfolding. *Nature* **325**, 763.

Paladini, A. A., & Weber, G. (1981). Absolute measurements of fluorescence polarization at high pressure. *Rev. Sci. Instrum.* **52**, 419.

Ramsay, G. D., & Eftink, M. R. (1994a). Analysis of multi-dimensional spectroscopic data that monitors the unfolding of proteins. *Meth. Enzymol.* **240**, 615.

Ramsay, G. D., & Eftink, M. R. (1994b). A multi-dimensional spectrophotometer for monitoring thermal unfolding transitions of macromolecules. *Biophys. J.* **66**, 516.

Royer, C. A., Hinck, A. P. Loh, S. N., Prehoda, K. E., Peng, X., Jonas, J., & Markley, J. L. (1993). Effect of amino acid substitutions on the pressure denaturation of staphylococcal nuclease A by fluorescence and nuclear magnetic resonance spectroscopy. *Biochemistry* **32**, 5222.

6

Pressure-Jump Relaxation Kinetics of Unfolding and Refolding Transitions of Staphylococcal Nuclease and Proline Isomerization Mutants

GEDIMINAS J. A. VIDUGIRIS, RAJ THOMAS, and CATHERINE A. ROYER

We present here the first report of the pressure dependence of pressure-jump relaxation kinetics for protein folding transitions. We have studied the relaxation kinetics for the unfolding/refolding of wild-type staphylococcal nuclease and have found that the relaxation kinetics observed at high pressure are much slower than those observed by pH or denaturant jumps at atmospheric pressure. This indicates that these processes have large, positive values for the activation volumes, most likely stemming from exclusion of solvent from a transition state that is less well packed than the native state. We examined the pressure-jump relaxation kinetics of three single-site mutations in nuclease that lead to alterations in the interactions between the two domains of the protein and changes in the equilibrium constant for isomerization of the lysine-116 to proline 117 peptide bond away from the cis form that predominates in the wild-type enzyme. At comparable pressures, the relaxation times for these mutants were significantly shorter than those observed for the wild type, indicating lower values of the activation volumes. We propose that these mutations cause a decrease in the cooperativity of the unfolding of the two domains, leading to a decrease in the degree of solvent exclusion at the rate-limiting step.

The mechanism by which a particular amino acid sequence determines the fold and stability of globular proteins remains one of the most interesting and important unresolved issues in biophysical chemistry. The approaches to increasing our understanding of this phenomenon typically have involved perturbation of the proteins by chemical means or by temperature extremes. The equilibrium or time-dependent responses to these perturbations are then monitored (using a spectroscopic signal, activity, or some other observable) to extract the energetic or kinetic aspects of the unfolding or refolding transitions. Another means of perturbing the system is to modify the protein itself, either chemically or by site-directed mutagenesis, and to assess the effects of modification on the equilibrium or kinetic folding or refolding profiles. This approach has generated a great deal of information about small globular proteins that denature reversibly. Certain general aspects of the process, such as hydrophobic collapse, proline isomerizations, secondary structure nucleation, and subunit assembly, have been elucidated. However, the basis for the specificity of the final folded structure, and the particular folding pathway(s) that are followed, remain elusive.

Clearly, to advance our understanding of protein folding, we must bring to bear every conceivable biophysical methodology, as each provides particular structural and energetic information. High pressure represents the other thermodynamic variable that can be used to perturb chemical systems. Its particular advantage is that it modifies only the volume, and not the heat or chemical content, of the system. As such, it is often considered to be a more straightforward perturbation technique from a thermodynamic point of view, although more complex experimentally. Moreover, the derivative of the Gibbs free energy change of unfolding with respect to changing pressure is the volume change of the reaction. Thus, high-pressure unfolding experiments yield the difference between the volume of the protein-solvent system when the protein is unfolded versus when it is folded.

High-pressure equilibrium unfolding and refolding studies were initiated on a number of protein systems in the 1970s. Among these was the seminal study of the high-pressure unfolding of metmyoglobin published by Zipp & Kauzmann (1973). Pressure effects on proteins have been reviewed by Heremans (1982), Weber & Drickamer (1983), Silva & Weber (1993), and Royer (1994). Typically, proteins such as lysozyme, myoglobin, and chymostrypsinogen do not exhibit appreciable unfolding below 4 kbar. Virtually every type of spectroscopy has been coupled to high-pressure studies of proteins, including absorption, fluorescence, NMR, FTIR, Raman, and, recently, circular dichroism (reviewed by Royer, 1995). These spectroscopic observables generally indicate that the high-pressure denatured state of proteins is quite similar in structure (or lack thereof) to the denatured states of proteins observed at high temperature, low pH, or high concentrations of chemical denaturants such as urea or guanidine hydrochloride. Thus, for all practical purposes, it would appear that the same type of transition in the structure of single-chain globular proteins occurs upon application of pressure as under other denaturing conditions. Thus, comparison of the thermodynamic parameters derived from high-pressure experiments can be compared with those obtained by other means.

Since the application of pressure eventually leads to protein denaturation in all systems investigated to date, the volume change associated with the unfolding equilibrium is necessarily negative. That is to say, the protein-solvent system occupies a smaller volume when the protein is in the unfolded state than it does when it is in its native, folded conformation. These negative volume changes for unfolding are sufficiently large that many proteins become denatured below 10 kbar. These values, typically near 50–100 ml mol, only represent approximately 0.5–2% of the hydrated molecular volume of the protein. The basis for these negative volume changes is not well understood. From strict calculations of hydrophobic hydration from data on specific volumes of transfer of liquid hydrocarbons, one would predict much larger negative values for the volume change of protein denaturation. However, unlike liquid hydrocarbons, proteins in their native conformations exhibit very small compressibilities, such that the application of pressure results not in the compression of the native globular form of the protein but, rather, in its unfolding, so as to increase the number of interactions between the protein molecule and the solvent. The conventional view is that these protein—water bonds are, on the whole, of shorter length than the sum of the lengths of the protein—protein and water—water bonds that they replace.[1] Moreover, it is thought that the elimination of free volume due to imperfections in the packing of the native structure upon unfolding and the exposure of buried protein surface to the solvent also contribute to the decrease in system volume observed upon unfolding. Electrostriction of buried ion pairs in the native protein upon their disruption and exposure to solvent also probably plays a role, at least for some protein systems. Thus, while it is most surely true that the negative volume charge for the unfolding of proteins is linked to an increased interaction of the protein with the solvent, the relative contributions of the various microscopic mechanisms (hydrophobic hydration, free volume, and electrostriction) have not been assessed and may differ among various protein molecules.

Staphylococcal nuclease has proven to be an excellent model system for high-pressure protein denaturation studies because its relatively low stability allows for full, reversible denaturation below 3 kbar of pressure, thus greatly simplifying the technical aspects of the experiment. Moreover, three-dimensional structures of the H124 and L124 wild-type proteins and a number of site-specific mutants are available (Hynes & Fox, 1991; Hodel et al., 1993; D. M. Trucksness & J. L. Markley, personal communication). A large number of unfolding studies, both equilibrium and kinetic, using chemical or heat perturbations, have been reported for this protein and its mutants (Shortle, 1986; Shortle et al., 1988; Shortle et al., 1990; Eftink et al., 1991a,b; Chen et al., 1992a, 1992b; Nakano et al., 1993). From a fluorescence spectroscopic point of view it is ideal, as it contains only one tryptophan residue, which has been characterized in great detail by a number of investigators (Brochon et al., 1974; Munro et al., 1979; Lakowicz et al., 1986; Eftink et al., 1991a,b; James et al., 1992; Royer et al., 1993). A ribbon diagram of nuclease is shown in Figure 6.1. The tryptophan residue, number 140, is the last residue resolved in the structure obtained from crystallographic studies (Hynes & Fox, 1991). Nonetheless, it has been demonstrated by ourselves and others (Brochon et al., 1974; Munro et al., 1979; Lackowicz et al., 1986; Eftink et al., 1991a; James et al., 1992; Royer et al., 1993) that the rotational mobility of this residue is very small. Its fluorescence emission spectrum is rather red, in comparison to trypto-phans that are buried in the protein structure, and thus it appears to be in a solvent-exposed, yet rigid, environment. Upon unfolding by virtually any perturba-tion method, including high pressure (Eftink et al., 1991b; Royer et al., 1993), the emission intensity of the tryptophan residue decreases as a result of both static

---

1. However, see Chapter 3 (Prehoda & Markley) in this volume for an alternate interpretation.

and dynamic quenching effects (Shortle, 1986; Eftink et al., 1991a; Royer et al., 1993; J. M. Beechem, personal communication). The spectrum also shifts to the red upon denaturation as a consequence of becoming more exposed to solvent, and one observes a large decrease in the fluorescence anisotropy due to an increase in local rotational mobility of the tryptophan residue upon loss of structure. Thus, unfolding can be monitored using a number of fluorescence observables, which at equilibrium have been shown to parallel unfolding profiles obtained from changes in circular dichroism (Shortle, 1986).

In recent studies, our group, in collaboration with those of John Markley and Jiri Jonas, have assessed the high-pressure equilibrium unfolding properties of single-site mutants of nuclease exhibiting altered *cis-trans* isomerization characteristics for particular prolyl bonds (Royer et al., 1993). Three of the mutations involve perturbations of the interactions between two loops connecting the α-helical with the β-sheet domains of the protein. At the top of the 115-119 loop in the α-helical domain is a proline residue (P117). In the wild-type protein,[2] the peptide bond between lysine-116 and proline-117 is in the *cis* conformation (Fox et al., 1986; Evans et al., 1987; Alexandrescu et al., 1989; Hinck et al., 1990), which normally is less stable than the *trans* conformation for proline residues in native proteins. Changing aspartate-77 to alanine (D77A) in the loop of the β-sheet domain results in the abrogation of a salt bridge between aspartate-77 and a peptide amine group in the 115-119 loop in the α-helical domain (see Figure 6.1). Replacing glycine at position 79 with a bulkier serine residue also disrupts the interactions between the two loops, as does replacing phenylalanine-76 by valine, presumably as a consequence of alterations in the packing of the hydrophobic core of the loop from the β-sheet domain. Disruption of the interactions between these loops loosens the structure and allows for the prolyl peptide bond at position 116-117 to shift toward the *trans* state, 50% for H124L + G79S and 100% for H124L + D77A and H124L + F76V (Alexandrescu et al., 1989, 1990). All of these mutants are less stable than the H124L wild-type protein. However, because the H124L change results in a stabilization, the H124L + G79S and H124L + F76V mutants exhibit approximately the same stability as the H124 wild-type protein, while the H124L + D77A mutant is less stable than the H124 wild type. The stabilities of these mutant proteins obtained from the fits of the high-pressure profiles correlated with their stabilities as assessed by alternative perturbations (that is, high temperature and chemical denaturants). The volume changes obtained from the high-pressure denaturation data (Royer et al., 1993) of the H124L + D77A and H124L + G79S mutants were similar to that of the H124 wild-type protein (near −85 ml mol for unfolding), whereas the volume change observed for the H124L + F76V mutant was significantly smaller, near −50 ml mol. This difference was interpreted as resulting from a change in the packing of the hydrophobic core. None of these proteins exhibited a dependence of the volume change on pH between

---

2. All mutants were in the background of the wild-type protein produced by the Foggi strain of *Staphylococcal aureus* (denoted H124L) which has a lysine at position 124 instead of histidine. Since the L124 wild type from Foggi is more stable than the H124 wild type produced by the V8 strain of *Staphylococcal aureus*, we could not observe denaturation in our apparatus. Thus, all pressure experiments on wild type were done on the H124 wild type, not the L124 wild type.

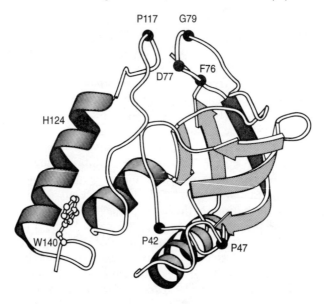

Figure 6.1. Ribbon diagram of staphylococcal nuclease taken from the structure determined by Hynes & Fox (1991). Sites of relevant mutations are as noted. Three of the proline residues known to contribute to native-state heterogeneity through their prolyl peptide bonds are shown. Note in particular proline 117. Also noted is tryptophan 140, which provides the observable signal in our experiments. (The MOLSCRIPT program [Kraulis, 1992] was used to generate the figure.)

pH 7 and 3.5, indicating that electrostriction contributes little to the volume change of unfolding for this particular set of proteins. In these previous studies, we also assessed the pressure effects on a series of mutants in which proline residues at positions 117, 47, and 42 were substituted by glycine. All of these mutants were more stable to pressure denaturation than the (H124) wild-type protein, and thus pressure-induced denaturation was only observed at pH 3.5 and below. Under these conditions, the apparent volume change for unfolding was much smaller in absolute value (10–20 ml mol) than that observed for the H124 wild-type and the loop mutants. In carrying out the experiments on the equilibrium unfolding properties of these mutants, we noticed that the equilibration time was unusually long. This prompted us to characterize the time dependence of the unfolding after a pressure jump to either higher (unfolding) or lower (refolding) pressure for the H124 wild-type and the less stable loop mutants in the H124L background.

RESULTS

The pressure-jump relaxation kinetic profiles for the unfolding and refolding of H124 wild-type nuclease at pH 5.5 can be seen in Figures 6.2a–b. Upon rapid increase of pressure, the tryptophan intensity decreases, indicating that the folding equilibrium has been perturbed toward the unfolded state. Upon rapid decrease of pressure, the opposite occurs, and the transition is 100% reversible, as indicated

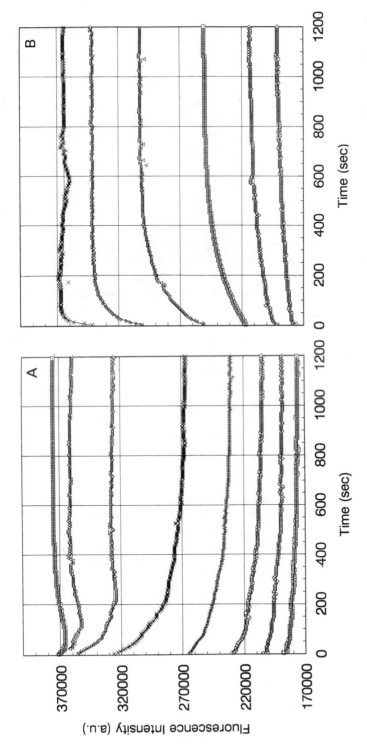

Figure 6.2. (**A**) Positive pressure-jump and (**B**) negative pressure-jump intrinsic tryptophan fluorescence intensity profiles as a function of time for staphylococcal nuclease H124 wild type. In (A), final pressures are 0.65, 1.02, 1.33, 1.70, 1.96, 2.17, 2.38, and 2.57 kbar from the top of the figure to the bottom; and in (B), final pressures were 0.50, 1.05, 1.39, 1.72, 2.01, and 2.26 from the top of the figure to the bottom. Data were acquired with a 5-second time base at 21 °C in 10 mM bis-Tris, pH 5.5.

by the values of the intensity upon return to atmospheric pressure. What is immediately apparent from these pressure-jump profiles (Figure 6.2) is that the time scale of the unfolding is much longer (near 10 minutes) than the time scale of unfolding by pH or denaturant jumps (about 1 second) (Chen et al., 1992a, 1992b; Nakano et al., 1993). Careful examination of the data in Figures 6.2a–b reveals that there is no burst phase; thus, the entire unfolding or refolding profile for each pressure jump is observed on this long time scale.

Another immediately apparent observation is that the average relaxation times appear to become longer at higher pressures for both the positive and negative pressure jumps. This indicates that either the forward or backward rate constant (or both) is decreasing as a function of pressure. At higher pressures, a single exponent satisfactorily describes the relaxation. Because our data acquisition and analysis are ongoing, in this preliminary report we will not present values for the relaxation times. However, we point out that simple inspection of the plots indicates that in the limit where refolding predominates (at pressures below the transition midpoint) in either the positive or negative pressure jumps, the relaxation times are much more sensitive to pressure than in the high-pressure limit where the unfolding rate predominates. This would indicate that a larger positive activation volume is associated with the refolding transition than with the unfolding transition, and that pressure has only a small effect on the rate of unfolding.

In the positive pressure jumps at lower pressures the transition is multiphasic, exhibiting a decrease in fluorescence intensity at short times (< 200 seconds), followed by a regaining of intensity at longer times. In the negative pressure-jump profiles at lower pressures, this complex behavior is less obvious. A double exponential decay model was required to fit the relaxation profiles for positive pressure jumps at low pressures. Thus, it would appear that an intermediate is observed, implying at least two steps in the unfolding model at these low pressures. Moreover, the biphasic nature of the profiles in this low-pressure range indicates that the quantum yield of the intermediate is larger than that of the wild-type or the denatured state.

In Figures 6.3a–d are plotted positive pressure-jump relaxation profiles near the midpoint pressure for each unfolding transition for the H124 wild-type and the H124L + D77A, H124L + F76V, and H124L + G79S mutants. Since the stabilities of the H124L + G79S and H124L + F76V mutants were nearly identical to that of the H124 wild-type, the pressures corresponding to the midpoint of the transition were also quite similar, near 1.25 kbar. The H124L + D77A mutant was much less stable than the others, and its midpoint relaxation profile, therefore, was obtained at 700 bar. What is immediately obvious is that the relaxation times for these mutants are much shorter than that of the wild type. The H124L + D77A mutant exhibits the fastest relaxation profiles, while the H124L + G79S and H124L + F76V profiles are intermediate between those of the H124L + D77A mutant and the wild type.

DISCUSSION

The data in the preceding section represent, to the best of our knowledge, the first studies of the pressure dependence of pressure-jump relaxation kinetics of protein unfolding and refolding transitions. Pryse et al. (1992) recently reported pressure-jump studies on the unfolding of cytochrome c at acid pH; however, these studies did not include the pressure dependence of the relaxation to equilibrium and thus provide no information on the activation volumes associated with unfolding/refolding transitions. The most startling result of the present studies is that at high pressure, the pressure-jump relaxations occur on very long time scales (10 minutes), compared to those observed using other perturbation techniques at atmospheric pressure (1 second). In our laboratory, the unpublished results of Teresa Fernando on the high-pressure unfolding of *trp* repressor also revealed extremely long (up to 2 hours) time scales for relaxation, whereas those observed at atmospheric pressure are on the order of 1 second, at the most. We have not examined enough different proteins to reach a general conclusion, but it would appear from this limited data set that either the unfolding or the refolding transition step involves a large positive activation volume.

Although a full explanation of the complex profiles at low pressures awaits more experimentation and analysis, it is clear from these preliminary results that the general effect of pressure is to slow the relaxation process significantly. In a simple protein-folding equilibrium, at atmospheric pressure, the refolding rate is much larger than the unfolding rate; hence, the protein is stable in its native form. Upon application of pressure, the protein unfolds, so either one or both of these rates must change. Assuming that only the refolding rate is decreased by pressure, then as one proceeds through the pressure-unfolding transition, the apparent relaxation time would increase as it tends toward the invariant slower unfolding rate. Alternatively, pressure could affect primarily the unfolding rate; however, in this case, one would observe a decrease in relaxation time as pressure is increased, whereas we observe the opposite. Finally, pressure could affect both rates; in this case, the refolding rate would have to decrease substantially more with pressure than that of unfolding, implying a significantly larger activation volume for the refolding transition than for the unfolding transition. Our observation is that relaxation times in both the positive and negative pressure-jump profiles increase with increasing pressure and exhibit a much larger pressure dependence at pressures below the midpoint, where the refolding rate predominates. Thus, it would appear that the bulk of the effect of pressure on the protein-folding equilibrium resides in a slowing of the rate of refolding. Such a result implies a large activation volume for the refolding process. Nonetheless, even at higher pressures the relaxation times continue to increase, indicating a small but significant activation volume for the unfolding process as well.

The differences in the density of protein bound and bulk water, as discussed in the opening text, give rise to the positive volume change for protein-folding equilibrium. This means that the protein-solvent system occupies a larger volume when the protein is in its native, folded conformation than when it is unfolded. By analogy, we ascribe the basis for the positive activation volume to the increase in volume associated with the exclusion of water from the protein surface upon collapse of the chain. In Chapter 3 of this volume, Prehoda and Markley provide an alternative view of the basis for the volume changes in protein folding. According to these authors, hydrophobic hydration makes a positive contribution to the equilibrium volume change of unfolding that is compensated for by a large decrease in the specific volume upon unfolding. In either case, in the transition state the chain would have collapsed to exclude solvent, yet it would not have achieved the high packing density of the native state—that is, it would present

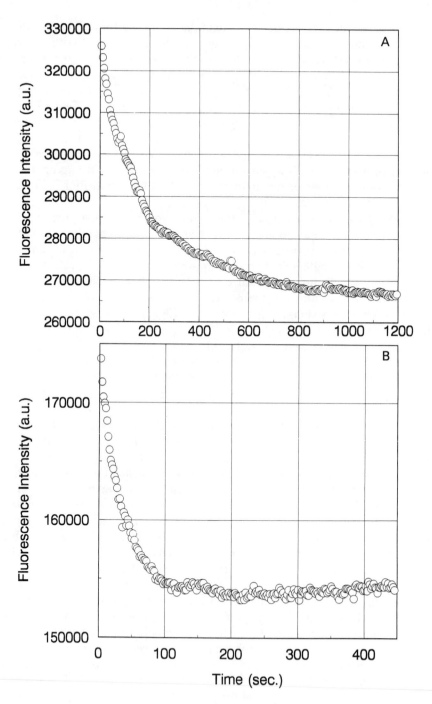

Figure 6.3. Pressure-jump unfolding fluorescence intensity profiles as a function of time obtained at the pressure for 50% dissociation for (**A**) H124 wild type; (**B**) H124L + G79S.

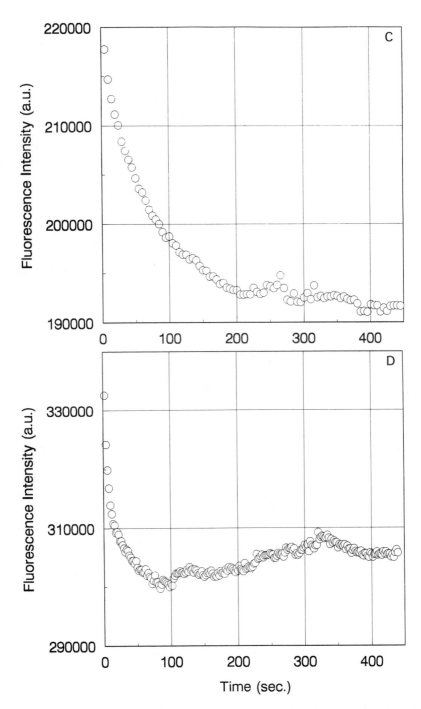

Figure 6.3. (**C**) H124L + F76V; and (**D**) H124L + D77A variants of staphylococcal nuclease. Data were obtained as in Figure 6.2. Note the difference in time scale for the wild type as compared with the three mutants.

many more small packing defects. These attributes are reminiscent of the so-called molten globule intermediate in protein folding. Such intermediates are also referred to as compact denatured states because reasonable amounts of secondary structure have formed and hydrophobic collapse (and thus exclusion of solvent) has occurred. However, these intermediates lack the specific tertiary structural inter-actions of the native form. Recent volumetric measurements of RNase A folding indicate that the refolding process involves an intermediate state of larger volume than the denatured state (Ybe & Kahn, 1994).

The data obtained for the particular mutant proteins we have examined remain to be fully analyzed and interpreted. Nonetheless, the large changes observed in the mutant relaxation profiles lead us to believe that the activation volumes for these mutants will be much lower than those observed with the wild-type protein. The mutations results in uncoupling of the two domains of nuclease by abrogation of critical interactions between the α-helical domain containing the reporter tryptophan residue and the largely β-sheet domain. The faster pressure-jump relaxations and their smaller pressure dependence observed for these mutants may indicate that the rate-determining step is the uncoupled folding of the α-helical domain alone. Because this domain is much smaller than the whole protein, it should exclude less solvent upon folding, and the activation volume associated with its transition state could be significantly smaller. A more detailed discussion of the studies of these mutants will be presented elsewhere.

Only at the very lowest pressures were the refolding relaxations complex. Thus, whatever the nature of the intermediate in the pathway, this intermediate phase does not contribute signifcantly to either folding or unfolding at higher pressures, where the bulk of the equilibrium unfolding occurs. Complex refolding kinetics for denaturant and pH jumps have been observed for nuclease at atmos-pheric pressure (Chen et al., 1992a, 1992b; Nakano et al., 1993). However, the large amplitude relaxation times obtained from those data were faster than 1 second. The longest relaxation time observed in those studies (near 1 to 10 minutes), depending on temperature and final conditions, exhibited an amplitude of approximately 12% of the signal change. By proline substitution mutations, this slow folding phase was convincingly demonstrated to correspond to a prolyl peptide bond isomerization. If the complex behavior observed at low pressures in the present work were due to prolyl peptide bond isomerization, then one would expect, as observed by Nakano et al. (1993), that replacement of the proline residue by glycine would result in elimination of the intermediate step. We are currently investigating the pressure-jump relaxation kinetics for a series of mutants in which the proline residues at positions 117, 42, and 47 have been replaced singly or in tandem by glycine.

ACKNOWLEDGMENTS: The authors thank Professors John L. Markley and Dexter Northrup for helpful discussions and past and present members of the Markley group for supplying the proteins used in this study. In particular, we are grateful to Dagmar M. Truckses, Kenneth E. Prehoda, Andrew P. Hinck, and Stuart N. Loh. We also acknowledge NSF MCB-9218461 to C.A.R. and NIH GM35976 to J.L.M.

REFERENCES

Alexandrescu, T. A., Ulrich, E. L., & Markley, J. L. (1989). Hydrogen-1 NMR evidence for three interconverting forms of staphylococcal nuclease: effects of mutations and solution conditions on their distribution. *Biochemistry* **28**, 204–211.

Alexandrescu, A. T., Hinck, A. P., & Markley, J. L. (1990). Coupling between local structure and global stability of a protein: mutants of staphylococcal nuclease. *Biochemistry* **29**, 4516–4525.

Bronchon, J.-C., Wahl, P., & Auchet, J.-C. (1974). Fluorescence time-resolved spectroscopy and fluorescence anistropy decay of the *Staphylococcus aureus* endonuclease. *Eur. J. Biochem.* **41**, 557–583.

Chen, H. M., Markin, V. S., & Tsong, T. Y. (1992a). pH-Induced folding/unfolding of staphylococcal nuclease: determination of kinetic parameters by the sequential-jump method. *Biochemistry* **31**, 1483–1491.

Chen, H. M., Markin, V. S., & Tsong, T. Y. (1992b). Kinetic evidence for microscopic states in protein folding. *Biochemistry* **31**, 12369–12375.

Eftink, M. R., Greyczynski, I., Wiczk, W., Lackzo, G., & Lakowicz, J. R. (1991a). Effects of temperature on the fluorescence intensity and anisotropy decays of staphylococcal nuclease and the less stable nuclease-con A-S28G mutant. *Biochemistry* **30**, 8945–8953.

Eftink, M., Ghiron, C., Kautz, R. A., & Fox, R. O. (1991b). Fluorescence and conformational studies of staphylococcal nuclease and its mutants, including the less stable nuclease-concanavalin A hybrids. *Biochemistry* **30**, 1193–1199.

Evans, P. A., Dobson, C. M., Fox, R. O., Hatfull, G., & Kautz, R. A. (1987). Proline isomerism in staphylococcal nuclease characterized by NMR and site directed mutagenesis. *Nature* **329**, 266–268.

Fox, R. O., Dobson, C. A., & Evans, P. A. (1986). Multiple conformations of a protein demonstrated by magnetization transfer NMR spectroscopy. *Nature* **320**, 192–194.

Heremans, K. (1982). High pressure effects on proteins and other biomolecules. *Annu. Rev. Biophys. Bioeng.* **11**, 1–21.

Hinck, A. P., Loh, S. N., Wang, J., & Markley, J. L. (1990). Histidine 121 of staphylococcal nuclease: correction of the $H^{\delta 2}$ ^1H NMR assignment and reinterpretation of the role this residue plays in conformational heterogeneity of the protein. *J. Am. Chem. Soc.* **112**, 9031–9034.

Hodel, A., Kautz, R. A., Jacobs, M. D., & Fox, R. O. (1993). Stress and strain in staphylococcal nuclease. *Protein Sci.* **2**, 838–850.

Hynes, T. R., & Fox, R. O. (1991). The crystal structure of staphylococcal nuclease refined at 1.7 Å resolution. *Proteins: Struct. Func. Genet.* **10**, 92–105.

James, E., Wu, P. G., Stites, W., & Brand, L. (1992). Compact denatured state of a staphylococcal nuclease mutant by guanidinium as determined by resonance energy transfer. *Biochemistry* **31**, 10217–10225.

Kraulis, P. J. (1991). MOLSCRIPT: a program to produce both detailed and schematic plots of protein structures. *J. Appl. Cryst.* **24**, 946–950.

Lakowicz, J. R., Laczko, G., Gryczynski, I., & Cherek, H. (1986). Measurement of subanosecond anisotropy decays of protein fluorescence using frequency domain fluorometry. *J. Biol. Chem.* **261**, 2240–2245.

Munro, I., Pecht, I., & Stryer, L. (1979). Subnanosecond motions of tryptophan residues in proteins. *Proc. Natl. Acad. Sci. USA* **76**, 56–60.

Nakano, T., Antonini, L. C., Fox, R. O., & Finck, A. L. (1993). Effect of proline mutations on the stability and kinetics of folding of staphylococcal nuclease. *Biochemistry* **32**, 2534–2541.

Pryse, K. M., Bruckman, T. G., Maxfield, B. W., & Elson, E. L. (1992). Kinetics and mechanism of folding of cytochrome *c*. *Biochemistry* **31**, 5127–5136.

Royer, C. A. (1995). The application of pressure to biochemical systems: What can we learn from the other thermodynamic variable? *Methods Enzymol.* **259**, 357–377.

Royer, C. A., Hinck, A. P., Loh, S. N., Prehoda, K. E., Peng, X., Jonas, J., & Markley, J. L. (1993). Effects of amino acid substitutions on the pressure denaturation of staphylococcal nuclease as monitored by fluorescence and nuclear magnetic resonance spectroscopy. *Biochemistry* **32**, 5222–5232.

Shortle, D. (1986). Guanidine hydrochloride denaturation studies of mutant forms of staphylococcal nuclease. *J. Cell. Biochem.* **30**, 281–289.

Shortle, D., Meeker, A. K., & Freire, E. (1988). Stability mutants of staphylococcal nuclease—large compensating enthalpy entropy changes for the reversible denaturation reaction. *Biochemistry* **27**, 4761–4768.

Shortle, D., Stites, W. E., & Meeker, A. K. (1990). Contributions of the large hydrophobic amino acids to the stability of staphylococcal nuclease. *Biochemistry* **29**, 8033–8041.

Silva, J., & Weber, G. (1993). Pressure stability of proteins. *Annu. Rev. Phys. Chem.* **44**, 89–113.

Weber, G., & Drickamer, H. G. (1983). The effect of high pressure upon proteins and other biomolecules. *Q. Rev. Biophys.* **16**, 89–112.

Ybe, J. A., & Kahn, P. C. (1994). Slow-folding kinetics of ribonuclease A by volume change and circular dichroism: evidence for two independent reactions. *Protein Sci.* **3**, 638–649.

Zipp, A., & Kauzmannm, W. (1973). Pressure denaturation of metmyoglobin. *Biochemistry* **12**, 4217–4228.

7

High-Pressure FTIR Studies of the Secondary Structure of Proteins

YOSHIHIRO TANIGUCHI and NAOHIRO TAKEDA

Infrared spectra of five globular proteins (bovine pancreas ribonuclease A, horse skeletal muscle myoglobin, bovine pancreas insulin, horse heart cytochrome c, egg white lysozyme) in 5% D_2O solutions (pD 7.0) were measured as a function of pressure up to 1470 MPa at 30 °C. According to the second-derivative spectral changes in the observed amide I band of the proteins, which indicate that the α-helix and β-sheet substructures of the secondary structures break dramatically into the random coil conformation, ribonuclease A and myoglobin are denatured reversibly at 850 MPa and 350 MPa, respectively. Lysozyme denatures partially and reversibly at 670 MPa, as shown by decrease in the α-helix and β-turn substructures, but no change occurs in the random coil and β-sheet substructures. The secondary structure of cytochrome c is not disrupted at pressures up to 1470 MPa, and partial transformation of the α-helix of insulin to random coil starts at 960 MPa. Hydrogen-deuterium exchange of protons on the amide groups in the protein interior is increased by external pressure and is associated with the pressure-induced protein conformational changes.

A number of studies on the effects of pressure on protein denaturation have been carried out using various high-pressure detection methods: ultraviolet absorbance spectroscopy (Brandts et al., 1970; Hawley, 1971), visible absorbance spectroscopy (Zipp & Kauzmann, 1973), fluorescence intensity spectroscopy (Li et al., 1976), polarization fluorescence spectroscopy (Chryssomallis et al., 1981), and enzyme activity assays (Taniguchi & Suzuki, 1983; Makimoto et al., 1989). These techniques have the great advantage of being applicable to pressure-induced reversible denaturation of proteins to identify the thermodynamic parameters, especially the volume change and compressibility of a protein in solution, because the experiments can be run under dilute conditions at a protein concentration of less than 0.05% w/v. Therefore, these data reflect the intramolecular phenomena of reversible pressure changes and provide the volume changes accompanying the denaturation of proteins, which are due to the difference in partial molal (specific) volume between the native and denatured proteins in solution. The isothermal compressibility related to the pressure dependence of the partial molal volume of a protein in solution is an important factor in understanding the structural stability and flexibility of a protein in a solution at high pressure (Taniguchi & Suzuki, 1983; Gekko & Noguchi, 1979; Gekko & Hasegawa, 1986). The mechanism of pressure denaturation can be explained by these two thermodynamic parameters (Taniguchi & Suzuki, 1983).

The thermodynamic data to not provide any direct information about the secondary structure, α-helix and β-sheet, of proteins. However, Raman and infrared spectroscopy are sensitive to conformational changes of both peptide backbone and side chains. Raman spectroscopic studies revealed that the secondary structures of trypsin and elastase (Heremans & Heremans, 1989) are stabilized up to 300 MPa and those of lysozyme (Remmele et al., 1990) up to 218 MPa, but at 550 MPa an irreversible denaturation and precipitation of lysozyme occurred in solution. High-pressure FTIR spectroscopy of protein solutions has been carried out at pressures up to 3000 MPa. An irreversible denaturation of chymotrypsinogen (Wong & Heremans, 1988) was found to be induced at 760 MPa, and this converted the α-helix and β-sheet substructures into a random coil. E. coli methionine repressor protein undergoes a rearrangement of α-helix segments into the β-sheet structure at 1800 MPa, and after the pressure has been decreased to 0.1 MPa, the β-strands reconvert into less ordered α-helix or random segments (Wong et al., 1989). Recent results from vibrational spectroscopic studies of proteins indicate that characteristic changes in secondary structure accompany irreversible pressure-denaturation reactions. However, these results do not provide an understanding of the mechanism of pressure denaturation from the viewpoint of protein molecular structure or the physical properties of proteins in solution. In the study reported here, we investigated the pressure-induced secondary structural changes in five globular proteins by means of high-pressure FTIR spectroscopy. We discuss the relationship between the molecular structures of these proteins and their physical properties in solution.

The strong amide I band in the infrared spectra of proteins has been widely used for determining secondary structure (Susi, 1969). This band occurs due to the in-plane $C≡O$ stretching vibration which is weakly coupled with the $C—N$ stretching and in-plane $N—H$ bending vibrations; it is located in the frequency band of 1600–1700 cm^{-1}. The maximum peak of the amide I band for various secondary substructures occurs at different frequencies: about 1655 cm^{-1} for α-helix, 1640 and 1688 cm^{-1} for β-sheet, and 1645 cm^{-1} for random coil. A globular protein molecule contains several segments with different conformational substructures, and the amide I band in the infrared spectrum of a globular protein usually appears as a broad band with several maxima. The broad amide I band is best analyzed by reference to the second derivative of the original spectrum. The change in the relative intensity of each of these maxima can be used to monitor the change in the conformational substructures of protein molecules. Infrared spectroscopy has the additional advantage of providing a way to monitor hydrogen/deuterium (H/D) exchange of the amide protons as a means to investigate changes in protein structure. It is generally accepted that the H/D exchange rate is extremely rapid for the labile protons of skeletal amide groups and side-chain amide groups located on the surface of proteins and freely exposed to bulk water. The H/D exchange rates are much slower for protons buried in the interior of proteins involved in internal hydrogen bonding. Rapid H/D exchange in chymotrypsinogen takes place over a period of 20 minutes at 0.1 MPa (Wong & Heremans, 1988). The number of rapidly exchanging protons increases with increasing pressure, and at higher pressures, rapid H/D exchange is completed within about 1 or 2 hours (Wong, 1990). Therefore, spectral measurements are taken at least 1 hour after the application of pressure to avoid the effects of rapid H/D exchange on the infrared spectrum.

Figure 7.1 shows second-derivative spectra containing the amide I band of RNase A in D_2O at various pressures, and Figure 7.2 shows the pressure dependence of the wavenumber of the amide I band of RNase A. H-containing amides give rise to amide I bands at 1638, 1657, 1678, and 1688 cm^{-1}, respectively, due to β-sheet, α-helix, β-turn, and β-sheet substructures. D-containing amides give rise to corresponding amide I bands at 1630, 1650, 1665, and 1680 cm^{-1}. The relative intensity of the 1638 (H) cm^{-1} component band is the strongest, which indicates that the β-sheet conformation of the amide groups predominates over other structures in RNase A. These infrared results are consistent with the generally accepted band assignments associated with the secondary structure of RNase A summarized in Table 7.1 (Olinger et al., 1986; Haris et al., 1986; Yamamoto & Tasumi, 1988, 1991). No significant changes in the infrared spectrum occur in the pressure range below 510 MPa. At 700 MPa, the pressure-induced H/D exchange is complete. Although the relative intensity of the deuterated amide I band for α-helix does not change, bands corresponding to the β-sheet conformation are shifted to lower frequencies. These changes are associated with an increase in the intensity of the deuterated amide I′ band at 1470 cm^{-1} at the expense of that of the band at 1550 cm^{-1}. At 780 MPa, the strong band at 1631 cm^{-1} and the weak bands at 1653 and 1680 cm^{-1} decrease dramatically due to the ordered structure in the α-helix and β-sheet (seen in the spectrum at 700 MPa of RNase A). All bands suddenly disappear in the spectrum at 1050 MPa due to the ordered structures. Surprisingly, as the pressure is increased beyond 1050 MPa, the sol state of the protein solution changes into a gel state, with no changes in the infrared spectrum observed at 1050 MPa. This sol-gel transformation is completely reversible. After the pressure has been released, the spectrum of the completely deuterated protein is different from the original spectrum of partially deuterated protein because the amide groups of RNase A have become completely deuterated at 1280 MPa. Consequently, the changes in the infrared spectrum strongly suggest that contributions from the random coil substructures increase in pressure-denatured RNase A at the expense of contributions from α-helix and β-sheet substructures. Appreciable changes were observed at high pressure in the infrared signals at 1610 cm^{-1} due to the tyrosine amino acid residues of RNase A. The band at 1610 cm^{-1} splits into two bands at 1609 and 1615 cm^{-1} as pressure is increased from 510 to 780 MPa (Figure 7.2). It is assumed that these changes in the infrared spectrum correspond to those detected in high-pressure ultraviolet spectra of RNase A (Brandts et al., 1970).

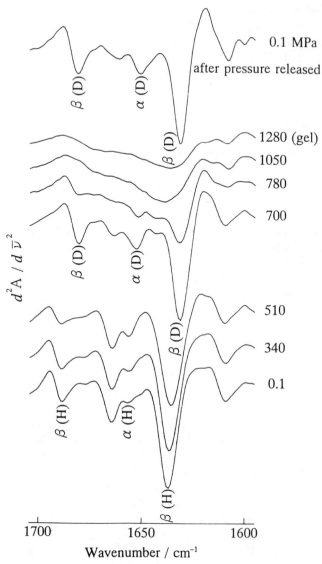

Figure 7.1. Second-derivative IR spectra of the amide I′ band of RNase A in D_2O at various pressures (pD 7.0, 30 °C).

The second-derivative infrared spectrum of myoglobin (Taniguchi & Takeda, 1992) at 0.1 MPa shows only the amide I bands of the deuterated amide groups of α-helix and β-turn substructures. Most of the labile protons of the skeletal amide groups and side-chain amide groups located on the surface of myoglobin are deuterated at atmospheric pressure. As a consequence, we cannot observe any protons of the amide I band for β-sheet and only a few protons of the α-helix substructure in the protein interior. The α-helix and β-sheet substructures of myoglobin are maintained in the pressure range of 0.1–130 MPa, but at 350 MPa,

Table 7.1. Amide I band assignments in D_2O and H_2O for ribonuclease A

	Band position (cm^{-1})								
	In D_2O						In H_2O		
	Completely (D)				Partially (H/D)				
Reference	TT[a]	OHJB[b]	HLC[c]	YT[d]	TT	HLC	OHJB	HLC	YT
Tyrosine	1609		1609		1609	1609		1614	1613
	1631	1632	1631	1633	1638	1637	1631	1631	1631
β-sheet							1641	1640	1641
	1681	1676	1680	1682	1688	1688	1687	1689	1689
α-helix	1650	1651	1651	1653	1650	1655	1655	1658	1658
					1657				
β-turn	1663	1662	1663	1662	1665	1664	1667	1667	1665
Undefined	1673	1643		1612	1677	1677	1678	1681	1674
				1618					1681
				1639					
				1646					
				1672					

[a] Second-derivative method, 50 mg/ml, pD 7.0 at 30 °C (this work).
[b] Deconvoluted method, 50 mg/ml, pD 6.0 (Olinger et al., 1986).
[c] Second-derivative method, 50 mg/ml, pD 6.6 at 20 °C (Haris et al., 1986).
[d] Deconvoluted method, 100 mg/ml, pD 7.0, 0.1 M NaCl at 28 °C (Yamamoto & Tasumi, 1988, 1991).

the ordered α-helix and β-sheet substructures exhibit a sharp, two-state transition to random coil. As pressure is increased to 960 MPa, we observe partial precipitation of myoglobin. After the pressure on the sample myoglobin has been returned to atmospheric pressure, the infrared spectrum shows only the deuterated amide I band of the protein, indicating only partial reversibility.

No significant changes in the high-pressure infrared spectrum of insulin (pD 3.0) and cytochrome c (pD 7.0) take place at external pressures up to 1320 and 1470 MPa, respectively, except for H/D exchange reactions and changes in the side-chain bands near 1610 cm^{-1}. H/D exchange in the component bands of the substructures takes place in the pressure range between 210 and 730 MPa for insulin, and between 350 and 510 MPa for cytochrome c. The side-chain band at 1612 cm^{-1} of insulin and cytochrome c splits into the 1610 cm^{-1} and 1620 cm^{-1} bands in the pressure range of 210–550 MPa.

The second-derivative infrared spectrum of lysozyme at atmospheric pressure shows component bands at 1639, 1655, 1668, 1675, and 1687 cm^{-1} due to the protonated amide I band of the amide group segments in the β-sheet, α-helix, β-turn, and β-sheet substructures, respectively. Pressure-induced H/D exchange takes place at 100 MPa for the β-turn conformation and at 460 MPa for the α-helix and β-sheet substructures. At higher pressure, the structure of lysozyme changes into the random coil form. By 670 MPa, the pressure-induced redistribution detected by the infrared spectrum has been completed. Maxima in the amide I band observed at 1630 and 1645 cm^{-1} are due to the deuterated random coil structure and the deuterated β-sheet substructure, respectively. The changes reported by the

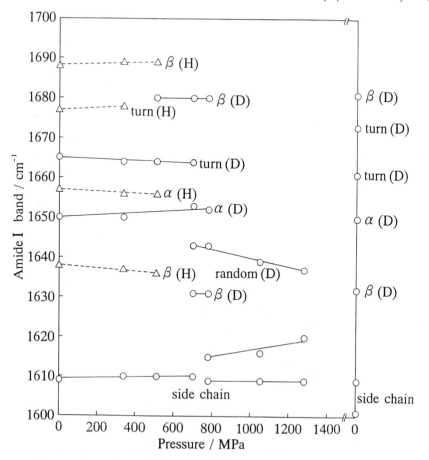

Figure 7.2. Pressure dependence of the wavenumber of the amide I' band of RNase A in D_2O (pD 7.0, 30 °C).

infrared spectrum of lysozyme at 1090 MPa are irreversible and remain after the pressure on the sample has been released. These results coincide with those seen in high-pressure Raman spectra (Heremans & Wong, 1985).

Table 7.2 summarizes the physical parameters that describe the pressure-induced secondary structural changes observed in the five globular proteins we studied: these are the most important factors needed for understanding the mechanism of pressure denaturation. The results of high-pressure infrared spectroscopy reveal that the H/D exchange reaction of the protons in protein molecules is commonly completed at a lower pressure than that at which the secondary structure is abolished by compression. Three mechanisms exist for H/D exchange in proteins: the local unfolding mechanism, the solvent penetration mechanism, and the regional melting mechanism (Kossiakoff, 1982). The penetration mechanism, by which some molecules diffuse to buried exchange sites by collective small amplitude fluctuations or by mobile defects in the packing of the protein interior,

Table 7.2. Physical properties describing the pressure denaturation of proteins in solution (pD 7.0)

Protein	v^a ml g^{-1}	$\beta_t{}^a$ 10^{-4} MPa^{-1}	$\delta V/V^a$ (%)	pI^b	H/D exchange pressure		Denatured pressure (MPa)	
Ribonuclease A	0.704	0.548	0.38	9.6	α-helix	700	850	(rever.)
					β-sheet	510	850	
Myoglobin	0.747	1.31	0.51	8.1–8.2	α-helix	130	350	(rever.)
Insulin	0.742	1.34	0.35	5.3–5.8	α-helix	730		native
					β-sheet	730		
Cytochrome c	0.725	0.427	0.34	10.1	α-helix	910		native
					β-sheet	910		
Lysozyme	0.712	0.773	0.43	11.0–11.4	α-helix	670	550^d	(irrever.)
					β-sheet	360		

[a] K. Gekko & Y. Hasegawa (1986).
[b] Isoelectric point.
[c] pD 3.0.
[d] K. Heremans & P. T. T. Wong (1985).

takes place cooperatively at the same time that pressure induces changes in the secondary structure of the protein molecules. The experimental results do not indicate a cooperative relationship between H/D exchange and pressure denaturation of the secondary structure of the proteins. Pressure-induced changes in the vibrational mode of the infrared spectra due to the side chain of protein molecules were observed below the pressure at which the secondary structure of proteins undergoes the transition from the α-helix to the random coil. The pressure-induced increase in H/D exchange and spectral changes attributed to protein side chains suggest that hydrogen exchange occurs by the local unfolding or regional melting mechanism.

We have already reported that the mechanism for pressure denaturation is related to the difference between the partial specific volumes of the native and denatured proteins in solution at high pressure (Taniguchi & Suzuki, 1983). Each partial specific volume is described by two factors—the partial specific volume (v_o) and the partial isothermal compressibility (β_t). Therefore, v_o and β_t values constitute a criteria for assessing pressure denaturation. However, as shown in Table 7.2, the order of the partial specific volume and partial isothermal compressibilities of the native proteins at 0.1 MPa does not agree with the order of pressures needed to induce protein denaturation. Surprisingly, there is a good correlation between the volume changes ($\delta V/V$) and the amount of pressure needed to induce the denaturation of proteins. That is, the larger the volume fluctuation, the lower the pressure for the protein denaturation. Myoglobin, which has the highest volume fluctuation ($\delta V/V = 0.51\%$), denatures at the lowest pressure (350 MPa). Insulin and cytochrome c, have the smallest volume fluctuation ($\delta V/V = 0.35$ and 0.34%, respectively) and do not denature at the highest pressure studied (1470 MPa). Consequently, it is clear from the present results that the volume fluctuation of a protein molecule is closely related to the pressure-induced changes in its conformational structure.

REFERENCES

Brandts, J. F., Oliveira, R. J., & Westort, C. (1970). Thermodynamics of protein denaturation: effect of pressure on the denaturation of ribonuclease A. *Biochemistry* **9**, 1038.

Chryssomallis, G. S., Torgersen, P. M., Drickamer, H. G., & Weber, G. (1981). Effect of hydrostatic pressure on lysozyme and chymotrypsinogen detected by fluorescence polarization. *Biochemistry* **20**, 3955.

Gekko, K., & Hasegawa, Y. (1986). Compressibility-structure relationship of globular proteins. *Biochemistry* **25**, 6563.

Gekko, K., & Noguchi, H. (1979). Compressibility of globular protein in water at 25 °C. *J. Phys. Chem.* **83**, 2706.

Haris, P. I., Lee, D. C., & Chapman, D. (1986). A Fourier transform infrared investigation of the structure differences between ribonuclease A and ribonuclease S. *Biochim. Biophys. Acta* **874**, 255.

Hawley, S. A. (1971). Reversible pressure-temperature denaturation of chymotrypsinogen. *Biochemistry* **10**, 2436.

Heremans, L., & Heremans, K. (1989). Pressure effects on the spectra of proteins: pressure-induced changes in the conformation of lysozyme in aqueous solutions. *J. Mol. Struct.* **214**, 305.

Heremans, K., & Wong, P. T. T. (1985). Pressure effects on the spectra of proteins: pressure-induced changes in the conformation of lysozyme in aqueous solutions. *Chem. Phys. Lett.* **118**, 101.

Kossiakoff, A. A. (1982). Protein dynamics investigated by the neutron diffraction-hydrogen exchange technique. *Nature* **296**, 713.

Li, T. M., Hook, T. W., Drickamer, H. G., & Weber, G. (1976). Plurality of pressure-denatured forms in chymotrypsinogen and lysozyme. *Biochemistry* **15**, 5571.

Makimoto, S., Nishida, H., & Taniguchi, Y. (1989). Pressure effects on thermal inactivation of Taka-amylase A. *Biochim. Biophys. Acta* **996**, 233.

Olinger, J. M., Hill, D. M., Jakobsen, R. J., & Brody, R. S. (1986). Fourier transform infrared studies of ribonuclease H_2O and ^2H_2O solutions. *Biochim. Biophys. Acta* **869**, 89.

Remmele, R. L. Jr., McMillan, P., & Bieber, A. (1990). Raman spectroscopic studies of hen egg-white lysozyme at high temperatures and pressures. *J. Protein Chem.* **9**, 475.

Susi, H. (1969). Infrared spectra of biological macromolecules and related system. In *Structure and Stability of Biological Macromolecules*, ed. S. N. Timasheff & G. L. D. Fasman. New York, Marcel Dekker. 575.

Taniguchi, Y., & Suzuki, K. (1983). Pressure inactivation of α-chymotrypsin. *J. Phys. Chem.* **87**, 5185.

Taniguchi, Y., & Takeda, N. (1992). Pressure-induced secondary structure of proteins studied by FT-IR spectroscopy. In *High Pressure and Biotechnology*, ed. C. Balny et al. London, Colloque INSERM/John Libbey Eurotext, **224**: 115.

Taniguchi, Y., & Takeda, N. (1994) The pressure-induced riversible changes in secondary structure of ribonuclease A: an FT-IR spectroscopic study. In *Basic and Applied High Pressure Biology*, ed. P. B. Bennett & R. E. Marquis. Rochester, NY, University of Rochester Press, 147.

Wong, P. T. T. (1990). FT-IR spectroscopic studies of the structure of proteins under high pressure. In *Proc. 2nd International Meeting on High Pressure Biology*, ed. J. Drouet, J. J. Risso, & J. C. Rostain. City, publisher, section 41, 1.

Wong, P. T. T., & Heremans, K. (1988). Pressure effects on protein secondary structure and hydrogen deuterium exchange in chymotrypsinogen: A Fourier transform infrared spectroscopic study. *Biochim. Biophys. Acta* **956**, 1.

Wong, P. T. T., Saint Girons, I., Guillou, Y., Cohen, G. N., Bârzu, O., & Mantsch, H. H.

(1989). Pressure-induced changes in the secondary structure of the Escherichio coli methionine repressor protein. *Biochim. Biophys. Acta Report* **996**, 260.

Yamamoto, T., & Tasumi, M. (1988). Infrared studies on thermally-induced conformational changes of ribonuclease A by the methods of difference-spectrum and self-deconvolution. *Can. J. Spectro.* **33**, 133.

Yamamoto, T., & Tasumi, M. (1991). FT-IR studies on thermal denaturation processes of ribonuclease A and S in H_2O and D_2O solutions. *J. Mol. Struct.* **242**, 235.

Zipp, A., & Kauzmann, W. (1973). Pressure denaturation of metmyoglobin. *Biochemistry* **12**, 4217.

8

High-Pressure NMR Studies of the Dissociation of Arc Repressor and the Cold Denaturation of Ribonuclease A

XIANGDONG PENG, JERSON L. SILVA, JING ZHANG,
LANCE E. BALLARD, ANA JONAS, and JIRI JONAS

We begin this article with a brief discussion of the specialized high-resolution NMR instrumentation developed for high-pressure studies of biochemical systems. We then present the potential for the unique information content of high-pressure NMR spectroscopy as illustrated by the results of two NMR studies performed recently in our laboratory.

Different denatured states of Arc repressor are characterized by one-dimensional (1D) and two-dimensional (2D) NMR. Increasing pressure promotes sequential changes in the structure of Arc repressor: from the native dimer through a predissociated state to a denaturated molten globule monomer. A compact state (molten globule) of Arc repressor is obtained in the dissociation of Arc repressor by pressure, whereas high temperature and urea induce dissociation and unfolding to less structured conformations. The presence of NOEs (Nuclear Overhauser Enhancement) in the β-sheet region in the dissociated state suggests that the intersubunit β-sheet (residues 6–14) in the native dimer is replaced by an intramonomer β-sheet. Changes in 2D NMR spectra prior to dissociation indicate the existence of a predissociated state that may represent an intermediate stage in the folding and subunit association pathway of Arc repressor.

The cold denaturation study of ribonuclease A has shown that high pressure can be utilized not only to perturb the protein structure in a controlled way but also to lower the freezing point of aqueous protein solutions substantially. As a result, one can access subzero temperatures and carry out cold denaturation studies of proteins. The results of the NMR study of the reversible cold denaturation are compared with the heat and pressure denaturation of bovine pancreatic ribonuclease A.

High-resolution NMR spectra of complex molecules in the liquid phase usually exhibit a great deal of structure and yield a wealth of information about the molecule. Therefore, it is not surprising that multinuclear high-resolution Fourier transform NMR spectroscopy at high pressure represents the most promising technique in studies of the pressure effects on biochemical systems (Jonas & Jonas, 1994). The high information content of the various advanced NMR techniques, including 2D NMR techniques such as NOESY, COSY, and ROESY, have yet to be fully exploited in high-pressure NMR experiments. Recent technological advances have resulted in the development of superconducting magnets capable of attaining a high homogeneity of the magnetic field over the sample volume, so that even without sample spinning one can achieve high resolution. It is quite remarkable that with the current instruments one can achieve an NMR line width of 1.2 Hz for sample diameters of 10 mm at a proton frequency of 300 MHz (H_o = 7.05 Tesla). At the same time, Fourier transform techniques make all these high-resolution experiments much easier to perform at high pressure than was possible with classical continuous wave (CW) techniques. Thus, the ability to record high-resolution NMR spectra on dilute spin systems opened an exciting new direction for high-pressure NMR spectroscopy dealing with pressure effects on biochemical systems (Jonas et al., 1988).

In our laboratory, we have used high-resolution, high-pressure NMR spectroscopy to investigate various problems dealing with simple molecular liquids (Jonas, 1982), and we have predicted the potential of this high-pressure technique in studies of biological systems (Wilbuy and Jonas, 1975). However, progress has been relatively slow in the application of high-pressure NMR to biochemical systems because the technique presents major experimental difficulties and requires the development of specialized equipment. The early studies in other laboratories (Wagner, 1980; Morishima, 1987) used the capillary techniques introduced by Yamada (1974), which have two inherent limitations for the investigation of biochemical systems: a limited pressure range (maximum pressures of 1.5 to 2 kbar) and a very small sample volume—usually a 1 mm-wide sample cell. This, of course, markedly decreases the sensitivity and precludes the use of dilute biochemical samples, which are very often necessary to prevent artifacts due to aggregation. Only very recently have we succeeded in developing high-pressure NMR probes (Jonas et al., 1993) that allow high-resolution and high-sensitivity experiments on biochemical systems.

Several features of NMR probe design are essential for biochemical applications of the high-pressure NMR technique: (a) high resolution, (b) high sensitivity, (c) wide pressure and temperature ranges, (d) large sample volume, (e) reliable RF (Radio frequency) feedthroughs, (f) contamination-free sample cell, (g) ease of assembly and use, and (h) suitability for superconducting magnets. Figure 8.1 shows a schematic drawing of the high-pressure vessel used for NMR studies with the 7.04 Tesla superconducting magnet. The vessel is made of titanium alloy (RMI 6A1-2Sn-4Zr-6Mo) and is used for pressures up to 9.6 kbar. The top and bottom of the vessel are closed by high-pressure plugs with C-seals and tightened by drivers. The high-pressure vessel is equipped with a cooling and/or heating jacket through which the cooling/heating fluid is circulated (temperature range: −30 to 90 °C). Since the use of C-seals, (commercially available from EG & G, Inc.)

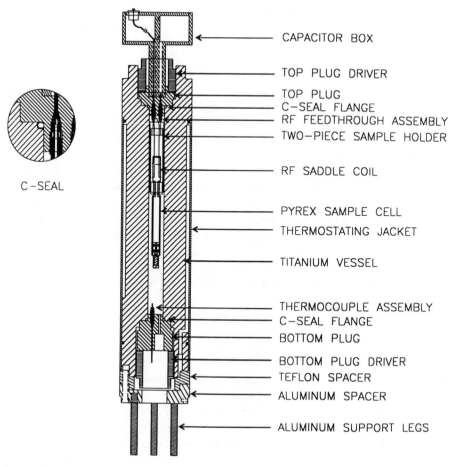

C-SEAL

CAPACITOR BOX

TOP PLUG DRIVER

TOP PLUG
C-SEAL FLANGE
RF FEEDTHROUGH ASSEMBLY
TWO-PIECE SAMPLE HOLDER

RF SADDLE COIL

PYREX SAMPLE CELL
THERMOSTATING JACKET

TITANIUM VESSEL

THERMOCOUPLE ASSEMBLY
C-SEAL FLANGE
BOTTOM PLUG

BOTTOM PLUG DRIVER
TEFLON SPACER
ALUMINUM SPACER

ALUMINUM SUPPORT LEGS

Figure 8.1. Schematic drawing of titanium-alloy high-pressure vessel used in the wide-bore 7.04 Tesla superconducting magnet (maximum pressure 9.6 kbar). The C-seal is shown in more detail in the inset.

is not widespread, we want to point out that they have many advantages over the classical Bridgman Everdur-lead-Everdur seals. The C-seal works well with both liquid and gaseous pressurizing media and is very easy to use. In addition, it is important that it can be reused many times. For studies of biochemical systems, we use a piston-type glass sample cell (Figure 8.2) to prevent contamination of the sample. The movement of the piston accommodates the volume changes due to compression. Several other designs of sample cells used in our laboratory are also shown in Figure 8.2.

We will use two examples to illustrate the new information that one can obtain by combining the high-resolution NMR technqiues with our high-pressure instrumentation. The first example deals with the pressure dissociation of the Arc repressor (Peng et al. 1993; Peng et al. 1994), and the second focuses on the cold denaturation study of ribonuclease A (unpublished results).

The results of our high-resolution NMR studies of the pressure-induced

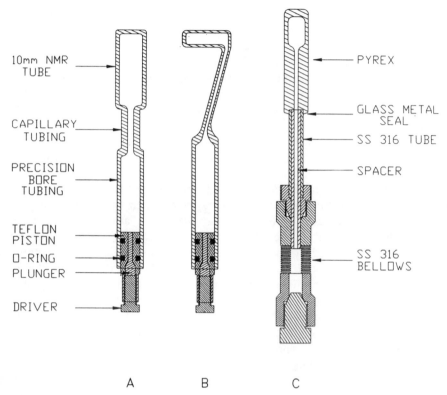

10mm NMR TUBE

CAPILLARY TUBING

PRECISION BORE TUBING

TEFLON PISTON

O-RING

PLUNGER

DRIVER

PYREX

GLASS METAL SEAL

SS 316 TUBE

SPACER

SS 316 BELLOWS

A   B   C

Figure 8.2. Types of sample cells used in the high-pressure NMR studies: (**A**) 10 mm, piston-type sample cell; (**B**) a sample cell use δ for ^{2}H NMR studies of model membranes; (**C**) an 8-mm sample cell with metallic bellows.

dissociation and denaturation of Arc repressor illustrate that pressure allows a more controlled, less drastic perturbation of proteins than either temperature or chemical denaturation allows. In fact, as we show here, pressure leads to "compact denatured states" of Arc repressor, a phrase introduced by Dill and Shortle (1991). Arc repressor is a small DNA-binding protein of 53 amino acid residues (Mr = 13,000) that is dimeric in solution. It has been reported that Arc, Mnt. and Met repressors belong to a new class of sequence-specific DNA-binding proteins that use a β-sheet as the DNA-binding motif, in contrast to other structural motifs of DNA-binding proteins, such as the helix-turn-helix, zinc-finger, and leucine-zipper. A tertiary structure model for Arc repressor has been proposed by Breg et al. (1989; 1990) on the basis of homology between Arc repressor and *E. coli* Met repressor and 2D NMR data. Silva et al. (1992) have used high-pressure fluorescence techniques to establish that the Arc repressor dimer reversibly dissociates into monomers by increase of pressure at constant protein concentration or by dilution at constant pressure. The dissociated monomer is compact and has properties characteristic of a molten globule.

Our study of Arc repressor (Peng et al. 1994) had several goals: (1) to use 1D and 2D NMR techniques to determine how different the pressure-dissociated

Arc repressor monomer is from the monomer forms obtained by thermal or urea denaturation; (2) to provide experimental evidence for the existence of a pressure-induced predissociated state of Arc repressor, as was suggested by our preliminary NMR experiments (Peng et al., 1993); (3) to partially characterize the structure of the predissociated state and the molten globule monomer state of Arc repressor; and, (4) to show in a general way that pressure leads to a more easily controlled and less drastic perturbation of protein structure than does either thermal or chemical denaturation.

A very important factor in the study of protein folding behavior is the investigation of the unfolding of proteins induced by various physicochemical perturbations such as temperature, pressure, pH, and denaturants. As one of the fundamental thermodynamic variables, high pressure can be utilized not only to perturb protein structure in a controlled way but also to substantially lower the freezing point of aqueous protein solutions by taking advantage of the high-pressure phase behavior of water (Jonas, 1982). Therefore, by using high pressures one can obtain not only pressure-denatured proteins but also cold-denatured proteins in aqueous solution. Cold denaturation—that is, the denaturation process of proteins by low temperatures—is thought to be a general property of all globular proteins (Privalov, 1990; Antonino et al., 1991); however, there has been scant experimental evidence for cold denaturation due to the fact that cold denaturation of proteins in aqueous solution usually is observed only at subzero temperatures at neutral pH. Different approaches have been utilized to prevent freezing of protein solutions, including the use of cryosolvents (Hatley & Franks, 1989), denaturants (Privalov, 1990), emulsions in oil (Franks & Hatley, 1985), and supercooled aqueous solutions (Tamura et al., 1991; Hatley & Franks, 1986). In our study of ribonuclease A (RNase A) (unpublished results), we wished to illustrate the advantages of using high-pressure, high-resolution ^{1}H NMR techniques to study the reversible cold-, heat-, and pressure-denaturation of proteins. Our study had the following objectives: (1) to investigate the pressure effects on the unfolding of RNase A: (2) to provide additional proof that cold denaturation is a general property of globular proteins; and (3) to compare the structures of cold-, heat-, and pressure-denatured RNase A.

RNase A is a pancreatic enzyme that catalyzes the cleavage of single-stranded RNA. This single polypeptide protein consists of 124 amino acid residues and has a molecular mass of 13.700. It has been used as a model for protein folding for many years (Brandts et al., 1970; Westmoreland & Matthews, 1973; Blume et al., 1978; Creighton, 1988; Biringer & Fink, 1988; Rothwarf & Scheraga, 1991) because it is small and stable and has a well-known crystal structure. The details of our study are given in the original references (Peng et al., 1993; 1994).

RESULTS AND DISCUSSION

Pressure Dissociation of Arc Repressor

The structure of Arc repressor based on the NMR data (Breg et al., 1990) contains an intertwined dimer, in which residues 8–14 of each subunit participate in the formation of an antiparallel β-sheet. The two α-helices in each subunit also have substantial intersubunit interactions. It can be argued that the intersubunit interactions are as important as the intrasubunit interactions for the maintenance of the native conformation of Arc repressor; therefore, it is conceivable that the monomers would lose most of their structure when separated from each other. Conformational changes in conjunction with the dissociation of Arc repressor were first observed on denaturation by guanidine (Bowie & Sauer, 1989). Large conformational changes were detected when the subunits were separated by hydrostatic pressure, but some residual structure remained (Silva et al., 1992).

A qualitative interpretation of the 1D ^{1}H NMR spectra both in the aromatic and aliphatic regions suggests that the conformation of the pressure-denatured Arc repressor is different from the thermal- and urea-denatured forms. It should be mentioned that we use the term *denaturation* for any major change in the conformation that leads to a disruption of the biological activity of a protein. In this case, it is legitimate to consider the dissociated state of Arc repressor as denatured, since monomeric Arc repressor loses its high-affinity binding to DNA (Brown et al., 1990; Silva et al., 1992). The denatured state of a protein is not necessarily similar to a random coil. Our spectra of the pressure-denatured Arc repressor appear more structured than the other denatured forms. At 70 °C or in 7 M urea, Arc repressor is completely denatured and has no secondary structure (Vershon et al., 1985) as determined by circular dichroism. According to Silva et al., (1992), under conditions comparable to those in our experiments, Arc repressor would begin to dissociate at about 1 kbar and completely dissociate above 3.5 kbar. Our spectra in the 1–5 kbar pressure range show substantial overlap and line broadening of many resonances. Similar spectral features have also been reported in the study of the molten globule state of guinea pig α-lactalbumin (Baum et al., 1989).

More specific information indicating that the pressure-dissociated monomer is different from the thermal or urea denatured states can be obtained by comparison of the 2D NOESY spectra in the H^α-H^α region where the NOE cross peaks in the β-sheet region (residues 8–14) are observed. Figure 8.3 compares the NOESY spectra in the H^α-H^α region for the native state, the pressure-induced monomer state, the thermally denatured state, and the urea-denatured state. The most important finding is that NOE cross peaks between H^α of Glu-9 and H^α of Arg-13 occur even in the pressure-denatured form, indicating the proximity of these two residues (Peng et al., 1993). By contrast, no NOEs are observed between H^αs of Glu-9 and Arg-13 in the NOESY spectra of Arc repressor at 70 °C, nor in the presence of 7 M urea. This observation directly proves that Glu-9 and Arg-13 are no longer close to each other in the thermally or chemically denatured states. As we mentioned in our preliminary note, we observed changes in 1D and 2D NMR spectra prior to pressure dissociation of the Arc repressor dimer and suggested the existence of a predissociated state. Subsequently, we carried out additional 1D and 2D NMR experiments to prove the existence of the pre-dissociated state. The relative changes in 1D spectra of both aromatic and aliphatic regions at 1 kbar with respect to 1 bar indicate that there must be some changes in the structure of Arc repressor prior to its dissociation. In this pressure range, there is no dissociation of 1 mM Arc repressor dimer according to the fluorescence emission and polarization experiments (Silva et al., 1992).

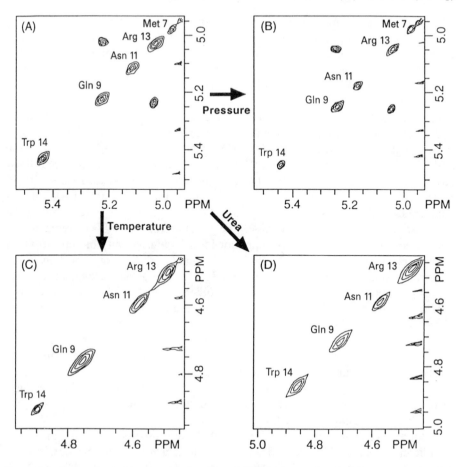

Figure 8.3. Comparison of an expanded region of NOESY spectra showing the H-H region in the β-sheet region of Arc repressor among the (**A**) native state, (**B**) pressure-induced molten globule state, (**C**) thermally denatured state, and (**D**) urea-denatured state.

Conclusive experimental evidence for the existence of a pressure-induced predissociated state of Arc repressor comes from the 2D NOESY experiments in the H^α-H^α region of the β-sheet residues (residues 8–14) performed at high pressure. Figure 8.4 shows the NOESY spectra in the H^α-H^α region in the pressure range from 1 bar to 5 kbar at 20 °C. As expected, at 1 bar we observe a strong cross peak between Glu-9 and Arg-13, which indicates the presence of the intermonomer β-sheet in the Arc repressor dimer as proposed by Breg et al. (1989; 1990). The increase in pressure to 500 bar does not produce any significant changes in the NOESY spectra, but drastic changes are evident at sites where, for example, an additional cross peak appears between Glu-9 and Trp-14 and between Met-7 and Glu-9. Essentially the same NOESY spectrum persists at 1.5 kbar, but at 2 kbar only the cross peak between Glu-9 and Arg-13 persists. This cross peak remains unchanged at pressures up to 5 kbar—that is, in the pressure regime in which Arc repressor dissociates into its molten globule monomers. This experimental evidence suggests that the conformation of the predissociated state could be intermediate between the antiparallel intermonomer β-sheet of native dimer (see Breg et al., 1990) and the intramolecular β-sheet of the molten globule monomer (Peng et al., 1993). Figure 8.5 shows the proposed conformations of the native dimer, predissociated state, and molten globule monomer of Arc repressor. The structures of the predissociated state and the molten globule monomer were obtained from molecular dynamics calculations and energy minimizations. The predissociated dimer may be related to the transition state in the protein folding theory described by Creighton (1988), corresponding to a high-energy distorted form of the native conformation. Further increases in pressure cause the dissociation of the predissociated state into molten globule monomers.

Additional information on structural changes during pressure dissociation of Arc repressor can be obtained from the chemical shift values of the resolved H^αs of the β-sheet residues (Glu-9, Asn-11, Arg-13, and Trp-14). It is well established that a relationship exists between the α-CH chemical shifts of individual residues and the protein secondary structure. Table 8.1 lists the chemical shift values of H^αs for four residues located in the β-sheet region of the native state, predissociated state, molten globule state, thermally denatured state, urea-denatured state, and

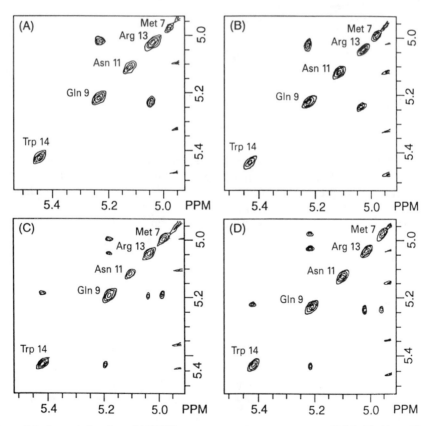

Figure 8.4. Expanded region of NOESY spectra at various pressures at 20 °C: (**A**) 1 bar, (**B**) 500 bar, (**C**) 1 kbar, (**D**) 1.5 kbar, (**E**) 2.5 kbar, (**F**) 3.5 kbar, and (**G**) 5.0 kbar.

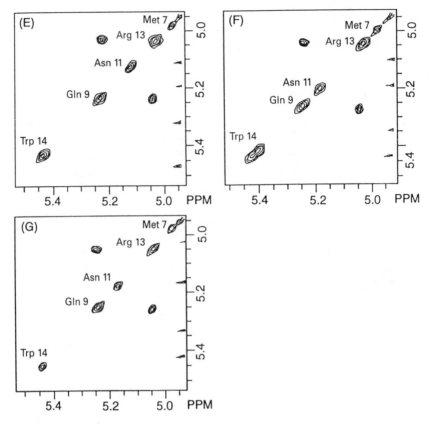

Figure 8.4. (Continued)

random coil state. Inspection of differences in these chemical shift values indicate that these values are slightly dependent on pressure but markedly dependent on temperature or the presence of urea. It can be seen from Table 8.1 that the chemical shift values for H^αs of those four residues in the molten globule state are much greater than the corresponding random coil values (Wüthrich, 1986), thus supporting the observation that these residues still maintain their β-sheet structure in the molten globule state (Wishart et al., 1992). However, the chemical shift values of Asn-11 in the thermally denatured state and in the urea-denatured state are less than the corresponding random coil value, indicating the disruption of the β-sheet in the thermally denatured state and the urea-denatured state.

Cold Denaturation of Ribonuclease A

RNase A has well-resolved histidine resonances which have been used to monitor the changes in protein structure. The four histidine proton resonances of RNase A, assigned according to literature (Blume et al., 1978), undergo chemical shift and intensity changes as pressure is increased from 1 bar (native state) to 5 kbar (denatured state) at 10 °C. All four native histidine resonances disappear at about 4 kbar, indicating that RNase A has become denatured. The most interesting

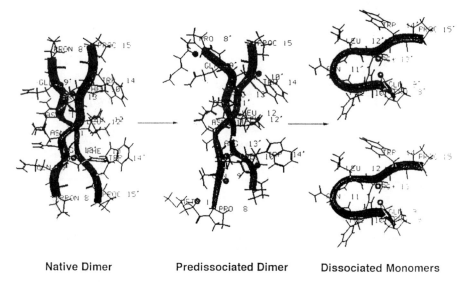

Native Dimer   Predissociated Dimer   Dissociated Monomers

Figure 8.5. The proposed β-sheet structure of Arc repressor in the native state, predissociated state, and dissociated molten globule state. The coordinates of the native dimer state were provided by R. Kaptein (University of Utrecht, The Netherlands).

feature in this pressure-induced unfolding process is that the histidine $H^{\varepsilon 1}$ resonances in the denatured state consist of two separated resonances (labeled D and D′ in Figure 8.6) which resemble the spectra of RNase A in the thermal refolding experiment by Blume et al. (1978). In this experiment, Blume et al. (1978) assigned the D′ resonance His $H^{\varepsilon 1}$ and proposed that it originates from a structural folding intermediate. In our pressure study, we observe similar spectra in the pressure-denatured state of RNase A. The intensity ratio of D to D′ is 3 to 1, suggesting that D′ is also from His-12. Therefore the structure of the pressure-denatured protein resembles the folding intermediate in the thermal refolding experiment. This result also indicates that pressure can be used to perturb the protein structure in a subtle way, retaining some organized structure.

We also carried out a variable temperature study in the temperature range from 0 to 50 °C to compare the structure of the pressure-denatured state with that

Table 8.1. Chemical shift values of α-protons of four residues in the β-sheet region in the various conformational states, of Arc repressor

Residue	Native	Predis (1 kbar)[a]	MG (4.5 kbar)[b]	Thermally-D	Urea-D	Random coil[c]
Gln 9	5.21	5.18	5.25	4.75	4.71	4.37
Asn 11	5.11	5.10	5.17	4.60	4.57	4.75
Arg 13	5.03	5.03	5.03	4.50	4.47	4.38
Trp 14	5.43	5.42	5.44	4.90	4.87	4.70

[a] Predissociated state.
[b] Molten globule state.
[c] Taken from Wüthrich (1986).

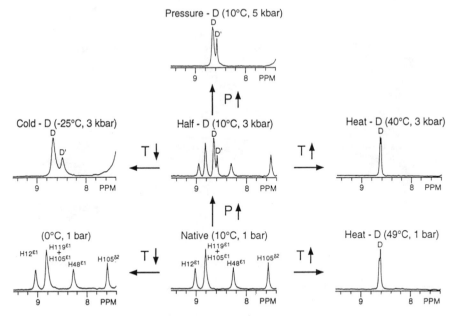

Figure 8.6. Comparison of the histidine region of ^{1}H NMR spectra among native, pressure-denatured, cold-denatured, and heat-denatured states of RNase A at pH 2.

of the heat-denatured state of RNase A. Our results are in good agreement with those of earlier thermal experiments (Westmoreland & Matthews, 1973; Blume et al., 1978) showing a doublet D from the denatured histidine resonances. Resonance D', seen in the folding intermediate and the pressure-denatured state, is absent. Although the heat-denatured protein has a more random structure (lack of D' resonance) than the pressure-denatured state, it is still not a random coil, as indicated by the doublet D instead of a singlet. The heat-denatured state of RNase A is observed at 49 °C at ambient pressure.

As we stated earlier, one of our goals was to take advantage of the pressure phase behavior of water to study the cold-denatured state of RNase A. It is difficult to obtain cold-denatured proteins by decreasing temperature below the freezing point of 0 °C at atmospheric pressure in aqueous solutions. However, one can easily obtain cold-denatured proteins by lowering the temperature well below 0 °C (−18 °C at 3 kbar in pure water) at high pressures.

To better understand the states of denatured RNase A under different denaturation conditions, we have compared the NMR spectra of cold-, heat-, and pressure-denatured RNase A in the aromatic and aliphatic regions. Figure 8.6 compares only the histidine regions of the NMR spectra of native, pressure-denatured, cold-denatured, and heat-denatured RNase A. At 10 °C, 1 bar RNase A is in the native state. As pressure is increased the protein begins to unfold. When the pressure reaches 3 kbar the protein is half-denatured, as evidenced by the appearance of denatured peaks D and D'. This half-denatured state is a pivotal state from which one can obtain the pressure-denatured state by increasing pressure to 5 kbar, the cold-denatured state by decreasing temperature to −25 °C, or the heat-denatured state by increasing temperature to 40 °C. The heat-denatured spectra at 3 kbar and 1 bar lack resonance peak D' while both pressure- and cold-denatured spectra show a D' peak in addition to the D peak. By analogy with the thermal refolding intermediate of Blume et al. (1978), the D' peak is attributed to the His-12 residue that is located in the N-terminal α-helix region. Clearly, the structures of pressure- and cold-denatured states are different from the heat-denatured state in that both the pressure- and cold-denatured states have some kind of residual helical structure around the His-12 residue, while such structures are absent in the heat-denatured states at both high and atmospheric pressures. Comparison of the spectra of cold- and pressure-denatured states reveals that the D' peak is closer to the D peak in the pressure-denatured spectrum than in the cold-denatured spectrum, indicating that the cold-denatured state is more structured than the pressure-denatured state in the N-terminal region.

The easy observation of the cold-denatured state of RNase A at high pressures clearly demonstrates the advantage of using high pressure to study cold-denatured proteins in aqueous solution. These results also provide supporting evidence that cold-denaturation of globular proteins is a general phenomenon. In a qualitative way, the comparison of aromatic and aliphatic regions of the NMR spectra of native, pressure-denatured, cold-denatured, and heat-denatured RNase A show that the resonance peaks are more dispersed in both the pressure- and cold-denatured spectra than in the heat-denatured spectra at 3 kbar and 1 bar, suggesting that the cold- and pressure-denatured states are, in general, more structured than the heat-denatured states.

We have recently developed a high-sensitivity, high-resolution NMR probe which can be used in NMR experiments at pressures up to 10 kbar. It will be very interesting to see how the structure of RNase A would change at pressures higher than 5 kbar. Detailed structural studies of the cold-, heat-, and pressure-denatured states of RNase A are in progress.

CONCLUSION

In addition to the work described here, our high-pressure high-resolution NMR studies of proteins (reviewed in Jonas & Jonas, 1994) to date have led to the following major conclusions: (1) different regions of a protein can have distinct reaction volumes, V, values for unfolding (Samarasinghe et al., 1992); (2) pressure can be used to attain subzero conditions which allow the study of the cold denaturation of proteins without the use of chemical additives (unpublished results); (3) different point mutants of a protein can have markedly different stabilities to pressure which do not directly correspond to their temperature stabilities (Royer et al., 1993); (4) temperature and pressure denaturation are distinct processes which lead to distinct denatured forms of proteins (Peng et al. 1994); and (5) the controlled and gentle unfolding and denaturation by pressure may stabilize molten globule or compact denatured states of proteins (Peng et al. 1993).

ACKNOWLEDGMENTS: This work was supported in part by the National Institutes of Health under Grant PHS 1R01 GM 42452.

REFERENCES

Antonino, L., Kantz, R. A., Nakano, T., Fox, R. O., & Fink, A. L. (1991). Cold denaturation and ^2H_2O stabilization of a staphylococcal nuclease mutant. *Proc. Natl. Acad. Sci. USA* **88**, 7715–7718.

Baum, J., Dobson, C. M., Evans, P. A., & Hanley, C. (1989). Characterization of a partly folded protein by NMR methods: studies on the molten globule state of guinea pig α-lactalbumin. *Biochemistry* **28**, 7–13.

Biringer, R. G., & Fink, A. L. (1988). Intermediates in the refolding of ribonuclease at subzero temperatures: 1. Monitoring by nitrotyrosine absorbance. *Biochemistry* **27**, 301–311.

Blume, A. D., Smallcombe, S. H., & Baldwin, R. L. (1978). Nuclear magnetic resonance evidence for a structural intermediate at an early stage in the refolding of ribnuclease A. *J. Mol. Biol.* **118**, 305–316.

Bowie, J. U., & Sauer, R. T. (1989). Identifying determinants of folding and activity for a protein of unknown structure. *Proc. Natl. Acad. Sci. USA* **86**, 2152–2156.

Brandts, J. F., Oliveira, R. J., & Westort, C. (1970). Thermodynamics of protein denaturation. Effect of pressure on the denaturation of ribonuclease A. *Biochemistry* **9**, 1038–1047.

Breg, J. N., Boelens, R., George, A. V. E., & Kaptein, R. (1989). Sequence specific ^1H NMR assignment and secondary structure of the Arc repressor of bacteriophage P22, as determined by two-dimensional ^1H NMR spectroscopy. *Biochemistry* **28**, 9826–9833.

Breg, J. N., van Opheusden, J. H. J., Burgering, M. J. M., Boelens, R., & Kaptein, R. (1990). Structure of Arc repressor in solution: evidence for a family of β-sheet binding proteins. *Nature* **346**, 586–589.

Brown, B. M., Bowie, J. U., & Sauer, R. T. (1990). Arc repressor is tetrameric when bound to operator DNA. *Biochemistry* **29**, 11189–11195.

Creighton, T. E. (1988). Toward a better understanding of protein folding pathways. *Proc. Natl. Acad. Sci. USA* **85**, 5082–5086.

Dill, K. A., & Shortle, D. (1991). Denatured states of proteins. *Annu. Rev. Biochem.* **60**, 795–825.

Franks, F., & Hatley, R. H. M. (1985). Low temperature unfolding of chymotrypsinogen. *Cryo-Letters* **6**, 171–180.

Hatley, R. H. M., & Fransk, F. (1986). Denaturation of lactate dehydrogenase at subzero temperatures. *Cryo-Letters* **7**, 226–233.

Hatley, R. H. M., & Franks, F. (1989). The effect of aqueous methanol cryosolvents on the heat- and cold-induced denaturation of lactate dehydrogenase. *Eur. J. Biochem.* **184**, 237–240.

Jonas, J. (1982). Nuclear magnetic resonance at high pressure. *Science* **216**, 1179–1884.

Jonas, J., & Jonas, A. (1994). High pressure NMR spectroscopy of proteins and membranes. *Annu. Rev. Biophys. Biomol. Struct.* **23**, 287–318.

Jonas, J. Xie, C.-L., Jonas, A., Grandinetti, P. J., Campbell, D., & Driscoll, D. (1988). High-resolution C^{13} NMR study of pressure effects on the main phase transition in 1-α-dipalmitoyl phosphatidylcholine vesicles. *Proc. Natl. Acad. Sci. USA* **85**, 4115–4117.

Jonas, J., Peng, X., Koziol. P., Reiner, C., & Campbell, D. M. (1993). High resolution NMR spectroscopy at high pressures. *J. Magn. Resn.* **102B**, 299–309.

Morishima, I., (1987). High pressure NMR studies of hemoproteins. In *Current Perspectives in High Pressure Biology*, ed. H. W. Jannasch, R. E. Marquis, & A. M. Zimmerman. New York, Academic Press, pp. 315–332.

Peng, X., Jonas, J., & Silva, J. L. (1993). Molten globule conformation of arc repressor monomers studied by high pressure (^{1}H) NMR spectroscopy. *Proc. Natl. Acad. Sci. USA* **90**, 1776–1780.

Peng, X., Jonas, J., & Silva, J. L. (1994). High pressure NMR study of the dissociation of Arc repressor. *Biochemistry* **33**, 8323–8329.

Privalov, P. L. (1990). Cold denaturation of proteins. *CRC Crit. Revs. Biochem. Mol. Biol.* **25**, 281–305.

Rothwarf, D. M., & Scheraga, H. A. (1991). Regeneration and reduction of native bovine pancreatic ribonuclease A with oxidized and reduced dithiothreitol. *J. Am. Chem. Soc.* **113**, 6292–6294.

Royer, C. A., Hinck, A. P., Loh, S. N., Prehoda, K. E., Peng, X., Jonas, J., & Markley, J. L. (1993). Effects of amino acid substitutions on the pressure denaturation of staphylococcal nuclease as monitored by fluorescence and clear magnetic resonance spectroscopy. *Biochemistry* **32**, 5222–5232.

Samarasinghe, S. D., Campbell, D. M., Jonas, A., & Jonas, J. (1992). High-resolution NMR study of the pressure induced unfolding of lysozyme. *Biochemistry* **31**, 7773–7778.

Silva, J. L., Silveira, C. F., Correia Jr., A., & Pontes, L. (1992). Dissociation of a native dimer to a molten globule monomer—effects of pressure and dilution on the association equilibrium of arc repressor. *J. Mol. Biol.* **223**, 545–555.

Tamura, A., Kimura, K., & Akasaka, K. (1991). Cold denaturation and heat denaturation of streptomyces subtilisin inhibitor ^{1}H NMR studies. *Biochemistry* **30**, 11313–11320.

Vershon, A. K., Youderian, P., Susskind, M. M., & Sauer, R. T. (1985). The bacteriophage P22 Arc and Mnt repressors. Overproduction, purification, and properties. *J. Biol. Chem.* **260**, 12124–12129.

Wagner, G. (1980). Activation volumes for the rotational motion of interior aromatic rings in globular proteins determined by high resolution at ^{1}H NMR at variable pressure. *FEBS Lett.* **112**, 280–284.

Westmoreland, D. G., & Matthews, C. R. (1973). Nuclear magnetic resonance study of the thermal denaturation of ribonuclease A: implications for multistate behavior at low pH. *Proc. Natl. Acad. Sci. USA* **70**, 914–918.

Wilbur, D. J., & Jonas, J. (1975). Fourier transform NMR in liquids at high pressure. *J. Chem. Phys.* **62**, 2800–2807.

Wishart, D. S., Sykes, B. D., & Richards, F. M. (1992). The chemical shift index—a fast and simple method for the assignment of protein secondary structure through NMR spectroscopy. *Biochemistry* **31**, 1647–1651.

Wüthrich, K. (1986). *NMR of Proteins and Nucleic Acids*. New York, Wiley.

Yamada, Y. (1974). Pressure-resisting glass cell for high pressure, high resolution NMR measurement. *Rev. Sci. Instr.* **45**, 640–642.

9

Exploring Structural, Functional, and Kinetic Aspects of Nucleic Acid–Protein Complexes with Pressure: Nucleosomes and RNA Polymerase

MAURO VILLAS-BOAS, ANA SEPULVEDA DE REZENDE, JERSON L. SILVA, ANNELIES ZECHEL, and ROBERT M. CLEGG

High pressure has provided us with new insights into two complex DNA-protein systems: nucleosomes and RNA polymerase. In spite of their complexity, we can derive new and useful information about them by coupling high pressure with a variety of other physical techniques and functional assays. These studies have shown clearly that multiple conformations of these large–molecular weight DNA-protein assemblies are present simultaneously in solution, although both molecular assemblies are generally considered to be single structures in most in vitro experiments. Considering the variety of different cellular situations encountered by nucleosomes and RNA polymerases, it is perhaps to be expected that evolution would select structures with flexible and multifarious conformations that possesses sufficient stability, rather than static, rigid, singular, and highly stable structures.

The molecular organization in the nucleus of a biological cell is extensive and involves intricate protein–protein and protein–nucleic acid interactions that are changing continually during the cell cycle. These dynamic activities in the nucleus are tightly coordinated with many extranuclear events throughout the cell. Highly organized molecular complexes involving multisubunit proteins (and higher order protein assemblies) interacting with the nucleic acid components are the rule rather than the exception in the nucleus (Alberts et al., 1983; Darnell et al., 1990; Lewin, 1994). For instance, chromosomes are organized in a structural hierarchy culminating in the metaphase state in which they are packed tightly together with proteins in a highly specific and economical manner that still largely eludes our understanding; the DNA of a eukaryotic cell is replicated with the help of a complex assembly of proteins; and information coded within the DNA sequence is transcribed with the assistance of multisubunit DNA-binding proteins, some acting as enzymes and others serving mainly as organizational and structural assistants to the catalytic process. Many important features of protein–nucleic acid (DNA and RNA) interactions have been elucidated in the last decade (Pabo & Sauer, 1992; Steitz, 1990), and exciting results have been obtained for single-protein molecules and dimers binding to DNA. Although we are a long way from understanding these interactions completely, enough aspects are known so that structural predictions are sometimes possible simply from the amino acid sequence. The task of determining structures at relatively high resolution becomes exceptionally difficult for more complex assemblies of protein and DNA. Nevertheless, certain components of these more complex DNA-protein, nucleic acid–nucleic acid, and protein–protein complexes can be separated or reconstituted in vitro, and these are amenable to structural and thermodynamic investigations. These dissociated and reconstituted systems often retain many of the functional features they possess when they are embedded in their natural surroundings as part of larger structures.

It is advantageous to use pressure as a structural perturbation for studying these complex molecular assemblies, and fluorescence is a convenient tool for measuring the molecular response. Thermodynamic studies of these structures are necessary to determine the stabilities of these complexes and their affinities for nucleic acids. The motives for applying high pressure to associating biological systems have been discussed previously (Gross & Jaenicke, 1994; Heremans, 1982; Weber & Drickamer, 1983). Pressure perturbation alters the intersubunit stability, arrangement, and ordering of many multisubunit protein assemblies at pressures less than 3000 bar (Weber & Drickamer, 1983); for this reason it provides an excellent method for probing the stability of these protein complexes in the presence and absence of DNA. High pressure has not been applied widely to DNA/RNA and DNA/RNA-protein associations as a tool to perturb the structures and their functions, in spite of the fact that it has been shown that high pressure has the potential to contribute significantly to our understanding of these structures (Macgregor, 1992; Robinson & Sligar, 1994; Royer et al., 1992, 1989; Villas-Boas et al., 1993). Fluorescence offers a sensitive way to observe the response of DNA and RNA molecules to high pressure (Benson et al., 1993; Cardullo et al., 1988; Clegg, 1992; Clegg et al., 1992, 1993; Guest et al., 1991b; Mergny et al., 1994; Morrison & Stols, 1993), and to observe and study the interactions of the different protein-DNA components (Allen et al., 1989; Guest et al., 1991a). We present pressure studies of two multisubunit protein-DNA systems: nucleosomes and E. coli RNA polymerase.

The internal structure and compartmentalization of the nucleus is currently a matter of debate, but much more is known about particular, separate components that can be separated, disassembled, and subsequently reassembled in vitro. Nucleosomes and E. Coli RNA-polymerases are composed of heterogeneous multicomponent protein structures, and they can be purified to a high degree and then reconstituted to form functional units. Nucleosomes are the lowest order structural elements of the chromosome; they involve extensive interactions between proteins such as histones and double-stranded DNA within the tightly organized hierarchy of chromosome packing (Thomas & Kornberg, 1975; van Holde, 1989; Wolffe, 1992). An intact core nucleosome contains a dimer of protein tetramers—that is, $(H2A-H2B-H3-H4)_2$ where H2A, H2B, H3, and H4 are the four histone subunits that form the protein assembly for the core of the nucleosomes. The DNA is wrapped around the octamer core, with an average of 200 base pairs per octamer.

The synthesis of RNA by transcription of DNA is one of the cardinal functions of a biological cell, and its importance has led to extensive research into its mechanism (Losick & Chamberlin, 1976). One of the best studied RNA polymerases is *E. coli* RNA polymerase. The core RNA polymerase enzyme is a tetramer of three subunits: α_2, β, and β'; an additional subunit, σ, is necessary for initiating synthesis at a promoter. When σ is present, the enzyme is called the holo RNA polymerase enzyme. In addition to the interprotein interactions between the protein subunits, histones and RNA polymerase interact strongly with DNA (and in the case of the RNA polymerase, also with RNA). We emphasize especially the kinetic response of these biological systems, (that is, the kinetics of the salt-induced conformational transition regions of the nucleosomes at high pressure and changes of the enzymatic activity of *E. coli* RNA polymerase that has been subjected to high pressure). These high-pressure studies provide evidence that nucleosomes and RNA polymerase exist in multiple conformations or states in solution despite their homogenous molecular composition and activities.

METHODOLOGY

Mono- and Polynucleotides

Nucleotides were purchased from Pharmacia (Freiburg, Germany). Poly[d(AT)] was purchased from Boehringer Mannheim. T7 DNA was grown in *E. coli*, and the DNA was separated by standard procedures or purchased from Sigma (St. Louis, Mo.). The T7 DNA was dialyzed into a buffer containing 10 mM Tris (pH = 8), 100 mM KCl, and 1 mM EDTA. The samples were then stored at 4 °C. Radioactively labeled nucleotides ($[^{14}C$ and $^3H]$-CTP and -UTP) were purchased from Amersham (Braunschweig, Germany).

Nucleosomes

Chromatin from chicken erythrocytes was partially digested with micrococcal nuclease, producing a series of oligomers of nucleosomes. The four core histones were then separated from the DNA and labeled with acrylodan dye (Prendergast et al., 1983; Villas-Boas et al., 1993). The DNA oligomers were separated according to size. To generate reconstituted nucleosomes at low salt, this DNA was combined with the purified, labeled histones in the presence of high concentrations of sodium chloride and 5 M urea; this solution was dialyzed against a series of buffers with progressively decreasing salt concentrations, in the presence of urea, until [NaCl] < 600 mM, whereupon dialysis was continued in the absence of urea. Solutions of these low-salt reconstituted nucleosomes were the starting stock solutions which were stored on ice. Core particles, mononucleosomes with a defined length of DNA, were prepared by treating the reconstituted, labeled nucleosomes with micrococcal nuclease to reduce the DNA to 146 base pairs. The size of the DNA was measured by electrophoresis (Noll & Kornberg, 1977). All results reported in this paper are for core particles.

RNA Polymerase

RNA polymerase from *E. coli* was prepared and purified according to the method of Burgess (Burgess, 1976; Burgess & Jendrisak, 1975). The protein was stored in 50% glycerol at $-80\,°C$, in which it remains stable for years without loss of activity. The radioactive incorporation assays of the RNA polymerase were carried out using either poly[d(AT)] or T7 DNA as templates.

T7 DNA Elongation Assay

The polymerase template functional assay of Chamberlin et al. (1979) was used to determine the rate of movement of the RNA polymerase along a T7 DNA template.

Poly[d(AT)] Assay

The dilution buffer for the enzyme was 10 mM Tris-HCl, pH 8.0, 1 mM MgCl$_2$, 1 mM β-mercaptoethanol, 0.05 mM EDTA pH 8.0, and 0.1 mg ml^{-1} BSA. The reaction mixture for the poly[d(AT)] assay was 40 mM Tris-HCl pH 8.0, 10 mM MgCl$_2$, 1 mM DTT, 0.2 mM EDTA, 0.2 mg ml^{-1} BSA, 400 µM ATP, 100 µM ^{14}C-UTP (20 cpm/pmol), and 100 µM poly[d(AT)]. The 100 µl assay mixture contained 0.3 µg enzyme. After incubation at 37 °C for variable lengths of time, the reaction was stopped with 1 ml of ice-cold quench solution, 0.1 M Na-pyrophosphate, and 3.5% perchloric acid. We added 25 µl of yeast RNA, 10 mg ml^{-1}, to aid the precipitation of the synthesized RNA. After cooling for 30 minutes in ice, the solutions were filtered on Whatman GF/C filter disks. The filters were dried and counted in the OPTI-FLUOR O scintillation cocktail from Packard Instrument Co. (Downers Grove, Ill.).

Labeling the H3 Histone with Acrylodan

The only cysteine residue in the core protein complex from chicken erythrocytes is located at position 110 of the H3 histones, and it is possible to label this amino acid specifically with sulfhydryl reactive fluorescence probes (Ashikawa et al., 1982b; Dieterich et al., 1977, 1979; Eshaghpour et al., 1980). These labeled, reconstituted histones were then used to carry out the high-pressure fluorescence experiments described here.

Pressure and Fluorescence Equipment

The fluorescence measurements in the pressure cell (Nova Swiss, Effretikon, Switzerland) were made on a fluorimeter constructed in our laboratory; many of the steady-state fluorescence spectra at 1 bar were taken with an SLM (8000S, Urbana, Ill.) fluorimeter. The data were collected with a Macintosh IIci computer (Apple) equipped with software (LabVIEW, National Instruments, Austin, Tex.) and an ADC data acquisition system (Lab-NB) from National Instruments (Austin, Tex.). The data analyses were carried out within the programming environments of LabVIEW and KaleidaGraph (Abelbeck Software). The high-pressure optical cell and hydraulic equipment were from Nova Swiss (Effretikon, Switzerland). The optical cell had been modified so the sample could be loaded from the top; the sample was contained in a quartz cuvette with a Kel-F plunger, held snugly in the quartz sample chamber with an O-ring, to transmit the pressure. The quartz sample cells (200 μl volume) that were placed in the optical HP cell are constructed at the MPI (Göttingen, Germany). The pressure fluid outside of this cuvette was ethanol. The high-pressure windows for excitation and emission were of sapphire (Nova Swiss, Effretikon, Switzerland). Measurements were made at 21 °C.

NUCLEOSOMES

Results and Discussion

Structure and Function of Nucleosomes

The nucleosome core structure is one of the most ubiquitous, conserved, and widespread biological structures known. With this multisubunit protein assembly, nature has apparently encountered and retained an optimal structure incorporating several seemingly incompatible characteristics: (1) the ability to assist the entire DNA of a cell to pack extremely tightly into a very compact, highly organized, ostensibly rigid structure (for example, in the metaphase chromosomes); (2) the capacity to remain flexible and dynamic and be able to respond swiftly to changing environmental situations (for example, during the process of transcription); and (3) the proficiency to assist as the basic building block in the hierarchical levels of highly organized structures (for example, during different phases of the cell cycle). To gain insight into the mechanisms by which these functions are realized, it is of great interest to know the free energy and various other thermodynamic parameters required to pass between different conformational states. Because the DNA is a highly negatively charged polyelectrolyte and the histones are highly positively charged polyelectrolytes, electrostatic forces play a major role in the stability of the protein-DNA complex. Therefore, another pivotal question is what role ions play in the transitions of the nucleosomes between their various canonical structures in solution. The major role that electrostatic forces play in the stability of the intact nucleosome becomes strikingly perspicuous when one examines the x-ray crystallographic structure of the nucleosome at 7 Å resolution (Finch et al., 1977; Richmond et al., 1984) and the recent structure of the histone octamer at 3.1 Å (Arents et al., 1991), which shows the probable complementary contacts of a "helix-strand-helix" repetitive motif of positive charges on the surface of the octamer with the negative charges on the DNA (Arents & Moudrianakis, 1993). It is, of course, most interesting to determine these variables under physiological conditions in which the nucleosome core structure is quite stable and does not easily change its structure in response to physical perturbations. In contrast to this apparent stability regarding external perturbations, nucleosomes can adjust their conformational state, and their degree of assembly, in response to molecular signals such as the presence of other protein complexes (transcription factors, polymerases, and so on.).

Labeling the H3 Histones with Acrylodan

The fluorescence ion titrations and the pressure measurements were carried out on core nucleosomes. We selected acrylodan as a fluorescence probe because this dye is exceptionally sensitive to the polarity of its environment (Bunker et al., 1993; Catalan et al., 1991; Macgregor & Weber, 1986; Torgerson et al., 1979; Weber & Farris, 1979). The basic chromophore of acrylodan is PRODAN, which does not have the SH reactive linker of acrylodan (Weber & Farris, 1979). The single cysteine group of the H3 histone[1] is located internally within the nucleosome particle (Finch et al., 1977; Richmond et al., 1984). Therefore, we reasoned that the dye would not be fully exposed to the solution in an intact nucleosome at intermediate salt concentrations. A red shift of the fluorescence spectrum is indicative of a more polar environment (Weber & Farris, 1979). A blue emission spectrum corresponds to acrylodan molecules that are less exposed to the polar solvent than those molecules that emit farther to the red. A red shift indicates an opening, or loosening, of the nucleosome structure, with exposure of the acrylodan to the polar solvent (see Figure 9.1). Absorbance spectroscopic measurements have shown that the H3 histones were essentially 100% labeled with acrylodan (data not shown).

The labeled nucleosomes respond in a characteristic fashion to variations in the salt concentration. There have been numerous reports concerning salt-dependent conformational changes of various nucleosome preparations (see van Holde, 1989, for a review). These structural transitions have been studied by a variety of physical methods, including fluorescence. We subjected our labeled nucleosomes to a wide range of salt concentrations to determine whether the acrylodan fluorophore is suitable as a sensitive probe for observing the low-salt ([NaCl] < 10 mM) and high-salt (0.4 M < [NaCl] < 0.7 M) conformational transitions that characterize unlabeled, intact, nucleosome preparations (Ashikawa et al., 1982a; Cary et al., 1978; Dieterich et al., 1979; Libertini & Small, 1982, 1987; Yager et al., 1989). These salt titrations demonstration that (1) reconstituted nucleosomes that have been labeled with acrylodan on the H3-SH group can associate in a defined manner with DNA; (2) the complex that is formed behaves as we expect for intact nucleosome structures; (3) the fluorescence properties of acrylodan are suitable for observing these transitions. Details of these fluorescence ion titrations will be published separately (Villas-Boas, et al., unpublished manuscript), and a preliminary report is available (Villas-Boas et al., 1993).

The simplest interpretation of our spectral results is that the salt concentration affects the relative population of two major structural conformations of the

---

1. Of all the subunits of the core structure (H2A, H2B, H3, and H4), only H3 has a cysteine.

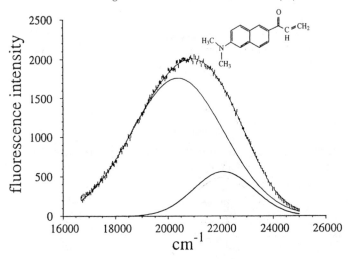

Figure 9.1. Emission spectrum of reconstituted nucleosomes labeled with acrylodan (see molecular structure shown), $\lambda_{ex} = 380$ nm. The spectrum has been transformed from the nm scale to the cm^{-1} mol^{-1} scale, and two Gaussian components have been fitted to the spectrum. The sample consists of core particles with 600 mM NaCl. The number of simultaneous Gaussian components, the position of the maximum peaks, the relative intensities of the various components, and the dependence of these spectral variables on the ion concentration depend in turn on the extent to which the DNA has been trimmed with nuclease, the homogeneity of the sample, and the presence of cosolvents. The spectra for 0.4–0.6 M salt in aqueous buffer are dominated by one Gaussian component. Whenever two Gaussian distributions were necessary, a minimum in the least-squares deviation of the data from the fit was always found when the Gaussian components were centered at 19,000–21,000 ml mol^{-1} and $\sim 22,000$ ml mol^{-1} (red component center $= 20,375 \pm 111$, half-maximum width $= 2490 \pm 56$; blue component center $= 22,111 \pm 38$, half-maximum width $= 1617 \pm 117$; red intensity/blue intensity $= 3.14$). The major features of the Gaussian fits that vary during the course of the salt titrations are (1) the number of Gaussian spectral components (one or two) required to fit the data, (2) the relative fraction of the two spectral Gaussian distributions, and (3) the center positions of the spectral components.

nucleosomes. Presumably this occurs because the apparent equilibrium constant coupling the two conformational states depends on the salt concentration. According to our paradigm for interpreting the position of the emission maxima (see two paragraphs above), we conclude that the presence of these two spectral components is in accord with an equilibrium between two major conformational states of the intact (that is, full occupancy by the histone subunits) nucleosomes. At physiological salt concentrations, the large majority of nucleosome core particles contribute to one Gaussian spectral component (see Figure 9.1). It is likely that a distribution of structurally related substates is hidden within each Gaussian component, especially for the lower energy component corresponding to the more exposed dye molecules, but for analyzing the static pressure measurements we will treat each distribution as a single spectral component.

Response of Labeled Nucleosomes to High Pressure

Static high-pressure experiments and kinetic measurements (slow pressure-jump) were carried out on the basis of results from the salt titrations. We restrict the discussion to the transition at higher salt concentration. In accordance with earlier studies that have shown no subunit dissociation at 600 mM salt and the fact that we have not observed any significant concentration dependence, we apply the simple monomolecular reaction scheme described in the next section by Eq. 1. Some pressure transitions involving multisubunit protein complexes do not show a concentration dependence, in spite of the fact that the overall reaction is multimolecular (Erijman & Weber, 1991; Ruan & Weber, 1989; Silva & Weber, 1988; Weber, 1986). Nevertheless, it has been shown by Burton et al. (1978) and others that at these ionic strengths there is no dissociation, and this agrees with our observations.

At ~ 100 mM NaCl (physiological conditions), the accessible pressures do not perturb the conformation of core nucleosomes significantly. Therefore, the ionic strengths of the solutions were adjusted to 0.4, 0.5, and 0.6 M NaCl to poise the nucleosomes in a state that is more responsive to pressure perturbation. The response of labeled nucleosomes to high pressure at these ionic strengths is shown in Figure 9.2. Even at these conditions of optimal ionic strength, it is only possible to drive the nucleosomes partially through the high-salt transition with 2000 bar at any single ionic strength. However, we can fit all the pressure transition curves simultaneously by assuming that the molar volume change, ΔV, of the molecular transition is the same at all three ionic strengths. This assumption is equivalent to assuming that the same transition mechanism and the same thermodynamic parameters controlling the transition pertain to all three ionic strengths. The result of such a fit is depicted in Figure 9.2, in which the solid lines are the simulations of the transition observed at each separate ionic strength calculated for $\Delta V = 55$ ml mol^{-1} nucleosome. It has been assumed that the transition is associated with a monomolecular equilibrium between two states, and that the equilibrium constant can be written as $K_{eq, [NaCl]_i} = (K_{eq, 1 bar})_{[NaCl]_i} \cdot \exp(p\Delta V/RT)$, where the variables have their conventional meaning: i refers to the ith concentration of NaCl. Thus, by fitting each of the transitions in Figure 9.2, we have determined not only ΔV, but also $(K_{eq, 1 bar})_{[NaCl]_i}$, which is the equilibrium constant for the monomolecular reaction representing the transition at 1 bar.

The natural logarithms of the equilibrium constants are plotted against ln[NaCl] as shown in Figure 9.3. We can interpret the plot according to the basic thermodynamic relation,

$$(\partial(\Delta G/RT)/\partial(\ln[NaCl])) = (\partial(\ln[(K_{eq, 1 bar})_{[NaCl]_i}]/RT)/\partial(\ln[NaCl])) = \Delta n,$$

where Δn is the difference in the number of moles of NaCl that must be added to solutions containing either the folded or the unfolded conformation of the nucleosomes in order to obtain a given value of the chemical potential (Privalov et al., 1969). The slope of the plot is $\Delta n = 6.8$.

Using the parameters of the linear fit for the data in Figure 9.3, we calculate that $[(K_{eq, 1 bar})_{[NaCl = 100 mM]}] = 1.7 \cdot 10^{-6}$. Thus, the nucleosomes are quite stable

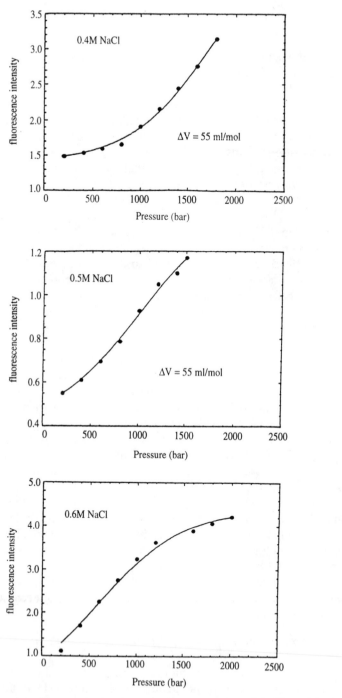

Figure 9.2. Pressure transition spectroscopic measurements taken at 0.4, 0.5, and 0.6 M NaCl. $\lambda_{ex} = 400$ nm, $\lambda_{em} = 520$ nm. The curves have been fitted globally to Eq. 3, giving $\Delta V = 55$ ml mol^{-1}. The same reaction parameters apply to all three fitted curves (solid lines).

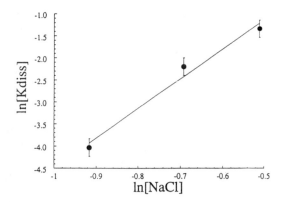

Figure 9.3. Plot of $\ln[(K_{eq,\,1\,bar})_{[NaCl]_i}]$ vs $\ln[NaCl]$. The fitted straight line has a slope of 6.8. See text for discussion.

in one conformation at physiological conditions, with only $1.7 \cdot 10^{-4}\%$ of the complexes in the unfolded state. This equilibrium constant corresponds approximately to a free energy change of $\Delta G = 7.9$ kcal mol^{-1} for the structural transition of core nucleosomes (at 300 mM salt $\Delta G = 5.9$ kcal mol^{-1}). Although most of the nucleosomes are in only one state at physiological conditions, the free energy difference between the two states is not large. The value of ΔG is consistent with the expectation that nucleosomes can adjust their equilibrium conformation without large changes in free energy to accommodate variable molecular situations, such as the binding of other proteins to the DNA or in response to cellular signals invoking particular functional cellular processes (such as transcription or cell division).

Thermal melting experiments coupled with exonuclease III (Simpson, 1979) and electronmicroscopic studies of the melting of native core particles (Poon & Seligy, 1980) and other exonuclease III experiments (Riley & Weintraub, 1978) have indicated that the end regions of the DNA (20–25 base pairs into the core particle) are especially labile. If we assume that every turn of the DNA contributes ~ 2 negative charges to the strong electrostatic attractions between the DNA and protein surface (McGhee & Felsenfeld, 1980), it is estimated that 8–9 electrostatic interactions are involved if both ends of the DNA dissociate from the octamer surface. This is in good agreement with our results. By assuming that 0.88 Na$^+$ ions are associated with every phosphate (Record et al., 1978) and that all ionic changes during the transition take place between the DNA and the octamer surface, we calculate that ~ 8 DNA-protein interactions are involved.

From our experiments we cannot ascribe the total ΔV explicitly to volume changes in the proteins, in the DNA, or in interactions between the DNA and proteins. However, if we were to attribute the total volume change only to the ionic interactions we determine by the analysis in Figure 9.3 (see last paragraph), an estimate of the volume change per ion contributing to the transition is ~ 55 ml$/\sim 8$ mol ions $= \sim 7$ ml mol^{-1} ions. This is an unusually large value for singly charged ionic species, as would be expected for the ionic interactions between the DNA phosphates and the lysines and glutamines on the surface of the histone core (Arents & Moudrianakis, 1993). This could be an indication that a ~ 20–25 ml mol^{-1} volume change should be attributed to rearrangements of the protein subunits, opening up part of their nonpolar surface contacts to the solution.

Pressure-Jump Kinetics

Pressure-jump kinetic measurements at different pressures in the transition indicate that the nucleosome structural transition (0.4–0.6 M NaCl) involves several intermediate states. To remain within the linear region required by relaxation kinetics, each ΔP was only 200 bar. If the perturbation is sufficiently small, the kinetic process can be analyzed as a sum of exponentials. Otherwise it is difficult to express analytically the time dependence of the progress of the reaction (Poland & Scheraga, 1966). Three series of measurements were made at NaCl concentrations of 0.4, 0.5, and 0.6 M. The first histones to dissociate completely from the DNA are H2A and H2B, and this occurs between 0.8 and 1.0 M NaCl (Burton et al., 1978); only at extremely low nucleosome concentrations does this dissociation occur earlier (Lilley et al., 1979). The progress of the reactions was followed by fluorescence, exciting at 400 nm and observing the fluorescence at 520 nm (see Figure 9.4). Corion (Holliston, Mass.) filters with width at half maximum = 10 nm were used to select both wavelengths. The progress curves were fitted to one or two exponentials.

The mean relaxation time of the faster process was ~ 10–20 seconds and did not change significantly within each series of experiments at any ionic strength. This rapid time is essentially the time required to adjust the pressure, and its inclusion was required to fit properly the complete time course of the kinetic curves. The amplitude of this faster component is probably partially due to small changes in the optical instrumentation and in the spectral properties of the dye molecule provoked by the pressure change. The faster amplitude is significantly smaller than the amplitude of the slower process, and it is not certain whether this rapid signal is related to a kinetic process of the transition; no obvious

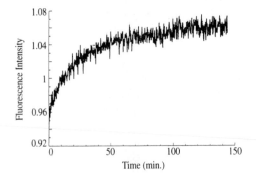

Figure 9.4. Example of a kinetic fluorescence measurement. [NaCl] = 0.5 M, the pressure jump = 200 bar, and the starting pressure was 1 kbar. Excitation = 400 nm, emission = 520 nm. The data have been fitted to a biexponential equation. The slow time constant is 25.6 min, and the fast time constant is 1.1 min.

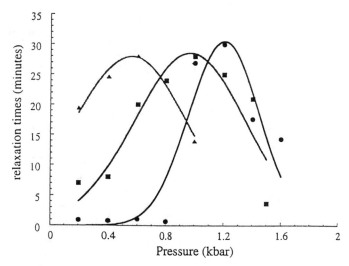

Figure 9.5. Plot of the mean relaxation time τ versus the pressure before a pressure jump. Each τ distribution has been fitted with a Gaussian curve to distinguish the data at different ionic strengths. Circles, squares, and triangles refer to 0.4, 0.5, and 0.6 M [NaCl], respectively. Enough time was alloted before every jump for equilibration. Temperature, 21 °C.

correlation was apparent. A rapid component following the pressure change was sometimes apparent outside the transition region.

The slow kinetic process displays a definite pattern (Figure 9.5). The plots of the mean relaxation times τ reveal several striking features: (1) the times do not increase or decrease monotonically, but display a pronounced maximum; (2) the maximum times are long in the center of the transition, $\tau = 25$–30 minutes; (3) the mean relaxation times undergo pronounced variations within each pressure transition curve (τ can vary up to a factor of 30); (4) the mean relaxation times within each series vary rapidly over a fairly narrow pressure range (we have approximated the distributions as Gaussians to demonstrate their symmetrical appearance); and (5) the static pressure values corresponding to the τ maxima are strongly dependent on the ionic strength. As expected, the total amplitudes of the kinetic curves (not shown) are well correlated with the first derivatives of the static pressure curves shown in Figure 9.2. The spectrum of the amplitudes is somewhat broader than the spectrum of the mean relaxation times at each ionic strength, but the amplitudes also display pronounced maxima, as expected when passing through a transition.

The behavior of the slow mean relaxation time described in the last paragraph is incompatible with the one-step reaction shown in Eq. 1.

$$A \underset{k_{-1}}{\overset{k_1}{\rightleftharpoons}} B \tag{1}$$

For such a one-step monomolecular mechanism, the reciprocal of the relaxation time, $\tau^{-1} = k_1 + k_{-1}$ (Eigen & De Maeyer, 1963) and the relaxation time would increase or decrease monotonically with the pressure. This is not consistent with the data shown in Figure 9.5. Such a narrow, pronounced distribution of τs with a maximum in the pressure dependence is difficult to reconcile with a simple mechanism consisting of only a few steps.

On the other hand, the spectrum of mean relaxation times is reminiscent of a cooperative transition in which kinetic processes become very slow in the center of the transition (Crothers, 1964; Poland & Scheraga, 1966; Schwarz, 1965). Such behavior is often an indication that a multitude of possible states exists in the transition region and that the number of significantly populated states often increases dramatically at the center of the transition region. Extended times may be required for such a large number of reaction species to reestablish an equilibrium following a perturbation; the slowest dynamic processes occur near the middle of the transition. The number of states that might be involved in the structural transition of the neuclesomes and the details of a reaction mechanism cannot be discussed at this point, but it is clear from our kinetic results that the mechanism of the nucleosome structural transition is more complex than a transformation between only two singular states.

Therefore, we tentatively suggest a description of the structural transition to account for the equilibrium and kinetic data. The reaction scheme, intended as a phenomenal representation and not as a quantitative kinetic model, rationalizes the following four results: (1) for the core structure at $[NaCl] < 30\ mM$ and $[NaCl] > 600$ mM, the fluorescence spectra can be decomposed into two distinct Gaussian distributions; (2) nucleosome preparations with DNA longer that 146 basepairs and nucleosome preparations containing polynucleosomes[2] (data not presented) show the same two spectral distributions even more distinctly; (3) the equilibrium pressure experiments could be interpreted satisfactorily in terms of a single-step conformational change, with a van't Hoff pressure dependence, and the parameters derived from such an analysis are reasonable; and (4) in spite of (1), (2) and (3), the kinetic data demand a complex reaction scheme:

$$\{A\} \underset{\langle k_{-1}\rangle}{\overset{\langle k_1 \rangle}{\rightleftharpoons}} \{B\} \tag{2}$$

Here the curly brackets $\{\ \}$ denote that the letters represent an ensemble of nucleosome structures with distinct conformations but similar spectral properties. The angular brackets $\langle\ \rangle$ denote that the observed kinetic constants are apparent values, coupling many different reactant species with product species, and weighted in an as yet undetermined fashion. There must be dynamic coupling between "reactant" $\{A\}$ species themselves and "product" $\{B\}$ species themselves. Analogous to a cooperative system, the number of populated species in the middle of the transition is expected to be greater (leading to longer times) than in the wings of the transition (where the shorter times correspond more closely to the individual reaction rates). The number of reactant and product species that are coupled directly and the values of the effective rate constant of Eq. 2 may depend on the position within the transition; presumably many pairs of reactant and

---

2. Polynucleosomes have more than one intact nucleosome particle attached to the same piece of DNA.

product species are not coupled directly. A mechanistic description of complex kinetic processes—such as nucleation phenomena, phase transitions, and polymer conformational transitions (Metiu et al., 1987)—can become extremely sophisticated, and our limited data do not justify an extended, quantitative treatment. However, although we cannot yet propose a detailed, quantitative model of the transition, our kinetic results with pressure jumps show that the transition is more complex than has often been assumed previously.

The additional complexities manifest in the kinetic experiments mean that we must approach the thermodynamic parameters derived above with caution. They do not correspond to a single reaction step of the transition. However, at each position in the transition, the time course of the one relaxation process that clearly corresponds to the dynamics of the transition can be represented very well by a single exponential component. Apparently, the two populations of structures that we observe in the decomposition of the fluorescence spectra relax to new fractional occupancies as though the mechanism involved only a single kinetic process. The maximum of the mean relaxation time at the center of the transition might result from the fact that at the center of the transition a maximum number of interactions must be made and destroyed on the path from one general structural population to the other, and this requires a longer time. This interpretation is in accord with the structure of intact nucleosomes, which are known to be composed of numerous protein-protein and DNA-protein contacts; thus, establishing a new distribution of intermolecular interactions will take longest where the number of configurations available to the nucleosome structure is greatest.

Conclusion

Our study has led us to the following conclusions: (1) The free energy difference ΔG between two major populations of nucleosome structures present at physiological conditions is only ~ 6–8 kcal mol^{-1} per nucleosome; (2) the apparent volume change ΔV between these two populations of conformations is $\sim 55 \text{ ml mol}^{-1}$ per nucleosome; (3) the number of ions associated with this conformational transformation is ~ 7 ions per nucleosome, corroborating the view that electrostatic forces play a major role in the stability and geometry of the intact nucleosome structure; (4) the rates of the structural transition in response to small pressure perturbations display a pronounced minimum near the center of the pressure-induced transition, indicating that the structural transition involves multiple conformations (possibly intermediates) of the nucleosomes in the transition region. In spite of these complexities, the spectral data, together with many features of both the static and kinetic pressure experiments, indicate that the conformational space available to nucleosomes can be apportioned into two predominant populations. The existence of multiple, related molecular assemblies is highly probable, considering the complexity of the nucleosome molecular assembly. If the results of our high-pressure experiments on reconstituted core particles can be extrapolated to in vivo situations, they may illustrate the multistate, flexible, and dynamic nature of the core nucleosome structure. These are probably essential characteristics that enable the nucleosomes to conduct their multifarious functions in the nucleus of cells.

E. COLI RNA POLYMERASE

Results and Discussion

Transcription involves an extremely complex series of reactions, and, although the major phases of the overall reaction have been identified for some time, many details remain unknown. The transcription process can be divided into well-defined phases that represent major separate activities; many functions of transcription can be studied by reference to isolated, independent systems that represent these separate phases. The four major divisions of the transcription process are recognition, initiation, elongation, and termination. RNA polymerase must bind to DNA, search for and recognize a promoter, proceed through several stages of initiation, and eventually progress into the elongation phase of transcription in which it transcribes DNA to RNA in a processive manner until terminating at particular sequences of the template (with or without the help of other proteins). Each phase comprises many substeps, and many control processes influence the enzymatic activity in each phase. The major function of transcription is the synthesis of an RNA polymer during elongation. Recently, Chamberlin (1994) proposed that the enzyme periodically undergoes major deformations during elongation, and this implies that the subunits of the multisubunit enzyme move relative to each other. Major structural transitions also occur in the DNA template; there is good evidence that a locally melted bubble of DNA is created and that a hybrid is formed between the newly synthesized RNA strand and the template strand of the DNA. The size of the bubble, the length of the DNA-RNA hybrid, and whether these values are variable (Chamberlin, 1994; Krummel & Chamberlin, 1989, 1991) or constant (Gamper & Hearst, 1982; Yager & von Hippel, 1987, 1991) during elongation have become controversial issues. Proper protein-protein and DNA-protein interactions are necessary for the enzyme to function. For instance, the simple copresence of previously separated subunits is not sufficient for the formation of an active enzyme; the mixture must first be annealed, presumably to generate the correct juxtaposition of the subunits. And yet the transcribing complex must remain flexible and dynamic. The polymerase molecule must recognize—and respond to—different sequences and structures of the DNA template and the RNA transcript (for example, pausing and termination sites) and be able to read through locations on the DNA template occupied by DNA binding proteins (for instance, nucleosomes in eucaryotic cells). This ostensible dichotomy, high stability, and dynamic flexibility, seems to be a prerequisite for the correct functioning of this multisubunit enzyme and is reminiscent of the nucleosome structure discussed earlier in this chapter. The task of interpreting physical measurements on RNA polymerase is fraught with uncertainties because we have only a very vague picture of the structure of the intact enzyme and the relative juxtapositions of the subunits. On the other hand, the enzymatic activity of the enzyme is relatively easy to measure, and assays have been developed for the different phases of the transcription process. We have undertaken a high-pressure study with the aim of perturbing the intersubunit interactions of RNA polymerase. We employ (1) fluorescence, (2) high-pressure electrophoresis, and (3) radioactive transcription assays to probe the effect of high pressure on the activity of the enzyme.

Fluorescence Measurements at High Pressure on E. coli RNA Polymerase

By monitoring changes in the center of mass of the tryptophan protein fluorescence emission (excitation = 296 nm), we observe a structural transition that is concentration dependent and not completely reversible. The fluorescence shifts slightly to the red at the higher pressures; by plotting the extent of this shift against the pressure, an approximate sigmoidal curve is obtained with a midpoint at approximately 1.2 kbar (data not shown). If the major transition region of this spectral curve is analyzed as though the enzyme dissociates into two subunits, we estimate an approximate ΔV of ~ 150 ml mol^{-1} and a dissociation constant at atmospheric pressure of $\sim 10^{-9} M$. The value of the apparent dissociation constant may be relevant because it is known that the enzyme rapidly loses activity below approximately 1 nM concentration in the absence of DNA, presumably due to subunit dissociation. There have been no previous determination of ΔV, but the order of magnitude is not in disagreement with those determined for other subunit proteins that have large areas of surface contacts. The transition curve is displaced somewhat to higher pressures by the presence of either T7 DNA or poly[d(AT)], but the increase in stability is only marginal, and the transition is still easily observed without applying pressures over 2 kbar. Because the structure of the enzyme and the locations of the tryptophans are not known, it is not possible to arrive at any structural conclusions beyond the implication that the enzyme complex dissociates.

High-Pressure Electrophoresis of E. coli RNA Polymerase

If the intact enzyme is subjected to high-pressure electrophoresis in a native gel, the integrity of the multisubunit enzyme is destroyed and subunits can be observed with differing mobilities. These results are still preliminary, but the experiments definitely show that the multisubunit enzyme dissociates under pressure. We are continuing these studies, and they will be published separately. The results are in agreement with the spectroscopic findings discussed in the previous paragraph and support the conclusion that high pressure leads to at least partial dissociation of the subunits, as expected.

Inactivation of the Enzymatic Activity of E. coli RNA Polymerase with High Pressure

The most interesting effects of high pressure have been obtained by measuring the decrease in the activity of the polymerization reaction. Figure 9.6 shows an inactivation curve obtained with poly[d(AT)] as the template. Poly[d(AT)] is transcribed by either core (minus σ subunit) or holo (plus σ subunit) enzyme preparations.[3] Somewhat different pressure inactivation curves are obtained in the presence or absence of DNA (poly[d(AT)] or T7 DNA). In the presence of DNA,

---

3. The σ subunit is required only for the initiation of polymerization on a template with a promoter, and it dissociates from the elongation complex after approximately 10 ribonucleotides have been incorporated.

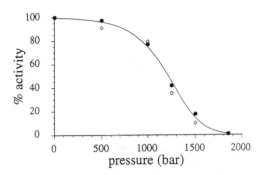

Figure 9.6. Inactivation of *E. coli* transcription at different pressures. RNA polymerase was subjected to the indicated pressure for 30 min and then assayed at atmospheric pressure as described in Methodology. The poly[d(AT)] template was present during application of pressure. The solid curve represents a fit to Eq. 2, which gives $\Delta V = 150$ ml mol^{-1}.

we can fit the curves best by a model that assumes no dissociation of the polymerase from the DNA at high pressure. The rate at which the pressure is increased or decreased does not affect the extent of inactivation. The curious thing about this pressure inactivation is that the fraction of activity retained depends on the pressure level, but not on the length of time the pressure has been applied.[4] The activity decrease is essentially the same whether the pressure has been applied for 20 minutes or 60 minutes (data not shown). Apparently, a certain fraction of the enzyme has become inactivated immediately upon the application of pressure, whereas the other fraction is not affected. We assume that the dissociated subunits undergo conformational changes at high pressure, rendering them unable to reassemble as an active enzyme molecule. This high-pressure sensitivity may be due to subtle differences in subunit conformations or to different subunit-subunit interactions. The important conclusion is that RNA polymerase molecules in solution are not homogeneous. Certain external perturbations such as pressure can discriminate between the individual molecular populations. Such a distribution of different RNA polymerase molecules is similar to other large multisubunit molecular complexes (Erijman & Weber, 1991; Ruan & Weber, 1989; Silva & Weber, 1988).

One would like to know which phases of the transcription process are inactivated by high pressure and what physical and enzymatic properties characterize those molecules that remain active subsequent to high-pressure treatment. Are those molecules that retain their activity damaged such that their rate of nucleotide incorporation per molecule has decreased, or are the molecules that are refractory to a certain level of high pressure even more active than the average value (before pressure), perhaps because they belong to a fraction of RNA polymerase molecules that were in better, more native conformations which are less vulnerable to high pressure? As we show now, neither of these instances appertains.

---

4. The activity of the enzyme decreases very slowly, at a rate similar to the rate of activity loss found for controls run at 1 atm pressure.

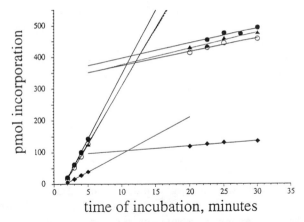

Figure 9.7. T7 elongation assay of *E. coli* RNA polymerase. Four sets of data are shown: a control reaction that has not been subjected to high pressure (○), and three reaction solutions that were subjected to 500 bar (●), 1000 bar (▲), and 1250 bar (◆) for 30 minutes. The first four points and the last four points were fitted with linear regression. The x-intercepts and the intersections of the lines define the beginning of transcription and the time required to attain the strong termination point, respectively. The latter are 10.5, 9.8, 10.6, and 9.3 minutes for 1, 500, 1000, and 1250 bar, respectively.

To address this problem, we employed an elongation assay (see Methodology) that detects the length of the RNA transcript as a function of time. This is a measure of the velocity of singular transcribing RNA polymerase molecules progressing along a T7 DNA template (Chamberlin et al., 1979). The polymerase was subjected to high pressure at 21 °C in the absence of T7 DNA and nucleotides. After returning to atmospheric pressure, the polymerase solution was kept on ice, and the assay was initiated by adding all four nucleotides and T7 DNA and warming the mixture to 30 °C.[5] After 1.8 minutes, heparin was added to arrest initiation of the RNA polymerase molecules that had not yet entered the elongation phase and 100 μl samples were withdrawn and assayed at the time points shown in Figure 9.7. At nucleotide number 7133, approximately 75–89% of the RNA polymerase molecules terminated, and only the remainder read through nucleotide 7133 and continued to transcribe. This is spontaneous termination, and it occurs at specific nucleotide sequences of many templates (Yager & von Hippel, 1987). The intersection between the initial and the later phase of the transcription event defines the time at which most of the RNA polymerase molecules have reached nucleotide number 7133 of the T7 template. The time it takes for the RNA polymerase molecules to travel from the promoter to the first termination point corresponds to this intersection. Figure 9.7 shows that the rate of nucleotide incorporation is identical for all the RNA polymerase

5. The T7 template has three strong promoters for *E. coli* RNA polymerase (A1, A2, and A3) located close to one other. The RNA polymerase/T7 DNA ratio is adjusted so that there is approximately a 1-to-1 ratio of polymerase to the single strongest promoter, A1.

molecules subject to different pressures. Thus, although the assay selected pressure-resistant molecules, those molecules still capable of transcribing have not experienced any modification of their catalytic elongation characteristics. In addition, the termination statistics remain the same as the control, and the initiation reaction cannot have been impaired significantly because initiation has not been noticeably retarded.

Conclusion

The picture that emerges from these pressure experiments is that a large ensemble of different RNA polymerase molecules is present in solution; these molecules probably differ in their subunit conformations and in their intersubunit interactions. Pressure can discriminate among the different structural species within the distribution. Higher pressures inactivate a larger fraction of RNA polymerase molecules, and the granularity of the distribution seems to be continuous (many different states). Nevertheless, despite the extended distribution in molecular characteristics defining the pressure sensitivity, once the refractory polymerase molecules become engaged in transcription, all fractions of the molecular distributions that have been subjected to different pressures have identical catalytic properties.

ACKNOWLEDGMENTS: R. M. C. is grateful to G. Weber for inviting him to his laboratory to become acquainted with high-pressure techniques, and for numerous discussions. We thank L. Erijman and C. Gohlke for discussions regarding the manuscript. This work has been supported in part by an international exchange grant from the European Economic Community (CtI*-CT90-0857) between R. M. C. and J. L. S; M. V.-B. and A. S. de R. also thank this organization for financial support and travel costs.

REFERENCES

Alberts, B., Bray, D., Lewis, J., Raff, M., Roberts, K., & Watson, J. D. (1983). *Molecular Biology of the Cell*. New York, Garland.

Allen, D. J., Darke, P. L., & Benkovic, S. J. (1989). Fluorescent oligonucleotides and deoxynucleotide triphosphates: preparation and their interaction with the large (Klenow) fragment of *Escherichia coli* DNA polymerase I. *Biochemistry* **28**, 4601–4607.

Arents, G., & Moudrianakis, E. N. (1993). Topography of the histone octamer surface: repeating structural motifs utilized in the docking of nucleosomal DNA. *Proc. Natl. Acad. Sci. USA* **90**, 10489–10493.

Arents, G., Burlingame, R. W., Wang, B.-C., Love, W. E., & Moudrianakis, E. N. (1991). The nucleosomal core histone octamer at 3.1 Å resolution: a tripartite protein assembly and a left-handed superhelix. *Proc. Natl. Acad. Sci. USA* **88**, 10148–10152.

Ashikawa, I., Nishimura, Y., Tsubol, M., & Watanabe, K. (1982a). Lifetime of tyrosine fluorescence in nucleosome core particles. *J. Biochem.* **91**, 2047–2055.

Ashikawa, I., Nishimura, Y., Tsubol, M., & Zama, M. (1982b). Micro-environment of the H3-H3 contact region of a nucleosome core particle, as revealed by a lifetime measurement of a fluorescent probe. *J. Biochem.* **92**, 1425–1430.

Benson, S. C., Mathies, R. A., & Glazer, A. N. (1993). Heterodimeric DNA-binding dyes designed for energy transfer: stability and applications of the DNA complexes. *Nucleic Acids Res.* **21**, 5720–5726.

Bunker, C. E., Bowen, T. L., & Sun, Y.-P. (1993). A photophysical study of solvatochromic probe 6-propionyl-2- (*n,n*-dimethylamino) naphthalene (PRODAN) in solution. *Photochem. Photobiol.* **58**, 499–505.

Burgess, R. R. (1976). Purification and physical properties of *E. coli* RNA polymerase. In *RNA Polymerase*, ed. R. Losick & M. Chamberlin. New York, Cold Spring Harbor Laboratory, p. 69.

Burgess, R. R., & Jendrisak, J. J. (1975). A procedure for the rapid, large-scale purification of *Escherichia coli* DNA-dependent RNA polymerase involving polymin P precipitation and DNA-cellulose chromatrography. *Biochemistry* **14**, 4634–4638.

Burton, D. R., Butler, M. J., Hyde, J. E., Philips, D., Skidmore, C. J., & Walker, I. O. (1978). The interaction of core histones with DNA: equilibrium binding studies. *Nucleic Acids Res.* **5**, 3643–3663.

Cardullo, R. A., Agrawal, S., Flores, C., Zzmecnik, P. C., & Wolf, D. E. (1988). Detection of nucleic acid hybridization by nonradiative fluorescence resonance energy transfer. *Proc. Natl. Acad. Sci. USA* **85**, 8790–8794.

Cary, P. D., Moss, T., & Bradbury, E. M. (1978). High-resolution proton-magnetic resonance studies of chromatin core particles. *Eur. J. Biochem.* **89**, 475–482.

Catalan, J., Perez, P., Laynez, J., & Garcia Blanco, F. (1991). Analysis of the solvent effect on the photophysics properties of 6-propronyl-2-(dimethylamino) nophthalene (PRODAN). *J. Fluorescence* **1**, 215–223.

Chamberlin, M. J. (1994). *Harvey Lectures* **88**, 1–21.

Chamberlin, M. J., Nierman, M. J., Wiggs, J., & Neff, N. (1979). A quantitative assay for bacterial RNA polymerases. *J. Biol. Chem.* **254**, 10061–10069.

Clegg, R. M. (1992). Fluorescence resonance energy transfer and nucleic acids. In *Methods in Enzymology*, ed. D. M. J. Lilley & J. E. Dahlberg. San Diego, Academic Press, pp. 353–388.

Clegg, R. M., Murchie, A. I. H., Zechel, A., Carlberg, C., Diekmann, S., & Lilley, D. M. J. (1992). Fluorescence resonance energy transfer analysis of the structure four-way DNA junction. *Biochemistry* **31**, 4846–4856.

Clegg, R. M., Murchie, A. I. H., Zechel, A., & Lilley, D. M. J. (1993). Observing the helical geometry of double-stranded DNA in solution fluorescence resonance energy transfer. *Proc. Natl. Acad. Sci. USA* **90**, 2994–2998.

Crothers, D. M. (1964). The kinetics of DNA denaturation. *J. Mol. Biol.* **9**, 712–733.

Darnell, J., Harvey, L., & Baltimore, D. (1990). *Molecular Cell Biology*, 2nd ed. New York, W.H. Freeman.

Dieterich, A. E., Axel, R., & Cantor, C. R. (1977). Dynamics of nucleosome structure studied by fluorescence. *Cold Spring Harbor Symp. Quant. Biol.* **42**, 199–206.

Dieterich, A. E., Axel, R., & Cantor, C. R. (1979). Salt-induced structural changes of nucleosome core particles. *J. Mol. Biol.* **129**, 587–602.

Eigen, M., & De Maeyer, L. (1963). Kinetics of the reaction between solvated electrons and water in ethylene-diamine. *J. Chem. Phys.* **39**(9), 2388–2389.

Erijman, L., & Weber, G. (1991). Oligomeric protein associations: transition from stochastic to deterministic equilibrium. *Biochemistry* **30**, 1595–1599.

Eshaghpour, H., Dieterich, A. E., Cantor, C. R., & Crothers, D. M. (1980). Singlet-singlet energy transfer studies of the internal organization of nucleosomes. *Biochemistry* **19**, 1797–1805.

Finch, J. T., Lutter, L. C., Rhodes, D., Brown, R. S., Rushton, B., Levitt, M., & Klug, A. (1977). Structure of nucleosome core particles of chromatin. *Nature* **269**, 29–36.

Gamper, H. B., & Hearst, J. P. (1982). A topological model for transcription based on unwinding angle *E. coli* RNA polymerase binary, initiation and ternary complexes. *Cell* **29**, 81–90.

Gross, M., & Jaenicke, R. (1994). Proteins under pressure: the influence of high hydrostatic pressure structure, function and assembly of proteins and protein complex. *Eur. J. Biochem.* **221**, 617–630.

Guest, C. R., Hochstrasser, R. A., Dupuy, C. G., Allen, D. J., Benkovic, S. J., & Millar, D. P. (1991a). Interaction of DNA with the Klenow fragment of DNA polymerase I time resolved fluorescence spectroscopy. *Biochemistry* **30**, 8759–8770.

Guest, C. R., Hochstrasser, R. A., Sowers, L. C., & Millar, D. P. (1991b). Dynamics of mismatched base pairs in DNA. *Biochemistry* **30**, 3271–3279.

Heremans, K. A. H. (1982). High pressure effects on proteins and other biomolecules. *Annu. Rev. Biophys. Bioeng.* **11**, 1–21.

Krummel, B., & Chamberlin, M. J. (1989). RNA chain initiation by *Escherichia coli* RNA polymerase: structural transitions of the enzyme in early ternary complexes. *Biochemistry* **28**, 7829–7842.

Krummel, B., & Chamberlin, M. J. (1991). Structural analysis of ternary complexes of *Escherichia coli* RNA polymerase: deoxyribonuclease I footprinting of defined complexes. *J. Mol. Biol.* **225**, 239–250.

Lewin, B. (1994). *Genes V*. Oxford, Oxford University Press.

Libertini, L. J., & Small, E. W. (1982). Effects of pH on low-salt transition of chromatin core particles. *Biochemistry* **21**, 3327–3334.

Libertini, L. J., & Small, E. W. (1987). Reversibility of the low-salt transition of chromatin core particles. *Nucleic Acids Res.* **15**, 6655–6664.

Lilley, D. M. J., Jacobs, M. F., & Houghton, M. (1979). The nature of the interaction of nucleosomes with a eukaryotic RNA polymerase II. *Nucleic Acids Res.* **7**, 377–399.

Losick, R., & Chamberlin, M., eds. (1976). *Regulatory Subunits of RNA Polymerase*. Cold Spring Harbor Monograph Series. Cold Spring Harbor, Cold Spring Harbor Laboratory.

Macgregor, R. B. (1992). Footprinting of *Eco*RI endonuclease at high pressure. *Biochem. Biophys. Acta* **1129**, 303–308.

Macgregor, R. B., & Weber, G. (1986). Estimation of the polarity of the protein interior by optical spectroscopy. *Nature* **319**, 70–73.

McGhee, J. D., & Felsenfeld, G. (1980). The number of charge-charge interactions stabilizing the ends of nucleosome DNA. *Nucleic Acids Res.* **8**, 2751–2769.

Mergny, J.-L., Boutorine, S., Garestier, T., Belloc, F., Rougée, M., Bulychev, N. V., Koshkin, A. A., Bourson, J., Lebedev, A. V., Valeur, B., Thuong, N. T., & Hélène, C. (1994). Fluorescence energy transfer as a probe for nucleic acid structures and sequences. *Nucleic Acids Res.* **22**, 920–928.

Metiu, H., Kitahara, K., & Ross, J. (1987). Statistical mechanical theory of kinetics of phase transition. In *Fluctuation Phenomena*, ed. E. W. Montroll & J. L. Lebowitz. Amsterdam, North Holland Personal Library, Elsevier Science, chap. 4.

Morrison, L. E., & Stols, L. M. (1993). Sensitive fluorescence-based thermodynamic and kinetic measurement hybridization in solution. *Biochemistry* **32**, 3095–3104.

Noll, M., & Kornberg, R. D. (1977). Action of micrococcal nuclease on chromatin and the location of H1. *J. Mol. Biol.* **109**, 393–404.

Pabo, C. O., & Sauer, R. T. (1992). Transcription factors: structural families and principles of DNA recognition. *Annu. Rev. Biochem.* **61**, 1053–1095.

Poland, D., & Scheraga, H. A. (1966). Kinetics of the helix-coil transition in polyamino acids. *J. Chem. Phys.* **6**, 2071–2090.

Poon, N. H., & Seligy, V. L. (1980). Substructure of thermally unfolded chromatin subunits or nucleos. *Exp. Cell Res.* **128**, 333–341.

Prendergast, F. G., Meyer, M., Carlson, G. L., Iida, S., & Potter, J. D. (1983). Synthesis, spectral properties, and use of 6-acryloyl-2-dimethylaminonaphthalene (Acrylodan): a thiol-sele polarity-sensitive fluorescent probe. *J. Biol. Chem.* **258**, 7541–7544.

Privalov, P. L., Ptitsyn, O. B., & Birshtein, T. M. (1969). Determination of stability of the DNA double helix in an aqueous medium. *Biopolymers* **8**, 559–571.

Record, M. T., Anderson, C. F., & Lohman, T. M. (1978). Thermodynamic analysis of ion effects on the binding and conformational equilibria of proteins and nucleic acids: the roles of ion association or release, screening, and ion effects on water activity. *Q. Rev. Biophys.* **11**, 103–178.

Richmond, T. J., Finch, J. T., Rushton, B., Rhodes, D., & Klug, A. (1984). Structure of the nucleosome core particle at 7 Å resolution. *Nature* **311**, 532–537.

Riley, D., & Weintraub, H. (1978). Nucleosomal DNA is digested to repeats of 10 bases by exo nuclease. *Cell* **13**, 281–293.

Robinson, C. R., & Sligar, S. G. (1994). Hydrostatic pressure reverses osmotic pressure effects on the specificity of *Eco*RI-DNA interactions. *Biochemistry* **33**, 3787–3793.

Royer, C. A., Rusch, R. M., & Scarlata, S. F. (1989). Salt effects on histone subunit interactions as studied by fluorescence spectroscopy. *Biochemistry* **28**, 6631–6637.

Royer, C. A., Ropp, T., & Scalata, S. F. (1992). Solution studies of the interactions between the histone core proteins and DNA using fluorescence spectroscopy. *Biophys. Chem.* **43**, 197–211.

Ruan, K., & Weber, G. (1989). Hysteresis and conformational drift of pressure-dissociated glyceraldehydephosphate dehydrogenase. *Biochemistry* **28**, 2144–2153.

Schwarz, G. (1965). On the kinetics of the helix-coil transition of polypeptides in solution. *J. Mol. Biol.* **11**, 64–77.

Silva, J. L., & Weber, G. (1988). Pressure-induced dissociation of brome mosaic virus. *J. Mol. Biol.* **199**, 149–159.

Simpson, R. T. (1979). Mechanism of a reversible, thermally induced conformational change in chromatin core particles. *J. Biol. Chem.* **254**, 10123–10127.

Steitz, T. A. (1990). Structural studies of protein–nucleic acid interaction: the sources of sequence-specific binding. *Q. Rev. Biophys.* **23**, 205–280.

Thomas, J. O., & Kornberg, R. D. (1975). An octamer of histones in chromatin and free in solution. *Proc. Natl. Acad. Sci. USA* **72**, 2626–2630.

Torgerson, P. M., Drickamer, H. G., & Weber, G. (1979). Inclusion complexes of poly-beta-cyclodextrin: a model for pressure effects upon ligand-protein complexes. *Biochemistry* **18**, 3079–3083.

van Holde, K. E. (1989). *Chromatin.* New York, Springer-Verlag.

Villas-Boas, Silva, J. L., & Clegg, R. M. (1993). Pressure studies on protein-DNA interactions. In *High Pressure Chemistry, Biochemistry and Materials Science*, ed. R. Winter & J. Jonas. Dordrecht, Kluwer, Academic, p. 579.

Weber, G. (1986). Phenomenological description of the association of protein subunits subjected to conformational drift: effects of dilution and of hydrostatic pressure. *Biochemistry* **25**, 3626–3631.

Weber, G., & Drickamer, H. G. (1983). The effect of high pressure upon proteins and other biomolecules. *Q. Rev. Biophys.* **16**, 89–112.

Weber, G., & Farris, F. L. (1979). Synthesis and spectral properties of a hydrophobic fluorescent: 6-propionyl-2-(dimethylamino)naphthalene. *Biochemistry* **18**, 3075–3078.

Wolffe, A. (1992). *Chromatin Structure and Function.* London, Academic Press.

Yager, T. D., & von Hippel, P. H. (1991). A thermodynamic analysis of RNA transcript elongation and termination in *Escherichia coli*. *Biochemistry* **30**, 1097–1118.

Yager, T. D., McMurray, C. T., & van Holde, K. E. (1989). Salt-induced release of DNA from nucleosome core particles. *Biochemistry* **28**, 2271–2281.

Yager, T. D., & von Hippel, P. H. (1987). Transcript elongation and termination in *Escherichia coli*. In *Escherichia coli and Salmonella typhimurium: Cellular and Molecular Biology*, vol. 2, ed. F. C. Neidhardt. Washington, D.C., American Society of Microbiology, pp. 1241–1275.

10

Pressure and Cold Denaturation of Proteins, Protein-DNA Complexes, and Viruses

JERSON L. SILVA, ANDREA T. DA POIAN, and DEBORA FOGUEL

The application of hydrostatic pressure provides a means of appraising inter-protein and intraprotein interactions isothermally and makes it possible to sample partially folded conformations. A number of proteins exhibit cold denaturation and cold dissociation. We have used the combined effects of pressure and low temperature to promote dissociation or denaturation of single-chain proteins, oligomers, protein-DNA complexes, and viruses. In this article, we summarize results that have biological relevance. The dissociation and denaturation of the hexameric protein, allophycocyanin, are accomplished only when the temperature is decreased to $-10\,°C$, indicating the entropic character of the folding and association reaction. The folding and dimerization of Arc repressor in the temperature range of 0–$20\,°C$ is also favored by a large positive entropy that counteracts an unfavorable positive enthalpy. On binding operator DNA, Arc repressor becomes extremely stable against denaturation. However, the Arc repressor-operator DNA complex is cold denatured at subzero temperatures under pressure. The entropy increases greatly when Arc repressor binds tightly to its operator sequence but not to a nonspecific sequence.

The dissociation and denaturation of icosahedral viruses by pressure and low temperature also have been studied. The procapsid shells of bacteriophage P22 only dissociate by pressure at temperatures below $0\,°C$. On the other hand, the monomeric coat protein is very unstable toward pressure. Cowpea mosaic virus (CPMV) dissociates only in the presence of $1.0\,M$ urea, at 2.5 kbar when the temperature is decreased to $-15\,°C$. At temperatures close to $-20\,°C$, partial denaturation is obtained even in the absence of urea. The assembly of CPMV is related to large and positive variations of enthalpy and entropy, making the assembly of ribonucleoprotein components an entropy-driven process. We conclude that protein folding, protein association, and protein-DNA recognition seem to need positive entropy to occur. We are facing a puzzle in which a final, apparently more ordered state is achieved, a state that paradoxically has more entropy.

In the last 20 years, several studies have described the cold denaturation of proteins (Brandts, 1964; Sturtevant, 1977; Privalov et al., 1986; Griko et al., 1988; Chen & Schellman, 1989, and as reviewed in Privalov, 1990). However, unlike thermal denaturation, cold denaturation is not well understood. One can fit the denaturation of a protein by high and low temperatures with a single equation that includes a term for the change in heat capacity (ΔC_p) (Griko et al., 1988; Privalov, 1990). This procedure seems not to be completely satisfactory because it uses the premise that the high- and low-temperature denatured forms have equivalent conformations. The combination of high pressure and low temperature appeared as an important tool for tackling this problem, especially because pressures in the range 0.5–2.5 kbar decrease the freezing point of water to negative values (Bridgman, 1964). At 2.0 kbar, the freezing point of water decreases to $-20\,°C$, providing the great advantage that one can study cold denaturation under equilibrium conditions in a manner different from other experimental approaches such as supercooling. Foguel et al. (1992) used this tool for the first time to study the dissociation and denaturation of phycobilisomes from cyanobacteria. Foguel and Weber (1995) have used the same approach to study the cold dissociation of allophycocyanin hexamers and the cold denaturation of allophycocyanin monomers. We have used the combined effects of pressure and low temperature to promote the dissociation or denaturation of single-chain proteins, oligomers, protein-DNA complexes, and viruses (Prevelige et al., 1994; Foguel & Silva, 1994; Da Poian et al., 1995).

Recently, Weber (1993) formulated a theory that attributes the entropy-driven character of protein associations to the large number of weak protein bonds rather than to solvent-solvent interactions. Weber's proposal takes into account the distribution of protein conformations that results from the dynamic nature of these macromolecules. The progressive application of hydrostatic pressure makes it possible to sample partially folded conformations. The effects of pressure on protein aggregates have been explained as differential bond compressibilities and appear to give a satisfactory account of the experimental observations (Silva & Weber, 1993). Recent NMR observations under pressure made it possible to detect the changes at the subunit interfaces that precede dissociation, and, in fact, the two-dimensional NOESY spectra of the predissociated state of Arc repressor revealed a decrease in the distance between several residues prior to dissociation (Peng et al., 1993, 1994) and the separation of other interactions that are already in closest approximation at atmospheric pressure. In the following section, we review recent experimental data on the combined effects of pressure and low temperature on several biological reactions.

PRESSURE AND LOW-TEMPERATURE EFFECTS ON PROTEIN-PROTEIN ASSOCIATION

Pressure changes the equilibrium between oligomers and subunits in such a way that the free energy of association of the subunits at atmospheric pressure in aggregates of known stoichiometry (dimers, trimers, and tetramers) can be obtained by extrapolation to 1 bar of the degree of dissociation obtained under increasing pressure. The effects of pressure on the interaction between segments of n separated subunits or between intramolecular segments are dictated by similar relations, which essentially describe the shift of the equilibrium by pressure in favor of the reactants that occupy the smaller volume (Le Chatelier principle). The Gibbs free energy and the equilibrium constant for either reaction will depend on the standard volume change of the reaction (ΔV) according to the relation

$$K_d(p) = K_{do} \exp(p\Delta V/RT) \tag{1}$$

where $K_d(p)$ and K_{do} are the equilibrium constants of association or folding at pressure p and atmospheric pressure, respectively. If we introduce the extent of reaction at pressure p, α_p, we deduce the following general equation for a dissociation or a denaturation process

$$\ln(\alpha_p^n/(1 - \alpha_p)) = p(\Delta V_p/RT) + \ln(K_{do}/n^n C^{(n-1)}) \tag{2}$$

where C is the total protein concentration and K_{do} is the dissociation or denaturation constant. In a denaturation process, $n = 1$, and the equilibrium does not depend on protein concentration. In dissociations of oligomers, n is the number of subunits, α is the degree of dissociation of the oligomeric protein, and C is the molar concentration of protein as oligomer. In these cases, the equilibrium between aggregates and subunits is expected to depend on the $n - 1$ power of the concentration. Equation 2 permits the calculation of the standard volume change in the reaction from the changes in the extent of reaction (α) with pressure, at constant concentration. We designate the volume thus obtained as ΔV_p.

In a dissociation process ($n > 1$), a change in protein concentration from C_1 to C_2 results in a parallel displacement (Δp) of the plot of $\ln K(p)$ versus p along the pressure axis. The volume change ΔV_C, calculated from the change in pressure, with concentration at a constant degree of dissociation, is given by

$$\Delta p = (n - 1)(RT/\Delta V_C) \cdot \ln(C_2/C_1) \tag{3}$$

The dependence on protein concentration furnishes an additional variable for the analysis of association equilibria, a possibility lacking in denaturation reactions.

From the dissociation constants at different temperatures, the standard entropy change on association is determined from a van't Hoff plot. The van't Hoff plot is based on a rearrangement of the Gibbs relation ($\Delta G = \Delta H - T\Delta S$), and shows the change in enthalpy and entropy from the slope (ΔH) and the intercept point on the ordinate ($-\Delta S$). The thermodynamic parameters of several aggregates obtained by these procedures are shown in Table 10.1. In all the cases studied, the predominant cause of the association is the increase in entropy of association. For a long time it has been maintained that the bulk of the entropy change in protein associations ought to follow the conversion of P—W bonds into W—W bonds (see Eq. 4), a result of special properties attributed to the hydrophobic bonds formed by apolar molecules with water. The alternative

Table 10.1. Thermodynamic parameters for the subunit association and folding of different proteins

Protein	Character of reaction	No. of subunits	$\Delta H°$ (kcal mol^{-1})	$T\Delta S°$ (kcal mol^{-1})
Yeast hexokinase[a] (1 °C, 1 bar)	Dissociation	2	17.00	38.00
β_2 subunit tryptophan[a] Synthase (4 °C, 1 bar)	Dissociation	2	17.70	28.40
Rhodobacter Rubisco[a] (15 °C, 1 bar)	Dissociation	2	13.90	26.80
Phosphorylase A Dimer[a] (1 °C, 1 bar)	Dissociation	2	4.00	17.00
Allophycocyanin[b] (0 °C, 1 bar)	Dissociation $(\alpha\beta)_3$ i $^3(\alpha\beta)$	3	72.00	93.40
Allophycocyanin[b] (0 °C, 2.2 kbar)	Dissociation $(\alpha\beta)$ i $\alpha + \beta$	2	14.28	20.36
Allophycocyanin[b] (0 °C, 2.2 kbar)	Denaturation $\alpha + \beta$ i $\alpha' + \beta'$	1	33.60	39.70
Arc repressor[c] (20 °C, 1 bar)	Dissociation/Denaturation	2	14.30	24.60

[a] Modified from Table 1 in Silva & Weber (1993).
[b] Foguel & Weber (1995).
[c] Foguel & Silva (1994).

hypothesis proposed by Weber (1993) is that the excess entropy arises from the conversion of the P—W bonds into the P—P bonds.

Several multimeric systems do not show dissociation at pressures up to 3 kbar, at room temperature, but can be dissociated by reducing the temperature below zero. Phycobilisomes from cyanobacteria dissociate from the photosynthetic membranes at high pressures and room temperature. High pressure at 0 °C elicits the disassembly of phycobilisomes into subunits, and lower temperatures result in denaturation of some of the phicobilisome components (Foguel et al., 1992). The allophycocyanin hexamer $(\alpha\beta)_3$, one of the phycobilisome components, shows complete dissociation into its subunits $(\alpha\beta)$ at a pressure of 2 kbar and 0 °C (Foguel & Weber, 1995). The dissociation of allophycocyanin was followed by changes in absorbance of the tetrapyrrolic chromophores, and by changes in the polarization and spectra distribution of its fluorescence. Since the application of pressure at room temperature alone did not promote appreciable dissociation of the hexameric protein, all the changes observed by the decrease in temperature under pressure have to be attributed to the additive perturbation of the equilibrium by low temperature and high pressure. Table 10.1 summarizes the parameters calculated for the allophycocyanin hexamer-dimer dissociation (Foguel & Weber, 1995).

The data indicate that, in common with other oligomeric proteins, the association of allophycocyanin has a large and positive enthalpy of association and a larger positive entropy change. Therefore, the association is opposed by the balance of bond energies (Weber, 1993), and the normal state of aggregation is the result of an entropy-driven reaction. Weber (1993) has proposed that the description of these association reactions can be simplified as reactions involving pairs of subunits

$$P—P + W—W \rightleftharpoons 2P—W \qquad (4)$$

where P—W represents two hydrated subunit interfaces, P—P the contacting protein interface, and W—W the pairs of water molecules released during association.

PROTEIN FOLDING

The folding of an expanded peptide chain into a globular structure is a more complicated phenomenon. At least two different processes are envisioned. The first is responsible for the formation of a globular structure in which bonds of the permanent dipole-induced dipole character with the water (P—W bonds) are replaced by apolar interactions (London dispersion forces) within the protein. In many respects, this reaction must evidently be similar to the association of globular subunits. The second process is a series of specific reactions that result in the formation of the different secondary structure based on hydrogen bonds (helices, β-sheets, etc.) with the protein.

Figure 10.1 shows a schematic representation of the two types of reaction for a hypothetical protein. The first reaction has a character similar to subunit dissociation and, in principle, its thermodynamic treatment can be simplified. As a simple approximation, the association of two folded protein subunits or two

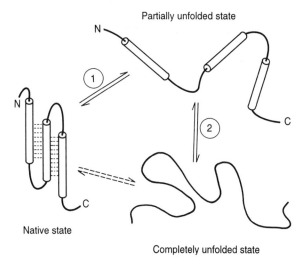

Figure 10.1. Schematic representation of the two main processes involved in protein folding: (1) formation of final tertiary structure; (2) formation of secondary structure elements. The dashed arrows represent a hypothetical cooperative transition (two-state) between the completely folded and the fully unfolded states.

protein domains to form a dimer or the final tertiary structure, respectively, may be considered as single reactions with each free energy of association determined by the difference in chemical potentials between products and reactants. To determine the free energy changes on folding, one has to determine the energy and entropy characteristics of the unfolded protein, and this must evidently vary according to the conditions of the experiments. The general assumption that a unique denatured state exists has prevented a rational examination of the folding process. It is quite evident that when the conformation is attained in concentrated urea or guanidine by dilution with water, or when a protein at high temperature or at high pressure is rapidly brought to room temperature and atmospheric pressure, different states result that may reach the folded condition by specific paths (Silva et al., 1992a; Peng et al., 1993, 1994; Prevelige et al., 1994). The pressure denatured protein represents the only case in which neither the temperature nor the solvent composition is altered in the process of generating the denatured or the native species (Silva, 1993). The hydrogen bonds differ little in their energy and compressibility, and one may expect that the pressure-denatured protein may not differ significantly from the native species in the energy of the hydrogen-bonded dispositions (Silva & Weber, 1993).

Only a very few observations have been made on the denaturation by hydrostatic pressure at or below room temperature. We regard these observations as showing the existence of considerable similarities between the folding and association reactions in that they are both driven by the intrinsic entropy and that they show, in van't Hoff plots, positive enthalpies of folding. The dissociation of allophycocyanin dimer ($\alpha\beta$) into subunits ($\alpha + \beta$) was followed by its concomitant denaturation (Foguel & Weber, 1995). As observed in the allophycocyanin dimer-hexamer association, the folding of the two peptide chains that form allophycocyanin is also an entropy-driven reaction. The van't Hoff plot for the equilibrium between $\alpha\beta$ and $\alpha + \beta$ at 2.4 kbar shows two distinct slopes. The one at higher temperatures was assigned to the dimer dissociation and the lower to the ensuing denaturation of the separated polypeptides. The extracted thermodynamic parameters are shown in Table 10.1. It is interesting to note that both the association and the folding processes are driven by entropy and opposed by a large positive enthalpy change.

An interesting model for studies of protein folding and dimerization is the Arc repressor protein. Arc repressor is a small, dimeric, DNA-binding protein ($M_r = 13,000$, monomer) that represses transcription from the P_{ant} promoter of *Salmonella* bacteriophage P22 (Sauer et al., 1983; Vershon et al., 1985; Knight et al., 1989; Breg et al., 1990). The Arc repressor dimer reversibly dissociates into partially folded subunits when pressure is applied (Silva et al., 1992a; Peng et al., 1993). The Arc monomer resulting from pressure dissociation has the characteristics of a molten globule (Ptitsyn, 1987): it is compact, with its nonpolar core partially exposed, and it has secondary structure (Silva et al., 1992a; Peng et al., 1993). Nuclear magnetic resonance studies show that the pressure-denatured Arc monomer contains some secondary structure, but most of the tertiary structure is lost (Peng et al., 1993), as expected for a molten globule conformation (Matouschek et al., 1990).

Pressure converts the DNA-binding Arc repressor protein from a native state to a denatured, molten globule state. Our data show that the folding and dimerization of Arc repressor in the temperature range of 0–20 °C is favored by a large positive entropy value (Foguel & Silva, 1994), so that the reaction proceeds in spite of an unfavorable positive enthalpy (Table 10.1).

DNA RECOGNITION

The mechanisms by which regulatory proteins recognize specific DNA sequences are not understood at the molecular level. DNA-binding proteins are responsible for many cellular functions, including transcription, replication, and restriction. The DNA-protein recognition process is the primary event in the cellular regulation of transcription (Brennan & Matthews, 1989; Steitz, 1990; Kim, 1992). Genetic activators and repressors constitute the main controls of growth, differentiation, and oncogenesis. Even though the three-dimensional structures of several repressor-operators are known, no unified rules for the various protein-DNA recognition motifs have emerged.

The magnitude of stabilization of the native Arc dimer is determined by the specificity of the protein-DNA interaction (Silva & Silveira, 1993). Operator DNA stabilizes the free energy of subunit association by about 4.4 kcal mol^{-1} of dimeric unit, whereas poly(dG–dC) stabilizes the subunit interaction by only 0.52 kcal mol^{-1}. This free-energy linkage is absent in a mutant Arc protein (PL8) that binds to operator and nonspecific DNA sequences with equally low affinity. We concluded that the coupling between DNA binding and the conversion of molten globule monomers to native dimer could account for the ability to recognize operator DNA (Silva & Silveira, 1993).

We have examined the basis for the stability of a protein-DNA complex using hydrostatic pressure and low temperature (Foguel & Silva, 1994). On binding operator DNA, Arc repressor becomes extremely stable against denaturation. However, the Arc repressor-operator DNA complex is cold denatured at subzero temperatures under pressure (Figure 10.2), demonstrating that the favorable entropy increases greatly when Arc repressor binds tightly to its operator sequence but not to specific sequences. We show how an increase in entropy may operate to provide the protein with a mechanism to distinguish between a specific and a nonspecific DNA sequence (Foguel & Silva, 1994).

Operator DNA prevents the dissociation of Arc to a partially folded monomer at room temperature (Silva & Silveira, 1993). This stabilization amounts to a decrease in the free energy of association of 4.4 kcal mol^{-1}. At -13 °C, Arc bound to operator DNA was completely denatured (Foguel & Silva, 1994). The Arc-operator DNA complex denatured as the temperature was decreased. At the lower pressures, lower temperatures were required to produce the same degree of denaturation of the complex. The van't Hoff plot of the data gave positive values for both entropy and enthalpy. The complex formed from Arc repressor and a nonspecific DNA, poly(dG–dC), was already 93% dissociated at 22 °C (Foguel & Silva, 1994). Table 10.2 shows the standard values of $\Delta G°$, $\Delta H°$, $\Delta S°$, and $T\Delta S$ (20 °C obtained in experiments using Arc bound to nonspecific DNA and Arc bound to operator DNA). It reveals that most of the stability in the specific complex arises from an increase in entropy. The enthalpy changes oppose the change in

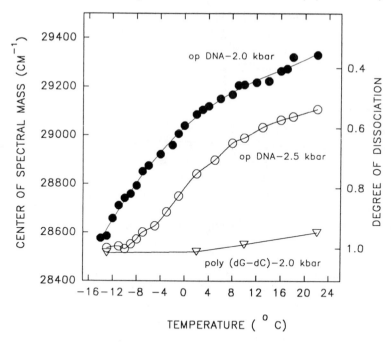

Figure 10.2. Plot of center of spectral mass (left axes) and of degree of denaturation (right axes) versus temperature: (●) = 1 µM Arc repressor −0.5 µM operator DNA at 2000 bar; (○) = 1 µM Arc repressor −0.5 µM operator DNA at 2500 bar; (▽) = 1 µM Arc repressor −13.8 mg mol⁻¹ of poly(dG–dC) at 2000 bar. The standard deviations (SD) are smaller than the symbols used (the SD values for the center of spectral mass in wavenumbers were between 5 and 10 cm⁻¹.

$\Delta G°$, reducing the increased affinity the repressor displays for the specific DNA sequence.

VIRUS ASSEMBLY

The coat protein subunits comprising the capsids of spherical viruses with $T > 1$ associate through the formation of nonequivalent bonding interactions (Rossmann & Johnson, 1989). Crystallographic studies have demonstrated that these non-equivalent interactions are due to conformational changes within the protein

Table 10.2. Thermodynamic parameters for the pressure stability of Arc dimer and Arc-DNA complexes

Complex	$\Delta H°$ (kcal mol⁻¹)	$\Delta S°$ (e.u.)	$T\Delta S°$ (20 °C) (kcal mol⁻¹)	$\Delta G°$ (20 °C) (kcal mol⁻¹)
Arc dimer ($p = 1$ bar)	14.3 ± 1.0	84.0	24.6 ± 1.1	−10.3
Arc-nonspecific DNA ($p = 1$ bar)	15.7 ± 1.2	89.9	26.3 ± 1.0	−10.6
Arc-operator DNA ($p = 1$ bar)	18.8 ± 1.4	93.9	27.5 ± 1.1	−8.7

subunits. Thus, chemically identical protein subunits take up unique position-dependent conformations during the process of assembly. The mechanism by which the plasticity required for successful assembly is coded into the folded conformation of the monomeric protein subunit is fundamental to the understanding of the viral assembly process.

Cold denaturation under pressure has been observed with the coat protein shells of bacteriophage P22 (Prevelige et al., 1994) and with the dissociation of the nucleoprotein particles of cowpea mosaic virus (Da Poian et al., 1994). The pressure stability of bacteriophage P22 coat protein in both monomeric and polymeric forms under hydrostatic pressure was examined using light scattering, fluorescence emission, polarization, and lifetime methodologies. The monomeric protein is very unstable toward pressure and undergoes significant structural changes at pressures as low as 0.5 kbar. These structural changes ultimately lead to denaturation of the subunit. Comparison of the protein denatured by pressure to that in guanidine hydrochloride suggests that pressure results in partial unfolding. In contrast to the monomeric protein subunit, when polymerized into procapsid shells the protein is very stable when pressure is applied and does not dissociate with pressure up to 2.5 kbar. However, under applied pressure the procapsid shells are cold labile, suggesting they are entropically stabilized (Prevelige et al., 1994).

Cowpea mosaic virus (CPMV) is the type member of the Comoviridae, a family of plant viruses that have a bipartite, single-stranded, positive-sense RNA genome. The two RNA molecules are encapsidated into separate virus particles. Empty shells are also formed in vivo, resulting in three different components that can be separated by gradient ultracentrifugation: the top (no RNA), the middle (smaller RNA), and the bottom (larger RNA) fractions. The stabilization conferred to the capsid by the RNA has been quantified by comparing pressure disassembly of empty and ribonucleoprotein particles at room temperature (Da Poian et al., 1994). Although the empty shell could be denatured by pressure in the presence of 1.5 M urea, pressure denaturation of the ribonucleoprotein particles was obtained only in the presence of 5.0 M urea. The denaturation of this virus under pressure is greatly facilitated by decreasing temperature to values below 0 °C. At room temperature, reversible pressure dissociation of both ribonucleoprotein particles is obtained only in the presence of 5.0 M urea, a concentration that by itself does not affect the virus structure. On the other hand, when the temperature is decreased to -15 °C, both ribonucleoprotein particles denature at 2.5 kbar in the presence of 1.0 M urea (Da Poian et al., 1995). At temperatures close to -20 °C, partial denaturation is obtained even in the absence of urea. Figure 10.3 shows the cold denaturation of bottom and middle CPMV components. The denaturation curves measured by fluorescence and light scattering were very similar for both components. The bottom component seemed to be slightly more stable in relation to low temperature than the middle one. The contributions of enthalpy (ΔH) and entropy (ΔS) for the free energy of association of CPMV are calculated from the cold-denaturation curves under pressure (Table 10.3). The entropy change is larger than the enthalpy change, making the assembly of ribonucleoprotein components an entropy-driven process. In addition, whereas the denaturation promoted by pressure and urea at room temperature is reversible,

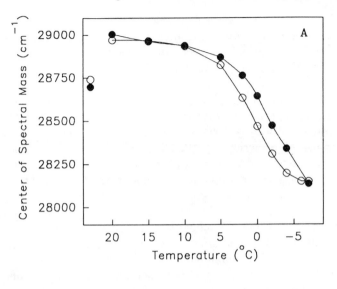

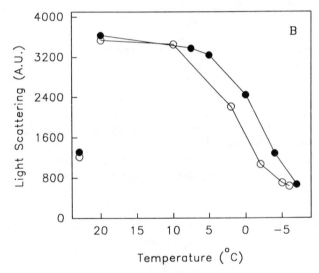

Figure 10.3. Disassembly of bottom and middle components. Cold-denaturation curves for the bottom (●) and middle (○) components of cowpea mosaic virus (CPMV) measured by (**A**) tryptophan fluorescence and (**B**) light scattering at 2500 bar, in the presence of 1.5 *M* urea. The isolated symbols correspond to the values after return to room temperature and atmospheric pressure.

virus particles denatured when temperature is decreased under pressure cannot reassemble. Bis-ANS binding data suggest that this irreversibility may be related to protein release from RNA, which probably does not occur under denaturating conditions at room temperature (Da Poian et al., 1995).

Table 10.3 summarizes the parameters obtained in the presence of different

Table 10.3. Enthalpy (ΔH) and entropy $(T\Delta S)$ contributions to the free energy of association[a] of CPMV bottom component

[urea] (M)	ΔH	$T\Delta S$	ΔG	$\Delta H/n$	$T\Delta S/n$	$\Delta G/n$
0	830.3	1458.5	−628.2	13.8	24.3	−10.5
0.5	1005.4	1627.9	−622.5	16.8	27.1	−10.4
1.0	1027.5	1616.3	−588.8	17.1	26.9	−9.8
2.5	1216.6	1799.2	−582.6	20.3	30.0	−9.7

[a] At 2.5 kbar, calculated for a temperature of 20 °C.

urea concentrations. If virus dissociation is facilitated by decreasing temperature, the assembly of the shells is expected to be an entropy-driven process. As found before for other protein aggregates (Ruan & Weber, 1988; Foguel et al., 1992; Erijman et al., 1993; Foguel & Weber, 1994), the association of CPMV is related to large and positive variations in enthalpy and entropy. The entropy change is larger than the enthalpy variation, indicating that maintenance of the assembled CPMV ribonucleoprotein particles is entropy driven. Our data suggest that low temperatures affect both protein-protein and protein-RNA interactions, showing that CPMV association can be seen as an entropy-driven process, and that protein binding to RNA may be essential to virus assembly.

DISCUSSION AND CONCLUSION

In proteins, the combination of hydrostatic pressure and low temperature appears able to produce graded effects that involve separation and denaturation of peptide chains. Application of pressure used to promote dissociation at subzero temperatures provide information on the thermodynamics of protein folding, DNA recognition, and virus assembly. The concept that observed entropy changes of dissociation or unfolding can be computed from the strengths of the protein-protein and protein-solvent bonds involved, and of their relative destabilization by pressure, should lead to an improved resolution of the origin of the thermodynamic parameters and perhaps to their estimation from structural information (Silva & Weber, 1993).

At the molecular level, the cold denaturation of proteins, protein-DNA complexes, and viruses indicates that these processes are accompanied by exposure to the solvent of a substantial proportion of nonpolar side chains. Against expectation, the interactions that lead to protein folding, DNA recognition, and virus assembly are entropy driven, which means that they are dominated by weak interactions (nonpolar) and a lower contribution of polar or H-bonding interactions, which would normally be expected to be driven by changes in ΔH. The cold-denatured capsid subunits of CPMV apparently expose the 14 tryptophan side chains of the large and small subunits and the hydrophobic regions that bind bis-ANS. Therefore, the virus particle can be pictured as a giant protein, and its constitutive capsid subunits as partly unfolded domains that, under specific conditions (chaperoned by RNA), undergo folding. These results may have direct implications for understanding the switches that control assembly and disassembly of viruses in vivo. Any cellular process that ultimately leads to disassembly of the virus should perturb the weak van der Waals interactions (hydrophobic bonds) that we show to be responsible for the stability of the virus.

All the processes described here lead to the apparent paradox that the London dispersion forces between nonpolar side chains would be responsible for the different types of biological recognition: folding, protein-protein interactions, specific DNA binding, and virus assembly. We are reluctant to accept this because it may imply a well-defined dynamic structure for these interactions and, therefore, the knowledge of the average molecular structure (given by x-ray or NMR structure) is not sufficient for comprehending the energetics of the system. In this article, we provide evidence that upon damping of the dynamics of the system by cooling, the structure is drastically perturbed. The "time structure" would restrict sampling in the conformational phase space. Although this in itself would impose some restriction of freedom, the specific available interactions would be enough to give a relatively high Boltzmann term.

Figure 10.4 shows a schematic representation of two of the processes reported here: DNA recognition by Arc repressor and assembly of cowpea mosaic virus. The model may oversimplify the problem, but it does give a picture of the

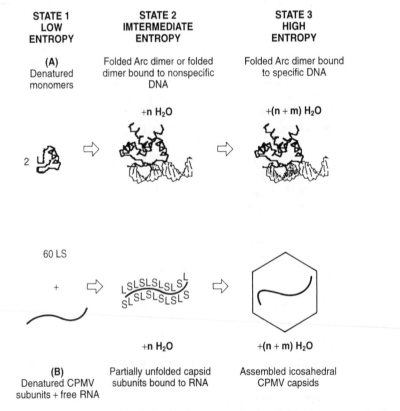

STATE 1 LOW ENTROPY	STATE 2 IMTERMEDIATE ENTROPY	STATE 3 HIGH ENTROPY
(A) Denatured monomers	Folded Arc dimer or folded dimer bound to nonspecific DNA	Folded Arc dimer bound to specific DNA

$+n\ H_2O$ $+(n + m)\ H_2O$

60 LS

$+n\ H_2O$ $+(n + m)\ H_2O$

| **(B)** Denatured CPMV subunits + free RNA | Partially unfolded capsid subunits bound to RNA | Assembled icosahedral CPMV capsids |

Figure 10.4. Schematic representation of (**A**) Arc-DNA recognition and (**B**) cowpea mosaic virus assembly.

importance of entropy-driven interactions in biological recognition. In the case of Arc repressor, at sufficiently high pressure and low temperature (or low-protein concentration), the protein is a partially folded monomer (denatured state), with only residual structure (Silva et al., 1992a; Peng et al., 1993, 1994) and low entropy. At atmospheric pressure and at high protein concentration either as a free dimer or binding a nonspecific DNA, it is more stable—at the expense of an increase in entropy. Its stability is further increased on binding the specific operator sequence (see state 3, Figure 10.4), once more at the expense of an increase in entropy. Solvent is displaced on folding and association of Arc repressor (Silva et al., 1992a; Oliveira et al., 1994), and the large change in specific volume in the dissociation-denaturation of Arc can be attributed to the interaction of buried amino acid side chains with the solvent as a consequence of denaturation. The formation of state 3 releases more solvent, which explains its entropy-driven character, as this solvent would not be displaced in nonspecific complexes. The release of these molecules of water is represented in Figure 10.4 by the letter m. Similarly, the assembly of CPMV is also represented in three stages: the state of lower entropy constitutes the separated subunits plus free RNA; it gives rise to a state of intermediate entropy on forming the nonicosahedral complex with RNA; and further increase in entropy elicits the formation of the $P = 3$ icosahedral structure. We also represent the release of molecules of water and formation of apolar interactions in virus assembly.

Several pieces of evidence indicate that apolar interactions are important for protein-DNA recognition. The three-dimensional structure of Arc-operator complex (Raumann et al., 1994) shows the formation of van der Waals contacts among thymine methyl groups with atoms from side chains in the β-sheet. Important apolar interactions have been found in the structure of other protein-DNA complexes such as in the zinc-finger glucocorticoid receptor (Luisi et al., 1991) and in the homeodomain bound to its homeobox (Kissinger et al., 1990). A large uncertainty is intrinsically related to the process of interaction based on apolar bonds; as pointed out by Cooper (1984), the more entropy they elicit, the lower the probability of resolving the structure at high resolution. Nevertheless, the few available structural data support our model that weak interactions are important for biological recognition. Our model supports the view that DNA recognition and virus assembly involve large conformational changes characterized mainly by a decrease in the contacts of nonpolar side chains with water—an increase in hydrophobic bonds. Recently, Weber (1993) has formulated a theory that attributes the entropy-driven character of protein associations to the large number of weak protein bonds rather than to solvent-solvent interactions. Weber's proposal takes into account the distribution of protein conformations that result from the dynamic nature of these macromolecules. The classical theory of the hydrophobic bond (Kauzman, 1959) and Weber's theory can equally explain the cold denaturation of proteins, protein-DNA complexes, and viruses. In both theories, the net result in the production of entropy-driven complexes is solvent displacement and formation of apolar interactions (London dispersion forces). However, Weber's theory assigns a more active participation of the weak inter-actions for the generation of the increase in entropy, indicating a higher degree of specificity for these interactions. In the coming years, we might face a renewed discussion of the participation of weak, nonpolar interactions in driving the specificity of folding, protein association, and protein–nucleic acid interaction. The combined use of hydrostatic pressure and low temperature will be useful for providing the experimental evaluation of apolar interactions.

The reduced stability of virus particles at high pressure and low temperature suggests that using these processes has a potential application for inactivating viruses of medical concern. We have found that several viruses are inactivated by pressures at room temperature (Silva et al., 1992b; Silva, 1993). The viruses that are stable against pressure at room temperature are inactivated when the temperature is decreased under pressure (unpublished results). These studies are important not only for suggesting a potential approach to the production of antiviral vaccines but also for the sterilization of biological products such as blood.

ACKNOWLEDGMENTS: This work was supported by grants from Conselho Nacional de Desenvolvimento Cientifico e Tecnologico (CNPq), Programa de Apoio ao Desenvolvimento Cientifico e Tecnologico (PADCT), and Financiadora de Estudos e Projetos (FINEP) of Brazil and by a grant from the European Economic Community to J.L.S.

REFERENCES

Brandts, J. F. (1964). The thermodynamics of protein denaturation. *J. Am. Chem. Soc.* **86**, 4291–4301.

Breg, J. N., van Opheusden, J. H. J., Burgering, M. J. M., Boelens, R., & Kaptein, R. (1990). Structure of Arc repressor in solution: evidence for a family of β-sheet DNA binding proteins. *Nature* **346**, 586–589.

Brennan, R. G., & Matthews, B. W. (1989). Structural basis of DNA-protein recognition. *TIBS* **14**, 286–289.

Bridgman, P. W. (1964). Water, in the liquid and five solid forms, under pressure. In *Collected Experimental Papers* Cambridge, Mass., 1: Harvard University Press 441–558.

Chen, B., & Schellman, J. A. (1989). Low temperature unfolding of a mutant of phage T4 lysozyme. *Biochemistry* **28**, 685–691.

Cooper, A. (1984). Protein fluctuation and the thermodynamic uncertainty principle. *Prog. Biophys. Mol. Biol.* **44**, 181–214.

Da Poian, A. T., Johnson, J. E., & Silva, J. L. (1994). Differences in pressure stability of the three components of cowpea mosaic virus: implications for virus assembly and disassembly, *Biochemistry* **33**, 8339–8346.

Da Poian, A. T., Oliveira, A. C., & Silva, J. L. (1995). Cold denaturation of an icosahedral virus: the role of entropy in virus assembly. *Biochemistry* **34**, 2672–2678.

Erijman, L., Lorimer, G. H., & Weber, G. (1993). Reversible dissociation and conformational stability of dimeric ribulose biophophate carboxylase. *Biochemistry* **32**, 5187–5195.

Foguel, D., & Silva, J. L. (1994). Cold denaturation of a repressor-operator complex: the role of entropy in protein-DNA recognition. *Proc. Natl. Acad. Sci. USA* **91**, 8244–8247.

Foguel, D., & Weber, G. (1995). Pressure-induced dissociation and denaturation of allophycocyanin at subzero temperatures. *J. Biol. Chem.*, **270**, 28759–28766.

Foguel, D., Chaloub, R. M., Silva, J. L., Crofts, A. R., & Weber, G. (1992). Pressure and low temperature effects on the fluorescence emission spectra and lifetimes of the photosynthetic components of cyanobacteria. *Biophys. J.* **63**, 613–622.

Griko, Y. V., Privalov, P. L., Sturtevant, J. M., & Venyaminov, S. Y. (1988). Cold denaturation of staphylococcal nuclease. *Proc. Natl. Acad. Sci. USA* **85**, 3343–3347.

Kauzmann, W. (1959). Some factors in the interpretation of protein denaturation. *Adv. Prot. Chem.* **13**, 1–63.

Kim, S.-H. (1992). B ribbon: a new DNA recognition motif. *Science* **255**, 1217–1218.

Kissinger, C. R., Liv, B., Martin-Blanco, E., Kornberg, T. B., & Pabo, C. O. (1990). Crystal structure of an engrailed homeodomain-DNA complex at 2.8 Å resolution: a framework for understanding homeodomain-DNA interactions. *Cell* **63**, 579–590.

Knight, K. L., Bowie, J. U., Vershon, A. K., Kelley, R. D., & Sauer, R. T. (1989). Arc and MNT repressors: a new class of sequence-specific DNA-binding proteins. *J. Biol. Chem.* **264**, 3639–3642.

Luisi, B. F., Xu, X. W., Otwinowski, Z., Freedman, L. P., Yamamoto, K. R., & Sigler, B. P. (1991). Crystallographic analysis of the interaction of the glucocorticoid receptor with DNA. *Nature* **352**, 497–505.

Matouschek, A., Kellis, J. T., Jr, Serrano, L., Bycroft, M., & Fersht, A. R. (1990). Transient folding intermediates characterized by protein engineering. *Nature* **346**, 440–445.

Oliveira, A. C., Gaspar, L. P., Da Poian, A. T., & Silva, J. L. (1994). Arc repressor will not denature under pressure in the absence of water. *J. Mol. Biol.*, **240**, 184–187.

Peng, X., Jonas, J., & Silva, J. L. (1993). Molten globule conformation of Arc repressor monomers by high pressure (^{1}H) NMR spectroscopy. *Proc. Natl. Acad. Sci. USA* **90**, 1776–1780.

Peng, X., Jonas, J., & Silva, J. L. (1994). High pressure NMR study of the dissociation of Arc repressor. *Biochemistry* **33**, 8323–8329.

Prevelige, P. E., King, J., & Silva, J. L. (1994). Pressure denaturation of P22 coat protein monomers and its entropic stabilization in the icosahedral shells. *Biophys. J.* **66**, 1631–1641.

Privalov, P. L. (1990). Cold denaturation of proteins. *Crit. Rev. Biochem. Mol. Biol.* **25**, 281–305.

Privalov, P. L., Griko, Y. V., Venyaminov, S. Y., & Kutyshenko, V. P. (1986). Cold denaturation of myoglobin. *J. Mol. Biol.* **190**, 487–498.

Ptitsyn, O. B. (1987). Protein folding: hypothesis and experiments. *J. Prot. Chem.* **6**, 277–293.

Raumann, B. E., Rould, M. A., Pabo, C. O., & Sauer, R. T. (1994). DNA recognition by β-sheets in the Arc repressor-operator crystal structure. *Nature* **367**, 754–757.

Rossmann, M. G., & Johnson, J. E. (1989). Icosahedral RNA virus structure. *Annu. Rev. Biochem.* **58**, 533–573.

Ruan, K., & Weber, G. (1988). Dissociation of yeast hexokinase by hydrostatic pressure. *Biochemistry* **27**, 3295–3301.

Sauer, R. T., Krovatin, W., DeAnda, J., Youderian, P., & Susskind, M. M. (1983). Primary structure of the immI immunity region of bacteriophage P22. *J. Mol. Biol.* **168**, 699–713.

Silva, J. L. (1993). Effects of pressure on large multimeric proteins and viruses. In *High Pressure Chemistry, Biochemistry and Material Sciences*, ed. R. Winter & J. Jonas. NATO ASI series. Dordrecht, Kluwer Academic Publishers, pp. 561–578.

Silva, J. L., & Silveira, C. F. (1993). Energy coupling between DNA binding and subunit association is crucial for the specificity of a DNA-protein interaction. *Prot. Sci.* **2**, 945–950.

Silva, J. L., & Weber, G. (1993). Pressure stability of proteins. *Annu. Rev. Phys. Chem.* **44**, 89–113.

Silva, J. L., Silveira, C. F., Correia, A., Jr., & Pontes, L. (1992a). Dissociation of a native dimer to a molten globule monomer: effects of pressure and dilution on the association equilibrium of Arc repressor. *J. Mol. Biol.* **223**, 545–555.

Silva, J. L., Peng, L., Glaser, M., Voss, E., & Weber, G. (1992b). Effects of hydrostatic pressure on a membrane-envelope virus: high immunogenicity of the pressure-inactivated virus. *J. Virol.* **66**, 2111–2117.

Steitz, T. A. (1990). Structural studies of protein–nucleic acid interactions: the sources of sequence-specific binding. *Q. Rev. Biophys.* **23**, 205–280.

Sturtevant, J. M. (1977). Heat capacity and entropy changes in processes involving proteins. *Proc. Natl. Acad. Sci. U.S.A.* **74**, 2236–2241.

Weber, G. (1993). Thermodynamics of the association and the pressure dissociation of oligomeric proteins. *J. Phys. Chem.* **97**, 7108–7115.

Weber, G., & Drickamer, H. G. (1983). The effect of high pressure upon proteins and other biomolecules. *Q. Rev. Biophys.* **116**, 89–112.

Vershon, A. K., Youderian, P., Susskind, M. M., & Sauer, R. T. (1985). The bacteriophage P22 Arc and Mnt repressors: overproduction, purification and properties. *J. Biol. Chem.* **260**, 12124–12129.

11

Sequence, Salt, Charge, and the Stability of DNA at High Pressure

ROBERT B. MACGREGOR JR., JOHN Q. WU,
and REZA NAJAF-ZADEH

The effect of pressure on the helix-coil transition temperature (T_m) is reported for the double-stranded polymers poly(dA)poly(dT), poly[d(A-T)], poly[d(I-C)], and poly[d(G-C)] and triple-stranded poly(dA)2poly(dT). The T_m increases as a function of pressure, implying a positive volume change for the transition and leading to the conclusion that the molar volume of the coil form is larger than the molar volume of the helix. From the change in T_m as a function of pressure, molar volume changes of the transition (ΔV_t) are calculated using the Clapeyron equation and calorimetrically determined enthalpies. For the double-stranded polymers, ΔV_t increases in the order poly[d(I-C)] < poly[d(A-T)] < poly(dA)poly(dT) < poly[d(G-C)]. The value of ΔV_t for the triple-stranded to single-stranded transition of poly(dA) 2poly(dT) is larger than that of poly[d(G-C)]. The magnitude of ΔV_t increases with salt concentration in all cases studied; however, the change of ΔV_t with salt concentration depends on the sequence of the DNA and the number of strands involved in the transition. In the model proposed to explain the results, the overall molar volume change of the transition is a function of a negative volume change arising from changes in the electrostatic interactions of the DNA strands, and a positive volume change due to unstacking the bases. The model predicted the direction of the change in the ΔV_t for several experiments. The magnitude of ΔV_t increases with counter ion radius, thus for poly[d(A-T)], ΔV_t increases in the series Na^+, K^+, Cs^+. The ΔV_t also increases if the charge on the phosphodiester groups is removed. The kinetics of the formation of double-stranded $(dA)_{19}(dT)_{19}$ in 50 mM NaCl are slowed approximately 14-fold at 200 MPa relative to atmospheric pressure. The implied volume of activation of $+37$ ml mol^{-1} in the direction of this change is also in agreement with the proposed model.

Abbreviations used: TrisHCl-Tris(hydroxymethyl)-aminomethane hydrochloride; EDTA-ethylenediaminetetraacetic acid.

The stability of double- and triple-stranded DNA helices in water around neutral pH depends on the base composition and sequence, as well as on the ionic strength of the solution. Each of these dependencies also defines how DNA interacts with water. It has been recognized since the early studies on the physical chemistry of DNA that its interactions with water play an important role in the relative stability of the various conformations (Chapman & Sturtevant, 1969; Clement et al., 1973; Wolf & Hanlon, 1975). More recently, there has been renewed interest in further characterizing water-DNA interactions in an effort to explain sequence-specific ligand interactions, to improve models of the dynamics of DNA structure, and to deepen the general understanding of this molecule.

The dependence of DNA stability on salt concentration and the strong dependence of ionic interactions on hydrostatic pressure led to the idea that the helix-coil transition of DNA might be accompanied by a significant volume change. In early studies, dilatometry and hydrostatic pressure techniques were employed to measure the volume changes of the helix-coil transition of natural sequence DNA. The molar volume of the helical form was found to be smaller than that of the coil form, (that is, the molar volume change for the helix-coil transition is positive, and the transition temperature T_m shifts to higher values with increasing hydrostatic pressure, although large volume changes were not recorded). Using DNA polymers from natural sources with varying base composition, Hawley and MacLeod (1977) analyzed the effect of base composition on the pressure effect and reported that the change in T_m with pressure increases with the GC content of the DNA. These investigators were also the first to report the increase in the volume change for poly[d(A-T)] with salt concentration (Hawley & MacLeod, 1974).

One of the most extreme cases of the effect of solvent conditions on the structure and stability of DNA is the B-Z transition of poly[d(G-C)] (Pohl & Jovin, 1972). The right-handed helical B form of this polymer can be converted to a left-handed helical Z form by adding to the solution a variety of agents that reduce the water activity (for example, inorganic salts, ethanol, and organic polyamines). Given the sensitivity of the stability of ionic interactions in water to changes in the hydrostatic pressure, and the salt dependence of the B-Z transition, it was not surprising that the equilibrium between the two forms of poly[d(G-C)] is sensitive to pressure, with elevated pressure favoring the more hydrated B form (Macgregor & Chen, 1990).

Buckin et al. (1989) published the results of a study using ultrasonic techniques to measure the apparent molar volume (ϕV_h) of the compressible water associated with several synthetic oligonucleotides with repetitive sequences. They observed large differences in the magnitude and sign of ϕV_h dependent on base composition as well as sequence. Thus, ϕV_h equals -14 ml mol^{-1} for poly[d(A-T)], -21 ml mol^{-1} for poly(dA)·poly(dT), and $+2 \text{ ml mol}^{-1}$ for double-stranded $d(CG)_4$. Double-stranded oligonucleotides composed of dAdT base pairs display similar sequence-dependent differences in ϕV_h. They attribute the differences in the apparent molar volumes to the strength of the interaction between the DNA and water, with more negative ϕV_h values being associated with stronger interactions. The rationale for this interpretation lies in the fact that stronger water-DNA interactions will lead to a higher apparent density (lower apparent molar volume) for the interacting water molecules.

Chalikian et al. (1994) have recently published new data on the molar compressibility and partial molar volumes of several natural and synthetic DNA sequences at low salt concentration. Their results indicate a minimum in the extent of hydration as measured by both the apparent volume and the apparent compressibility for sequences with approximately 50% AT content. The results seem to contradict some of the findings of Buckin et al. (1989); however, the difference in experimental conditions may account for the discrepancies.

The results outlined here suggested that it might be interesting to explore the effect of pressure on the helix-coil transition as a function of sequence and salt concentration. The availability of synthetic oligonucleotides and polynucleotides of definable sequence allows us to extend the earlier investigations of the volume difference of the helix and coil forms of DNA and assign the measured volumes to specific sequences.

RESULTS

Effect of Hydrostatic Pressure on the Transition of Polymers Containing Simple Alternating Sequences

Using an instrument we have recently described, which allows us to spectro-scopically monitor helix-coil transitions throughout a wide temperature range and pressures up to 300 MPa (Wu & Macgregor, 1993a), we have measured the change in the helix-coil transition temperature (T_m) as a function of sodium chloride concentration for poly[d(A-T)], poly(dA)poly(dT), poly[d(G-C)], poly[d(I-C)], and the triple-stranded to single-stranded transition of poly(dA)2poly(dT).

dAdT-Containing Polymers

Figure 11.1 shows the transition curves of poly[d(A-T)] and poly(dA)poly(dT) as a function of NaCl concentration at two pressures (Wu & Macgregor, 1993b). The data are represented as the extent of transition (θ) versus temperature. The transition curves remain highly cooperative and shift to higher temperatures at high pressure.

The helix-coil transition temperature (T_m) is defined as the temperature at which $\theta = 0.5$; the T_m of these two polymers is plotted as a function of pressure in Figure 11.2. The shift of T_m to higher temperatures with pressure (Figures 11.1, 11.2) indicates stabilization of the helical form by pressure. The dependence of T_m on pressure (dT_m/dP) increases with the NaCl concentration; this change is not the same for poly[d(A-T)] and poly(dA)poly(dT). Using the rate of change in T_m as a function of pressure, the molar volume change of the helix-coil transition (ΔV_t) was calculated using the Clapeyron equation ($dT_m/dP = T_m \Delta V_t/\Delta H_{cal}$), where P is pressure, ΔH_{cal} is the molar enthalpy of the transition measured calorimetrically, and T_m is the transition temperature at atmospheric pressure. Figure 11.3 shows ΔV_t as a function of the logarithm of the NaCl concentration for the two polymers. The enhanced stabilization of the helical form of DNA by

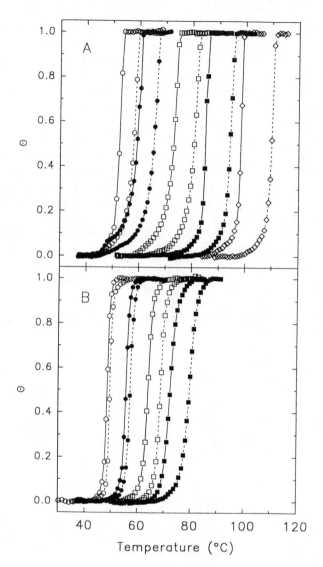

Figure 11.1. Thermal denaturation curves of poly(dA)poly(dT) and poly[d(A-T]) at different NaCl concentrations and pressures. (**A**) Poly(dA)poly(dT): ○-20 mM. ●-50 mM, □-200 mM. Poly(dA)$_2$poly(dT): ■-1 M, ◇-3 M. (**B**) Poly[d(A-T)]: ○-20 mM, ●-50 mM, □-200 mM, ■-1 M. Solid lines represent data acquired at atmospheric pressure (0.1 MPa), broken lines at 200 MPa. In addition to NaCl, the solutions contained 20 mM TrisHCl pH 8.8 (at 20 °C, atmospheric pressure), 0.1 mM EDTA. The concentration of the DNA was approximately 80 μM (base pairs). The parameter θ is calculated according to $\theta(T) = [A(T) - A_L(T)]/[A_H(T) - A_L(T)]$, where $A_H(T)$ and $A_L(T)$ are the temperature dependencies of the high and low-temperature base lines and $A(T)$ is the absorption at temperature T.

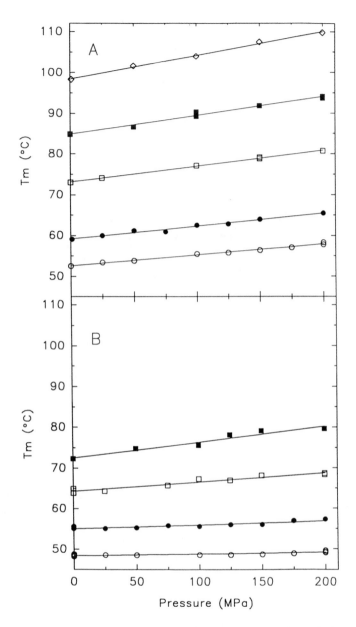

Figure 11.2. Helix-coil transition temperature (T_m) versus pressure for poly(dA)poly(dT) and poly[d(A-T)] at several NaCl concentrations. The lines are least-squares linear fits to the data. (**A**) Poly(dA)poly(dT): ○-20 mM, ●-50 mM, □-200 mM. Poly(dA)$_2$poly(dT): ■-1 M, ◇-3 M. (**B**) Poly[d(A-T)]: ○-20 mM, ●-50 mM, □-200 mM, ■-1 M.

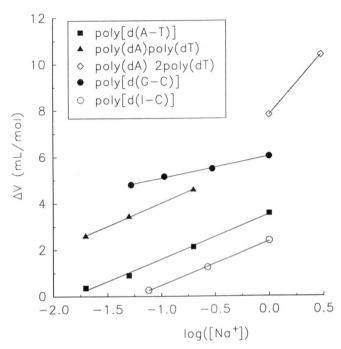

Figure 11.3. Molar volume change of the helix-coil transition (ΔV_t) versus NaCl concentration for poly(dA)poly(dT), poly[d(A-T)], poly(dA)2poly(dT), poly[d(G-C)], and poly[d(I-C)]. Except for poly[d(G-C)], all solutions contained 20 mM TrisHCl, pH 8.8 (20 °C, atmospheric pressure), 1 mM EDTA. Due to the large enthalpy of ionization of TrisHCl, the pH at the temperature of T_m was between 8.1 and 7.3. For poly[d(G-C)], the solutions contained 10 mM sodium phosphate, pH 7.6 (20 °C, atmsopheric pressure), 1 mM EDTA; the pH of the solution at the T_m is estimated to be ~6.8.

pressure ($dT_m/dP > 0$) is manifested as a positive ΔV_t—that is, the molar volume of the denatured coil form is larger than the molar volume of the helix form.

The data for the dAdT polymers, summarized in Table 11.1, illustrate several features about the pressure dependence of the stability of DNA. The molar volume of the helix-coil transition is sequence dependent, the ΔV_t of poly[d(A-T)] is smaller than that of poly(dA)poly(dT) by approximately 2.5 ml mol^{-1}. As mentioned previously, the partial molar volume of the water hydrating poly(dA)poly(dT) is smaller than that of poly[d(A-T)], presumably on account of the stronger interactions formed between water and poly(dA)poly(dT). The volume change of the transition reflects the difference in the extent of hydration between the helix and coil forms; thus, poly(dA)poly(dT) undergoes a larger change in hydration between the these two states. Because ΔV_t is positive, the water-DNA interactions are either weaker or less numerous in the coil form.

The ΔV_t valves of the two double-stranded dAdT polymers change linearly and in a parallel fashion with log[Na$^+$]. The linear change in ΔV_t with the logarithm of the salt concentration is indicative of a relation between this parameter (ΔV_t) and the release of counter ions during the transition to the coil form (Record et al., 1978). As discussed later in greater detail, it is not surprising

Table 11.1. Volume change of the helix-coil transition of poly(dA)poly(dT) and poly[d(A-T)] as a function of NaCl concentration

[NaCl] (M)	$100 \times dT_m/dP(°C/MPa)$	$T_m(°C)$	$\Delta H_{cal}(kJ \, mol^{-1})$	$\Delta V \, (ml \, mol^{-1})$
		Poly(dA)poly(dT)[a]		
0.02	2.49 ($\pm$ 0.10)	52.7	34.0	**2.60** ($\pm$ 0.13)
0.05	3.15 ($\pm$ 0.20)	59.2	36.3	**3.44** ($\pm$ 0.24)
0.20	3.86 ($\pm$ 0.08)	73.1	41.2	**4.59** ($\pm$ 0.15)
1.0	4.50 ($\pm$ 0.24)	85.1	62.2	**7.81** ($\pm$ 0.50)
3.0	5.80 ($\pm$ 0.22)	98.4	66.6	**10.4** ($\pm$ 0.60)
		Poly[d(A-T)]		
0.02	0.36 ($\pm$ 0.08)	48.4	32	**0.36** ($\pm$ 0.09)
0.05	0.93 ($\pm$ 0.17)	55.0	32	**0.90** ($\pm$ 0.20)
0.20	2.26 ($\pm$ 0.24)	64.3	32	**2.14** ($\pm$ 0.35)
1.0	3.86 ($\pm$ 0.46)	72.5	32	**3.57** ($\pm$ 0.62)

[a] 1 and 3 M NaCl are for the triple-stranded transition: poly(dA)$_2$poly(dT) → poly(dA) + 2 poly(dT).

that counter ion release would elicit a volume change due to the volume changes accompanying changes in ionic interactions in water arising from electrostriction. However, it is perhaps unexpected that the effect is linear in log[NaCl], but there is no theory now available to describe the behavior of polyelectrolytes as a function of hydrostatic pressure.

Poly(dA)poly(dT) disproportionates to form one mole of triple-stranded poly(dA)2poly(dT) and one mole of single-stranded ply(dA) at the two highest salt concentrations studied, 1 and 3 M NaCl. Although the melting behavior of this polymer is rather complex at salt concentrations between 0.2 and 1 M (involving thermally induced transitions between the double-stranded and triple-stranded forms), at the salt concentration used in these experiments the transition involves only triple- and single-stranded forms (Klump, 1988a). The ΔV_t for the triple strand–single-strand transition is larger than the ΔV_t of double-stranded poly(dA)poly(dT) by a factor of approximately 1.5 if the data are extrapolated to similar salt concentrations. Thus, it appears that the volume of the helix-coil transition is linearly dependent on the number of strands involved in the transition.

Poly[d(G-C)] and Poly[d(I-C)]

The thermodynamics of poly[d(G-C)] at atmospheric pressure are inconvenient to study because of the extreme stability of this polymer; in aqueous solutions containing more than 20 mM salt, the helix-coil transition temperature is above 100 °C. This has limited study of the factors important in stabilizing this polymer; however, such data would be interesting both for better understanding this polymer and for gaining insight into the possible mechanism of the B-Z transition. This conformational change from a right-handed helix to a left-handed helix must involve a complete opening of the base pairs as an intermediate state. We investigated the helix-coil transition at four NaCl concentrations below what is needed to induce the B-Z transition (Wu & Macgregor, 1994).

The data in Figure 11.3 show that the ΔV_t for this polymer varies from $+4.8$ to $+6.0$ ml mol^{-1} as the NaCl concentration changes from 52 mM to 1 M (Table 11.2). The magnitude of ΔV_t for poly[d(G-C)] is the largest of the double-stranded polymers studied, implying that, of these polymers, poly[d(G-C)] undergoes the largest change in hydration during the transition. As for the dAdT polymers, the ΔV_t changes linearly with the logarithm of the salt concentration. In calculating the volume change via the Clapeyron equation, calorimetrically determined enthalpies of Klump (1988b) were employed. A plot of the pressure dependence of the van't Hoff enthalpy revealed that the magnitude decreases with increasing salt concentration but is independent of pressure.

The salt dependence of the ΔV_t of poly[d(I-C)] is shown in Figure 11.3 and is summarized in Table 11.3. The calorimetric enthalpies used to calculate ΔV_t are taken from Rentzeperis et al. (1992). This polymer exhibits a very small positive volume change at the NaCl concentrations studied. The reason for the small magnitude of the ΔV_t is apparently due to the similarity of the hydration of the helix and coil forms; however, the lack of structural information regarding this polymer makes it difficult to ascribe this observation to any particular structural property. Although there are only three points, the ΔV_t displays a salt dependence similar to the three other double-stranded molecules.

As was found for the dAdT polymers, the ΔV also changes linearly with the logarithm of the salt concentration for poly[d(G-C)] and poly[d(I-C)], implying that the volume change is a function of the release of counter ions. However, the magnitude of the dependence of the ΔV_t on the salt concentration is a function of the composition of the polymer. The slope, $d\Delta V_t/d(\log[Na^+])$, is equal to approximately 2 for the double-stranded dAdT polymers and poly[d(I-C)], whereas $d\Delta V_t/d(\log[Na^+]) \approx 1$ for poly[d(G-C)]. The dependence of the value of $d\Delta V_t/d(\log[Na^+])$ on the polymer's sequence implies that the difference in hydration of the helix and coil forms is more sensitive to the salt concentration for poly(dA)poly(dT), poly[d(A-T)], and poly[d(I-C)] than for poly[d(G-C)]. That is, either the helical form is hydrated more extensively or the coil form is hydrated less extensively as the salt concentration increases, or both; furthermore, this effect depends on the base composition of the DNA. Increased hydration of the helical form at higher salt concentrations is physically unreasonable; thus, it seems likely that the salt dependence is related to decreasing water interactions with the coil form of the polymers. The dependence of this effect on the base composition remains to be explained.

Effect of Pressure on the Enthalpy of the Transition

Because dT_m/dP depends on both the enthalpy and the volume of the transition, it is possible that the pressure dependence of T_m illustrated in Figure 11.2 is also a function of the enthalpy change. Although we currently cannot directly measure ΔH_{cal} at high pressure, it is possible to determine the van't Hoff enthalpy of the transition (ΔH_{vH}) from our data. We have found that the enthalpy data are quite noisy if the melting curves are used directly to find ΔH_{vH}. To reduce error in the enthalpies, the θ versus temperature data are Fourier transformed, high-frequency components contributing less than 1% of the maximum amplitude are removed,

Table 11.2. Volume change of the the helix-coil transition of poly[d[(G-C)] as a function of sodium ion concentration

[Na$^+$] (M)	$100 \times dT_m/dP$ (°C/MPa)	T_m^a (°C)	ΔH_{vH}^b (kJ mol^{-1})	$\Delta V_{t,vH}^b$ (ml mol^{-1})	ΔH_{cal}^b (kJ mol^{-1})	$\Delta V_{t,cal}^b$ (ml mol^{-1})	Ratio[d]
0.052	4.51 (± 0.27)	106.5 (± 0.3)	1010	120 (± 8)	40.4[c]	4.80 (± 0.56)	25
0.107	4.79 (± 0.41)	110.1 (± 0.4)	930	116 (± 11)	41.3	5.16 (± 0.67)	23
0.30	5.01 (± 0.48)	114.7 (± 0.5)	840	109 (± 13)	42.6	5.50 (± 0.74)	20
1.0	6.41 (± 0.40)	115.8 (± 0.4)	670	111 (± 7)	36.6	6.03 (± 0.76)	17

[a] T_m values determined from extrapolation of a least squares fit of the data in Figure 11.2 to atmospheric pressure.
[b] Value at atmospheric pressure.
[c] Data of Klump (1988b). The error in ΔH_{cal} was assumed to be ± 4 kJ mol^{-1}.
[d] Ratio of the van't Hoff to calorimetrically determined enthalpies or volumes. Equals the length of the cooperative unit for the transition.

Table 11.3. Volume change of the helix-coil transition of poly[d(I-C)] as a function of NaCl concentration

[NaCl] (M)	$100 \times dT_m/dP$ (°C/MPa)	T_m(°C)	ΔH^a_{cal}(kJ mol^{-1})	ΔV_t(ml mol^{-1})[b]
0.075	0.28 (± 0.5)	51.9	30.5	**0.26** (± 0.47)
0.27	1.36 (± 0.09)	59.2	30.5	**1.25** (± 0.15)
1.0	2.64 (± 0.14)	63.2	30.5	**2.39** (± 0.27)

[a] Error in the enthalpy assumed to be 3 kJ mol^{-1}.
[b] All ΔV_t calculated using calorimetric enthalpy change.

and the data are back-transformed. The Fourier-filtered data are fitted to a fourth-degree polynomial which is then differentiated. As a check on the validity of this procedure, we find that the maximum of the differentiated curve corresponds within experimental error to the T_m measured directly from the transition curve. The half-width of differentiated transition ($d\theta/dT$ versus T) is used in calculating ΔH_{vH} as described by Marky and Breslauer (1987). Figure 11.4 shows that the van't Hoff enthalpy of poly[d(G-C)] does not depend strongly on pressure. The pressure-induced change in enthalpy for the other polymers is similar (Wu & Macgregor, 1993). In the absence of other data, we assume the pressure dependence of ΔH_{vH} to be representative of the pressure dependence of the calorimetrically determined enthalpy (ΔH_{cal}). Because $d\Delta H_{vH}/dP$ is small (Figure 11.4), we assume that enthalpy changes with pressure make a negligible contribution to the change

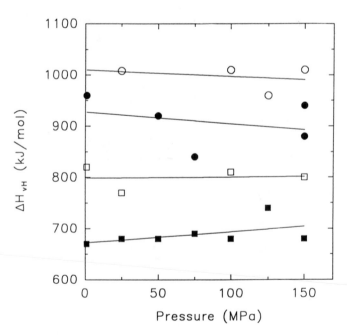

Figure 11.4. Van't Hoff enthalpy (ΔH_{vH}) versus pressure for poly[d(G-C)] in phosphate buffer (pH 7.6 at 20 °C, atmospheric pressure) at four NaCl concentrations: ○-52 mM, ●-107 mM, □-300 mM, ■-1 M.

in T_m with pressure and that the use of calorimetric enthalpies measured at atmospheric pressure will not introduce a significant systematic error in the calculation of the molar volume change.

A Model for Interpreting the Effect of Pressure on the Stability of DNA

These results lead us to propose that the ΔV_t can be written as the sum of two components: a volume change arising from changes in the electrostatic interactions of water and counter ions with the DNA (ΔV_e), and a volume change due to unstacking the bases (ΔV_{us}). The net volume change is given by $\Delta V_t = \Delta V_e + \Delta V_{us}$. Figure 11.5 is a schematic representation of the model. Describing the volume change in terms of two components is certainly a simplification; however, it provides a framework for designing other experiments.

The lower charge density of the coil form of DNA results in the association of fewer counter ions with this conformation than with the helix; thus, one consequence of the helix-coil transition is a net release of counter ions from the DNA. We hypothesize that the cations that move from the vicinity of DNA to the bulk solvent cause a negative volume change due to increased electrostriction of the water in their hydration spheres. Because there is no measurable difference in the inner hydration spheres of cations in the vicinity of DNA or in bulk solution, it is proposed that the volume change involves changes in the interaction of water molecules in the outer hydration sphere with the counter ions. The density of these outer-sphere water molecules is also larger than that of bulk water (Spiro et al., 1968); however, the difference is smaller than that between water molecules in the inner hydration sphere of the ion and bulk water. A large number of released cations each producing a small volume change will elicit the same effect as changes in the inner hydration sphere of the ion resulting from direct ionic binding and

Origin of the Volume Change Due to DNA Denaturation

Water electrostriction around released counter-ions: $\Delta V < 0$

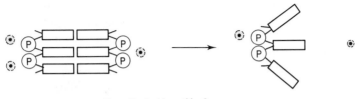

Base Unstacking: $\Delta V > 0$

Figure 11.5. Model of the molecular origin of the effect of hydrostatic pressure on the helix-coil transition temperature. Increasing the temperature to above T_m leads to a loss of base pairing, unstacking of the bases, and counter ion release due to the lower charge density of the single-stranded form. The speckled cricles are intended to depict the positively charged counter ions, and the difference in the size of the circles represents electrostriction effects in outer hydration spheres of the counter ion. Only one of the two resultant single strands is shown.

water release from the DNA structure. From our data, we cannot ascertain if the change is due to a difference in the number of water molecules in the outer hydration sphere or due to stronger interaction between the ion and the outer hydration sphere. Either mechanism would lead to a net decrease in the volume of the system.

According to the model, if ΔV_e is negative, then the volume change associated with unstacking the base pairs, ΔV_{us}, must be positive; that is, unstacking of the bases is accompanied by an increase in the volume of the system. Results from several studies involving model compounds are consistent with this proposal. In one set of experiments, pressure was shown to increase the stacking association of two aromatic molecules (flavin and tryptophan) tethered together by a neutral spacer (Visser et al. 1977). The authors concluded that disruption of the complex between the flavin and the trypotophan leads to a value change of $+5$ ml mol^{-1}. The volume change favors the disruption of noncovalent complexes of stacked nucleic acid bases in aqueous solution by pressure (Høiland et al., 1984); however, the opposite is true for the stacking of isolated, unbound nucleosides. In the latter case, perhaps the orientational constraints placed on the stack complex by the presence of the deoxyribose moiety plays an important role. Another example of the negative volume change that arises due to the formation of complexes of stacked aromatic molecules is given by the volume of intercalation of the aromatic phenanthroline moiety of ethidium between the bases of poly[d(A-T)] and poly[d(G-C)] (Macgregor et al., 1985, 1987). Although the volume of intercalation into poly(dA)poly(dT) is positive, the difference is attributed to the unusual conformation of the bases in this polymer (Marky & Macgregor, 1990).

The effect of hydrostatic pressure on the formation of π-π interactions between aromatic molecules is not simply a manifestation of the hydrophobic effect. Nonaromatic hydrocarbons frequently display an increase in aqueous solubility at elevated pressure; however, in general the effect is difficult to predict a priori (see, for example, Isaacs, 1981).

We therefore conclude from the present data and data on related model systems that the signs of the two components (ΔV_e and ΔV_{us}) are opposite; because the net volume change is small, the magnitudes of the electrostatic and stacking components must be similar. Based on the data of Visser et al. (1977), we estimate that each base pair contributes on the order of $+5$ ml mol^{-1} to the value of ΔV_{us}. Experiments described later in this Chapter, using uncharged methylphosphonate oligonucleotides, corroborate this estimate.

This model predicts that both solvent- and DNA-dependent parameters will influence the effect of pressure on the thermal stability of helical DNA. Changes in the solvent or DNA that lead to diminished importance of electrostriction are expected to cause ΔV_t to become more positive. Similarly, changes that give rise to better base stacking are predicted to increase ΔV_t. The model can also be tested from the point of view of helix formation instead of helix disruption by measuring the effect of pressure on the rate of recombination of single strands to form a double or triple helix. The following sections describe experiments carried out to further characterize the effect of pressure on DNA stability and the model.

Effect of Counter Ion Radius on ΔV_t

The extent to which water molecules in outer solvation spheres will undergo a density increase due to the transfer of counter ions from the vicinity of negatively charged DNA to the bulk solvent on account of the helix-coil transition depends on the charge density of the cation. In the series Na^+, K^+, Rb^+, and Cs^+, the charge density decreases by more than a factor of 5 between Na^+ and Cs^+ due to the relative sizes of the ions (Dean, 1985). Larger radius and lower charge density lead to weaker interactions between water molecules and the ion, and the water-ion interactions will be less dependent on the presence of a nearby anion. Thus, we would predict that the hydration of a larger ion would change less than that of a smaller ion with a high charge density, which, in turn, should mean that the electrostatic term, ΔV_e, should approach zero (that is, become less negative).

The molar volume of the transition for poly[d(A-T)] was measured in solutions of varying concetrations of KCl and CsCl, in addition to the measurements made in NaCl in the previous sections; ΔV_t is plotted as a function of the log([salt]) in Figure 11.6 (Najaf-Zadeh et al. 1995). The transition temperature of poly[d(A-T)] is independent of the nature of the ion at atmospheric pressure. Differences in the interactions between the water, salt, and DNA become apparent only with increasing pressure. In calculating the ΔV_t values, the calorimetric enthalpies of poly[d(A-T)] in NaCl solutions were employed.

At low salt concentrations, the prediction of the model is borne out: the

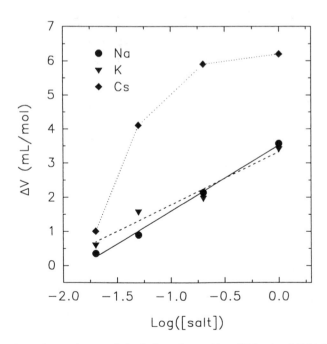

Figure 11.6. Molar volume change of the helix-coil transition (ΔV_t) of poly[d(A-T)] versus salt concentration for three different cations, Na^+, K^+, and Cs^+

magnitude of ΔV_t increases with ionic radius. With increasing salt concentration and Na^+ or K^+ as the counter ion, the observed volume changes become equal. This leveling-off effect implies that other factors dominate ΔV_t at these concentrations, and the difference in ionic radius of Na^+ and K^+ makes a negligible contribution. However, as the concentration of CsCl is increased, ΔV_t increases much more rapidly than in solutions containing the other two cations and then appears to level off at about 6.5 ml mol^{-1}. Thus, ΔV_t follows the predicted trend of increasing with ionic radius; however, the nonlinearity $d\Delta V_t/dP$ was unexpected.

In all except the lowest concentration of CsCl the change in T_m with pressure is also nonlinear; at high pressure dT_m/dP approaches zero, and this is especially evident in the 1 M CsCl solution. This is also in contrast to the behavior of all other systems studied. In 20 mM, CsCl dT_m/dP is linear within the resolution of our instrument. To calculate the molar volume change, the change of T_m with respect to pressure at the lowest pressures (0.1 and 50 MPa) was used. The nonlinearity of T_m with P implies that there is a larger compressibility for the transition of poly[d(A-T)] dissolved in aqueous CsCl than in any of the other solutions studied and that the compressibility increases with CsCl concentration.

The dependence of the equilibrium constant for the helix-coil transition on pressure can be expanded in terms of p

$$RT \ln \frac{K_p}{K_o} = \Delta V_t p - \left(\frac{\partial \Delta V_t}{\partial P}\right) p^2 + \cdots \tag{1}$$

where R is the gas constant, K_p and K_o refer to the equilibrium constant at pressure p and atmospheric pressure, and p_o (= atmospheric pressure) is assumed to be negligible with respect to pressure p. The second term on the right-hand side is related to the compressibility ($\beta = -1/V[\partial V/\partial p]$). The adiabatic compressibilities of many aqueous solutions of electrolytes have been compiled; for monovalent ions β_s decreases with electrolyte concentration. Furthermore, for the series NaCl, KCl, and RbCl, β_s is approximately independent of the type of ion (Schaafs, 1967). Assuming the trend in these data can be extrapolated to the behavior of CsCl in water, the nonlinearity of T_m with p is not due to differences in the compressibility of the solvent, but instead arises from changes in the interactions between poly[d(A-T)] and water.

Chain Length and DNA Stability at High Pressure

Bases located at the ends of double-stranded DNA molecules are less stable than those formed at positions distant from the ends. The base pairs at the ends are said to be frayed, and the extent of fraying depends on temperature and solvent conditions and whether AT or GC base pairs are involved. Fraying causes the bases to become somewhat unstacked, but it does not substantially alter the charge density of the helix. By studying the behavior of a series of oligonucleotides with the same sequence but differing lengths, the relative contribution of fraying to a particular parameter can be assessed. We have measured the effect of pressure on the stability of a series of oligonucleotides, $(dA)_n(dT)_n$, where $n = 11, 15, 19,$ and

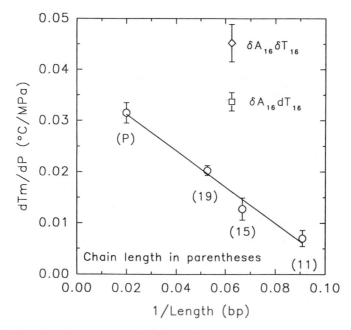

Figure 11.7. The effect of pressure on the helix-coil transition temperature as a function of the chain length of the oligonucleotide and oligonucleotide charge.

the polymer (Macgregor in press). The results are shown in Figure 11.7. The parameter, dT_m/dP, approaches zero with decreasing chain length. The cooperative unit was chosen as the chain length for the polymer.

The number of bases frayed at each end is independent of the chain length and, by decreasing the length, the number of bases in the helical form decrease and thus make less contribution to the observed pressure effect. If we assume that the contribution of fraying is approximately zero for the polymer and that ΔV_t is the volume change for a standard base pair distant from the ends, then we calculate that approximately one base pair is frayed at each end of the oligonucleotides. Further, the volume change for fraying at one end is approximately $+12$ ml mol^{-1} (end). The sign of this volume change is consistent with a decrease in the stacking of the bases at the frayed ends, as described in the model. The effect of pressure on T_m decreases with decreasing chain length because an increasing proportion of the base pairs are unstacked even in the native "helical" form. The loss of stacking interactions means that the initial native form has a larger volume than if all of the possible base pairs were formed; consequently, the volume difference between the helical and coil forms becomes smaller.

The Stability of Nonionic Oligonucleotide Helices at High Pressure

Oligonucleotides that have nonionic methylphosphonate groups between successive bases instead of a phosphodiester linkage can form stable helix structures with other methylphosphonate oligonucleotides (MPO) and with phosphodiester oligonucleotides (PDO). In certain cases, helices consisting of MPOs are more stable (higher T_m) than helices made from PDOs of the same sequence (Kibler-Herzog et al., 1990). MPOs and other covalently modified DNA structures have attracted interest recently because of their potential use in targeting specific DNA sequences in vivo via antisense interactions (Stein & Cheng, 1993).

Helices made from MPOs can be uncharged if both strands are MPO, or have one-half the charge of a conventional helix if one strand is an MPO and the other a PDO. The charge properties of MPO-containing helices, which are so different from standard DNA molecules, should have a significant effect on their interactions with cations. The change in the number of cations that interact with the helix and coil forms should be substantially reduced in the case of MPO:PDO duplexes, and MPO:MPO duplexes should have no ionic interactions in either helix or coil. A consequence of this is the insensitivity of the T_m of MPO:MPO helices to salt concentration.

If the difference in the extent of ionic interactions with the helix and coil forms is decreased significantly, then the model predicts that the volume change arising from changes in the electrostatic interactions with the DNA should vanish. Having lost this negative contribution to the volume change of the transition, the magnitude of ΔV_t is expected to increase. Figure 11.7 shows that the removal of charge on the DNA does, indeed, cause the value of dT_m/dP to increase.

Because the calorimetric enthalpy changes have not been measured, the data from this series of experiments are reported in terms of dT_m/dP instead of ΔV_t to avoid confusing volume changes calculated using van't Hoff enthalpies with the volume changes calculated with calorimetric enthalpy changes reported for DNA polymers. However, to verify that the differences observed in dT_m/dP were not due to enthalpy changes, the van't Hoff enthalpies were determined from the slope $(d\theta/dT)$ at the midpoint of the transition. The enthalpy values show no obvious trend as the pressure is changed, although as before, the ΔH_{vH} values are uncertain. The average value of ΔH_{vH} at all pressures for $(\delta A)_{16}(dT)_{16}$ equals 440 kJ mol^{-1} which, within experimental error, is equal to 470 kJ mol^{-1}, the average ΔH_{vH} for $(\delta A)_{16}(dT)_{16}$. The van 't Hoff enthalpy of the fully uncharged duplex $(\delta A)_{16}(\delta T)_{16}$, is equal to 220 kJ mol^{-1}, approximately half the value of the other two; the difference is perhaps due to an uncharacterized structural difference in the helix or coil form of $(\delta A)_{16}(\delta T)_{16}$ such as greater end fraying.

One important structural difference is that the methylphosphonate groups are a racemic mixture, whereas the phosphodiester groups of standard DNA are not chiral. Although other work has indicated the importance of the stereo-chemistry in the stability of helices formed by MPOs (Quartin & Wetmur, 1989; Summers et al., 1987), it appears from the similarity of the values of ΔH_{vH} for $(dA)_{15}(dT)_{15}$ and $(\delta A)_{16}(dT)_{16}$ that the stereochemistry is not too important in the values of ΔH_{vH}; however, until racemicly pure MPOs are available, the full ramifications of stereochemistry on the stability of helices formed from these oligonucleotides will remain unknown. With this caveat, the values of dT_m/dP apparently reflect changes in the molar volume of the transition. Thus, as predicted by the model, the volume increases when the charge on the DNA is removed. However, the increase of dT_m/dP for $(\delta A)_{16}(\delta T)_{16}$ is more than compensated for by a decrease in ΔH_{vH}; thus, the magnitude of ΔV_t increases in the following order, $(dA)_{16}(dT)_{16} < (\delta A)_{16}(\delta T)_{16} < (\delta A)_{16}(dT)_{16}$. From this apparent ordering, we infer that the molecular origin of the overall volume change involves more than the two factors examined in the model.

Effect of Pressure on the Denaturation/Renaturation Hysteresis of dA$_{19}$dT$_{19}$

To a first approximation, the model of the influence of pressure on the thermal denaturation of DNA accounts for all of the effects we have observed. However, in the results just presented, it has only been tested from the viewpoint of the transition from multiple-stranded DNA to single-stranded DNA by measuring dT_m/dP; a further test of the model involves describing the presure dependence of the reverse reaction, namely the formation of multiple- (double: or triple-) stranded helices from single strands. Because of the extreme stability of DNA helices, conventional techniques for studying their formation (for example, titration) are not feasible, especially at high pressure. We chose instead to measure the reformation of the helical structure at different pressures by cooling solutions of thermally denatured DNA. According to the model, the amount of helical structure in isolated single strands should increase with pressure because ΔV_{us} is greater than zero, and $\Delta V_{us} > \Delta V_e$. Single-stranded DNA has a certain amount of helical structure even at atmospheric pressure; the actual amount depends on the sequence of the DNA and solvent conditions. Pressure should decrease the fraction of random coil structure and increase the fraction of stacked bases, and regions along the length of DNA containing single-stranded structure should be less conformationally flexible (the persistence length increases). A consequence of the reduced flexibility is that it should be more difficult to form double- (or triple-) stranded helices.

The reformation of a helical structure starting from single strands can be monitored by carrying out the thermal denaturation measurement in reverse—that is, by measuring the absorption of the DNA-containing solution as it is slowly cooled, in which case the absorption is observed to decrease slowly as helix is formed. The renaturation curve will not be superimposable on the denaturation curve if the rate of reformation of the helix is slow relative to the cooling rate. Because we expect the single strands to become less flexible as the pressure is increased, the rate of reformation of helix is also expected to decrease. Therefore, we hypothesized that the hysteresis between the denaturation and renaturation curves should increase with pressure.

Figure 11.8 shows the result of carrying out denaturation/renaturation of dA$_{19}$dT$_{19}$ in 50 mM NaCl at three pressures. In this figure, the curves are paired, with one set for each of the three pressures. For each pair, the curve on the right (higher temperatures) is denaturation, and the curve on the left (lower temperatures) is renaturation. The heating and cooling rates were $\sim 0.3\,°$C/minute. The difference between the curves at their midpoints increases from 1 °C at atmospheric pressure to 3.5 °C at 200 MPa.

Knowing the heating rate and cooling rate, denaturation and renaturation curves such as those shown in Figure 11.8 can be used to calculate the rate constants for the reaction leading to helix formation between two single strands

$$2ss \underset{k_{-1}}{\overset{k_1}{\rightleftarrows}} ds \tag{2}$$

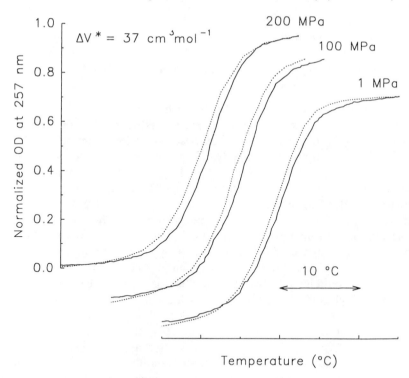

Figure 11.8. The denaturation/renaturation hysteresis of $(dA)_{19}(dT)_{19}$ in 50 mM NaCl, 20 mM TrisHCl, pH 8.8 (20 °C, atmospheric pressure) at three different pressures.

where k_1 and k_{-1} are the forward and reverse rate constants, and ds and ss refer to double- and single-stranded DNA. We define α, the fraction of strands present, as helices

$$\alpha = \frac{[ds]}{[ss]_o} \; ; \; [ss]_o = [ss] + 2[ds] \tag{3}$$

where $[ss]_o$ is the total concentration of strands. The rate of change of α with respect to temperature, the heating (or cooling) rate, and the rate equation for the reaction can be written as

$$\frac{d\alpha_{h,c}}{dT} = \left(\frac{dT}{dt}\right)^{-1}_{h,c} \{k_1[ss]_o(1 - 2\alpha_{h,c})^2 - k_{-1}\alpha_{h,c}\} \tag{4}$$

The subscripts h and c refer to the heating and cooling curves, T is the temperature, and t is time (Rougée et al., 1992). For any pair of heating and cooling curves, the forward and reverse rate constants can found using Eq. 4.

For the data shown in Figure 11.8, the forward rate constants vary from $6.2 \times 10^5 \, M^{-1}s^{-1}$ at atmospheric pressure to $0.43 \times 10^5 \, M^{-1}s^{-1}$ at 200 MPa, a greater than 14-fold reduction in the rate of the reaction. The rates at atmospheric pressure differ in the expected manner from rates measured under different conditions, with oligonucleotides of shorter chain length using temperature jump relaxation kinetics (Craig et al., 1971; Pörschke & Eigen, 1971). The value of k_{-1} is equal to 0.063 s^{-1} and is apparently independent of pressure.

Assuming a two-state transition, the pressure dependence of the rate constants is given by

$$k_1[ss]_0 = k_1^0[ss]_0 \exp\left[-\left(\frac{\Delta V^*}{RT}\right)(P - P_0)\right] \tag{5}$$

ΔV^* is the molar volume of activation for the transition, $R = 8.31$ ml MPa/(K mol), and T is the absolute temperature. The subscript o, refers to a standard reference state chosen to be atmospheric pressure.

On performing these calculations on the data in Figure 11.8, one obtains a molar volume of activation, ΔV^*, equal to $+37$ ml mol^{-1}. Thus, the volume is positive, in agreement with the hypothesis that elevated pressure leads to an increase in the amount of stacking in the single-stranded form. To form double-stranded DNA from the isolated strands, a nucleation complex must be formed, and the high-pressure data imply that the relative stability of the nucleation complex relative to that of the double strands is determined in part by the extent that the single strands form helical stacked structures. Extensive stacking of the *ss* forms decreases the probability that an initial nucleation complex between single strands will lead to formation of a helix. Thus, factors that modulate the amount of stacking in the single strands are expected to affect the rate of helix formation.

DISCUSSION

Over the last 30 years, the behavior of the helix-coil transition at elevated hydrostatic pressure has been studied for some natural sequence DNA and RNA molecules under a variety of salt conditions (Gunter & Gunter, 1972; Hughes & Steiner 1966; Weida & Gill, 1966; Hawley & MacLeod, 1974, 1977; Nordmeier, 1992). The volume changes of the helix-coil transition we have measured for simple repetitive sequence polymers define the upper and lower extremes of this parameter determined for these natural sequence polymers.

However, Marky and coworkers have used densitometry to observe a significantly larger slat dependence of the ΔV_t of poly(dA)-poly(dT) and poly(rA)-poly(dT) at 20 °C (Rentzeperis et al., 1993). At low ionic strength, they obtain molar volume changes similar in sign and magnitude to our values; however, they observe much larger volume changes at higher (100 mM) salt concentrations. Although the origin of this effect is not understood, it may be related to the temperature difference between the two measurements due to a temperature dependence of the molar volumes of one or both (helix, coil) of the states of the polymers. The difference between our data and that of Rentzeperis et al. (1993) emphasizes the importance of using both of these volume techniques for the characterization of DNA conformational forms. Mixing dilatometry yields model independent values for the molar volumes and can be used to measure the volume of helix formation at any temperature; however, these capabilities must be weighed against the necessity of mixing (not all DNA sequences can be used) and the extreme care required for the measurements (that is, extremely good temperature control). Determining the effect of pressure on the helix-coil transition curves has the advantage of relative experimental ease and the capability of being used for any sequence; the disadvantages are the reliance on a one-step model of the transition, the determination of the pressure effect at temperatures near the T_m, and the fact that only changes in volume can be measured.

Taken together, the results presented here and those of other laboratories highlight the fact that sequence-dependent volumetric changes occur upon formation of DNA helices. Clearly, the goal of this work is not the understanding of DNA at high pressure, per se. Experiments using elevated hydrostatic pressure and other volumetric measurements have as their goal the understanding of the thermodynamics and kinetics of DNA conformational changes in vivo. Interest in these types of measurements comes from the hypothesis that volume changes are related to differences in the interaction of the helix and coil forms of DNA with water and that hydration changes are crucial to DNA function in the cell. Future work will focus on investigating pressure effects on RNA and unusual DNA structures, elucidating the molecular origins of these changes, studying the role of volume changes in the kinetics of helix formation, and then using this knowledge to deepen our understanding of the processes involving DNA in vivo.

ACKNOWLEDGMENTS: This research was supported by the National Institutes of Health (GM 47358) and the Connaught Foundation.

REFERENCES

Buckin, V. A., Kankiya, B. I., Bulichov, N. V., Lebedev, A. V., Gukovsky, I. Ya., Chuprina, V. P., Saravazyan, A. P., & Williams, A. R. (1989). Measurement of anomalously high hydration of $(dA)_n(dT)_n$ double helices in dilute solution. *Nature* **340**, 321–322.

Chalikian, T. V., Sarvazyan, A. P., Plum, G. E., & Breslauer, K. J. (1994). Influence of base composition, base sequence, and duplex structure on DNA hydration: apparent molar volumes and apparent molar adiabatic compressibilities of synthetic and natural DNA duplexes at 25 °C. *Biochemistry* **33**, 2394–2401.

Chapman, R. E., Jr., & Sturtevant, J. M. (1969). Volume changes accompanying the thermal denaturation of deoxyribonucleic acid. *Biopolymers* **7**, 527–537.

Clement, R. M., Sturm, F., & Daune, M. P. (1973). Interaction of metallic cations with DNA: VI. Specific binding of Mg^{2+} and Mn^{2+}. *Biopolymers* **12**, 405–421.

Craig, M. E., Crothers, D. M., & Doty, P. (1971). Relaxation kinetics of dimer formation by self complementary oligonucleotides. *J. Mol. Biol.* **62**, 383–401.

Dean, J. A. (1985). *Lange's Handbook of Chemistry*, 13th ed. New York, McGraw-Hill. pp. 10–54.

Gunter, T. E., & Gunter, K. K. (1972). Pressure dependence of the helix-coil transition temperature for polynucleic acid helices. *Biopolymers* **11**, 667–678.

Hawley, S. A., & MacLeod, R. M. (1974). Pressure-temperature stability of DNA in neutral salt solutions. *Biopolymers* **13**, 1417–1426.

Hawley, S. A., & MacLeod, R. M. (1977). The effect of base composition on the pressure stability of DNA in neutral salt solution. *Biopolymers* **16**, 1833–1835.

Høiland, H., Skauge, A., & Stokkeland, I. (1984). Changes in partial molar volumes and isentropic partial molar compressibilities of stacking of some nucleobases and nucleosides in water at 298.15 K. *J. Phys. Chem.* **88**, 6350–6353.

Hughes, F., & Steiner, R. F. (1966). Effects of pressure on the helix-coil transitions of the poly(A)poly(U) system. *Biopolymers* **4**, 1081–1090.

Isaacs, N. S. (1981). *Liquid Phase High Pressure Chemistry.* New York, Wiley, pp. 115–117.

Kibler-Herzog, L., Kell, B., Zon, G., Shinozuka, K., Rahman, S. M., & Wilson, W. D. (1990). Sequence dependent effects in methylphosphonate deoxyribonucleotide double and triple helical complexes. *Nucleic Acids Res.* **18**, 3545–3552.

Klump, H. H. (1988a). Energetic of order/disorder transitions in nucleic acids. *Can. J. Chem.* **66**, 804–811.

Klump, H. H. (1988b). In *Biochemical Thermodynamics*, 2nd ed., ed. M. N. Jones. New York, Elsevier Science Publishers, pp. 100–144.

Macgregor, R. B., Jr. (1995). Chain length and oligonucleotide stability at high pressure. *Biopolymers*, in press.

Macgregor, R. B., Jr., & Chen, M. Y. (1990). Na^+-induced B-Z transition is positive. *Biopolymers* **29**, 1069–1076.

Macgregor, R. B., Jr., Clegg, R. M., & Jovin, T. M. (1985). Pressure-jump study of the kinetics of ethidium bromide binding to DNA. *Biochemistry* **24**, 5503–5510.

Macgregor, R. B., Jr., Clegg, R. M., & Jovin, T. M. (1987). Viscosity dependence of ethidium-DNA intercalation kinetics. *Biochemistry* **26**, 4008–4016.

Marky, L. A., & Breslauer, K. J. (1987). Calculating thermodynamics data for transitions of any molecularity form equilibrium melting curves. *Biopolymers* **26**, 1601–1620.

Marky, L. A., & Macgregor, R. B., Jr. (1990). Hydration of dAdT polymers: role of water in the thermodynamics of ethidium and propidium intercalation. *Biochemistry* **29**, 4805–4811.

Najaf-Zadeh, R., Wu, J. Q., & Macgregor, R. B., Jr. (1995). Effect of cations on the volume of the helix-coil transition of poly[d(A-T)]. *Biochim. Biophys. Acta* **1262**, 52–58.

Nordmeier, E. (1992). Effects of pressure on the helix-coil transition of calf thymus DNA. *J. Phys. Chem.* **96**, 1494–1501.

Pohl, F. M., & Jovin, T. M. (1972). Salt-induced cooperative conformational change of a synthetic DNA: equilibrium and kinetic studies with poly(dGdC). *J. Mol. Biol.* **67**, 375–396.

Pörschke, D., & Eigen, M. (1971). Cooperative non-enzymatic base recognition. *J. Mol. Biol.* **62**, 361–381.

Quartin, R. S., & Wetmur, J. G. (1989). Effect of ionic strength on the hybridization of oligodeoxynucleotides with reduced charge due to methylphosphonate linkages to unmodified oligodeoxynucleotides containing the complementary sequence. *Biochemistry* **28**, 1040–1047.

Record, M. T., Jr., Anderson, C. F., & Lohman, T. M. (1978). Thermodynamic analysis of ion effects on the binding and conformational equilibria of proteins and nucleic acids: the roles of ion association or release, screening, and ion effects on water activity. *Q. Rev. Biophys.* **11**, 103–178.

Rentzeperis, D., Kupke, D. W., & Marky, L. A. (1992). Differential hydration of homopurine sequences relative to alternating purine/pyrimidine sequences. *Biopolymers* **32**, 1065–1075.

Rentzeperis, D., Kupke, D. W., & Marky, L. A. (1993). Volume changes correlate with entropies and enthalpies in the formation of nucleic acid homoduplexes: differential hydration of A and B conformations. *Biopolymers* **33**, 117–125.

Rougée, M., Faucon, B., Mergny, J. L., Barcelo, F., Giovannangeli, C., Garestier, T., & Hélène, C. (1992). Kinetics and thermodynamics of triple-helix formation: effects of ionic strength and mismatches. *Biochemistry* **31**, 9269–9278.

Schaafs, W. (1967). *Landolt-Börnstein Tabellen und Zahlenwerte*, Berlin, Springer Verlag, 5:5–126.

Spiro, T. G., Revesz, A., & Lee, J. (1968). Volume changes in ion association reactions. Inner- and outer-sphere complexes. *J. Am. Chem. Soc.* **90**, 4000–4006.

Stein, C. A., & Cheng, Y.-C. (1993). Antisense oligonucleotides as therapeutic agents—is the bullet really magical? *Science* **261**, 1004–1012.

Summers, M. F., Powell, C., Egan, W., Byrd, R. A., & Wilson, W. D. (1987). Alkyl phosphotriester modified oligodeoxynucleotides: VI. NMR and UV spectroscopic studies of ethyl phosphotriester (Et) modified R_p-R_p and S_p-S_p duplexes, [d[GGAA(Et)TTCC]]$_2$. *Nucleic Acids Res.* **14**, 7421–7436.

Visser, A. J. W. G., Li, T. M., Drickamer, H. G., & Weber, G. (1977). Volume changes in the formation of internal complexes of flavinyltryptophan peptides. *Biochemistry* **16**, 4883–4886.

Weida, B., & Gill, S. J. (1966). Pressure effect on deoxyribonucleic acid transition. *Biochim. Biophys. Acta* **112**, 179–181.

Wolf, B., & Hanlon, S. (1975). Structural transitions of DNA in aqueous electrolyte solutions: II. The role of hydration. *Biochemistry* **14**, 1661–1670.

Wu, J. Q., & Macgregor, R. B., Jr. (1993a). Pressure dependence of dAdT polymers. *Biochemistry* **32**, 12531–12537.

Wu, J. Q., & Macgregor, R. B., Jr. (1993b). A temperature-regulated iso-hyperbaric spectrophotometer: construction and performance chanracteristics. *Anal. Biochem.* **211**, 66–71.

Wu, J. Q., & Macgregor, R. B., Jr. (1995). Pressure dependence of the helix-coil transition temperature of poly[d(G-C)]. *Biopolymers* **35**, 369–376.

12

Application of Pressure Relaxation to the Study of Substrate Binding to Cytochrome P-450$_{cam}$ versus Temperature, Pressure, and Viscosity

GASTON HUI BON HOA, CARMELO DI PRIMO,
ERIC DEPREZ, and PIERRE DOUZOU

The pressure-jump relaxation technique is a convenient and interesting means of studying rapid reversible reactions of biological systems. According to the change in reaction volume that accompanies a biochemical process, a rapid pressure change Δp produces a relative equilibrium shift $\Delta K/K$, which is given by $\Delta \ln K = \Delta K/K = -(\Delta V°/RT)\, \Delta p$, where $\Delta V°$ is the reaction volume change. If the pressure change has a very short transition time, then relaxation kinetic measurements near equilibrium are possible, allowing the elucidation of reaction mechanisms through the detection of eventual reaction intermediates and the characterization of elementary kinetic and thermodynamic parameters. Our reversible pressure-jump method described in this chapter is capable of producing a sharp pressure change of ± 20 MPa in less than 3 milliseconds allowing the determination of relaxation rates in the time range of several milliseconds to several minutes at any final pressure up to 400 MPa, and in any viscosity solution. This technique was employed to study the binding kinetics of camphor and its analogues to bacterial cytochrome P-450$_{cam}$ as functions of temperature, pressure, and viscosity. The results obtained are discussed in terms of conformational dynamics of the protein associated with the entry and the exist of water molecules and specific interactions of the substrate with the apolar residues in the active site of cytochrome P-450$_{cam}$.

The binding of ligands or substrates to proteins can exhibit multistate kinetic behavior similar to transient-state enzyme kinetics and isomerizations of proteins.

The underlying elementary reaction mechanisms can be elucidated by the use of rapid mixing techniques. Usually a reaction is initiated by mixing the reactants as rapidly as possible, and the approach to equilibrium is monitored. This method has been adapted to the study of enzyme reaction mechanisms under extreme conditions of temperature and pressure (Hui Bon Hoa & Douzou, 1973; Balny et al., 1984). However, this approach is limited by the deadtime, the large amount of sample required, and difficulties in using the apparatus to study viscous solutions, such as Schlieren effects, caused by incomplete mixing in flow experiments. Relaxation techniques overcome these problems by the application of a physical perturbation to a system already at equilibrium. In this way, kinetic experiments can be extended to relaxation times in the microsecond range (Eigen & De Mayer, 1962). There are two means of perturbing a system at equilibrium: temperature-jump and pressure-jump methods. In the temperature-jump method, rapid heating pulses are generated by microwaves, laser pulses, and electrical discharges (Rigler et al., 1974). The fast temperature rise then induces a rapid shift in the equilibrium constant of most biochemical processes which are characterized by significant reaction enthalpies. The pressure-jump method involves the application of a static pressure to a solution. The sharp change in pressure is usually produced by bursting a membrane (Goldstack et al., 1989). Because most biochemical processes are accompanied by changes in reaction volumes, a rapid pressure change, Δp, produces a relative equilibrium shift $\Delta K/K$ which is given by $\Delta \ln K = \Delta K/K = -(\Delta V^{\circ}/RT)\,\Delta p$, where ΔV° is the reaction volume change, R is the gas constant, and T is the absolute temperature. To obtain a relative shift, $\Delta K/K = 0.10$, for a reaction volume $\Delta V^{\circ} = 50$ ml mol^{-1} at a temperature $T = 300$ K, it is necessary to apply a pressure change of about 5 MPa. Similarly, the dependence of the rate constant k on the pressure is $\Delta \ln k = -(\Delta V^{*}/RT)\,\Delta p$, where ΔV^{*} is the activation volume change. According to Le Chatelier's principle, high pressure inhibits the rate or extent of reactions that result in an increase in the volumes of activated complexes or end products. Conversely, high pressure promotes reactions that decrease overall system volumes. These volumes changes generally differ in sign and absolute value, so that a change in pressure may have sharply different effects on different processes. In many cases, the volume changes are more likely to arise from changes in water density around proteins, ions, and substrates, or in the active site of proteins, than from changes in the volumes of the macromolecules themselves (Jaenicke et al., 1988, Somero, 1940).

Instead of using P-jump (pressure-jump) techniques that rely on the destruction of various rupture discs to obtain different bursting pressures, we have used a somewhat slower technique in which a valve is opened rapidly between the pressure cell and a reservoir chamber. The initial pressure of the cell can be adjusted to P_0, the valve closed and the reservoir chamber pressure was chosen to be equal to $P_0 \pm \Delta P$. The sharp change of pressure was obtained by quickly valving off the volume of the reservoir chamber into the high-pressure cell. The resulting pressure jump, typically 20 MPa, can be positive or negative, and the final pressure of the cell ranges from 0.1 to 400 MPa. This reversible P-jump technique is more convenient and permits the determination of relaxation kinetics as a function of pressure (during a positive- or negative-increment P-jump experiment). A plot of the logarithm of the measured rate constant versus pressure yields the activation volumes for the transition states. Our method can also be used to study the effect of viscosity on bimolecular reactions. The viscosity dependence of the rate constant most commonly used is of the form described by Kramers (Kramers, 1940): $k = (A/\eta) \exp(-(\Delta E^* + p\Delta V^*)/RT)$, where η is the viscosity of the solvent, ΔE^* is the activation barrier, and ΔV^* is the activation volume; A is a coefficient that describes the barrier shape (Larson & Kostin, 1980) and contains the activation entropy. We point out that the viscosity is temperature dependent: $1/\eta = \exp(\Delta E\eta/RT)$. Thus, the observed energy will be the sum $\Delta E_\eta + \Delta E^*$, where ΔE^*, where ΔE^* represents the barrier characteristic of the system measured at constant viscosity.

Preliminary kinetic studies were carried out on a bacterial enzyme, cytochrome P-450$_{cam}$ from *Pseudomoneas putida*. The three-dimensional structure is known (Poules et al., 1987). The enzyme is a monooxygenase which is responsible for the regio- and stereo-specific hydroxylation of camphor (Atkins & Sligar, 1989). The substrate-free ferric form of P-450 is low spin and hexacoordinated, with its Soret absorbance centered around 417–419 nm. The binding of camphor to the enzyne shifts the Soret absorption band to 390 nm, and results in changes in the spin state, from the low spin to the high spin, and the redox states of the heme becomes pentacoordinated. These effects are correlated with the activity of the enzyme and are of major importance in considering the mechanism of action of cytochrome P-450. In addition, relaxation times are slow (several milliseconds) following pH and potassium ion perturbations from equilibrium of camphor-bound ferric cytochrome P-450, as monitored by the absorbance (spin) change (Lange et al., 1979). This indicates that the spin transition might involve a conformational change. We have investigated by p-jump techniques, the kinetics of the interaction of camphor, and analogues of camphor with bacterial cytochrome P-450$_{cam}$. We have also examined the viscosity dependence of the relaxation rate constants of the camphor-bound enzyme complex.

EXPERIMENTAL PROCEDURES

Bacterial cytochrome P-450$_{cam}$ was generated and purified in the laboratory of S. G. Sligar in the University of Illinois, at Urbana-Champaign. The degree of purity of the protein was determined from the ratio of the absorbance at 392 and 280 nm. Optical titrations as a function of temperature and pressure were recorded on an Aminco DW2 spectrophotometer equipped with a cryothermostat and a high-pressure bomb (Hui Bon Hoa et al., 1982).

The P-jump apparatus consists of a high-pressure bomb with flat, sapphire windows polished to optical flat, and a high-pressure reservoir chamber in which a preset high pressure can be chosen, higher or lower than the pressure in the bomb. One extremity of the reservoir is connected to the bomb through a high-pressure air-operated valve; the other extremity of the reservoir is connected to a 700 MPa high-pressure generator by a manual valve. A quartz piezoelectric transducer (Kistler Co.) is mounted inside the bomb to record the time course of the pressure change and to trigger the scanning of a numerical oscilloscope (Lecroy). The high-pressure sample cell was a 400-µl Pyrex cuvette with an optical path of 0.5 cm. A teflon membrane separated the sample from the pressurizing fluid–pentane or heptane. Stable temperature control was obtained by surrounding the bomb with a copper jacket. When in operation, both valves are opened to set an initial pressure in the bomb. The air-operated high-pressure valve is then closed, and a pressure higher or lower than the pressure in the bomb is delivered to the reservoir. After the manual valve is closed, a fast positive or negative pressure jump, typically 10 or 20 MPa, is delivered by electrical switching of the air-operated valve. The total pressure change occurs in less than 3 milliseconds, as probed by the output signal change of the transducer and the absorbance change at 450 nm of a solution of paranitrophenol dye, which has an ionization volume of -11 ml mol^{-1}. Pressure-jump experiments were combined with the dual wavelength detection mode of the spectrophotometer. The first wavelength was centered on the maximum absorbance of the low spin species (417 nm) and the second wavelength on the isosbestic point (404 nm), or nonabsorbing region. D-camphor, d-fenchone, and 5-bromocamphor were purchased from Aldrich Chemical Co. The camphor-binding formalism used for analysis of kinetic data was a specified three-state model: camphor enters the low-spin species of P-450 which absorbs at 417 nm without any change of the spectrum. The first diffusion step is followed by a relaxation step which converts the protein from the low spin (LS) to the high-spin (HS) state, which absorbs at 392 nm. This second step should correspond to the specific binding of the substrate to the active set. The observed quantities are the spin equilibrium constant $K = [\text{HS}]/[\text{LS}]$, and the observed relaxation rate constant, k_{obs}

$$P_{ls} + C \; \underset{}{\overset{K_1^d}{\rightleftharpoons}} \; P_{ls} - C \; \underset{}{\overset{K_2}{\rightleftharpoons}} \; P_{hs} - C \tag{1}$$

$$k_{obs} = k_{-2} + k_2 (P + C)_f / K_1^d + (P + C)_f \tag{2}$$

where $P =$ P-450 and $C =$ camphor,

$$K_1^d = k_{-1}/k_1, \quad K_2 = k_2/k_{-2} \tag{3}$$

The dissociation constant, K_1^d, of the first step, is equal to the ratio between the rate constants for the exit (k_{-1}) and the entry (k_1) of camphor. The spin equilibrium constant of the second step is equal to

$$K_2 = k_2/k_{-2} \tag{4}$$

At saturation with the substrate, the overall relaxation rate constant is

$$k_{obs} = k_2 + k_{-2} \tag{5}$$

the sum of high spin and low spin relaxation rate constants. At low concentration of the substrate

$$k_{obs} = k_2 (P + C)/K_1^d + k_{-2} \tag{6}$$

The slope of the plot gives k_2/K_1^d and the intercept is k_{-2}. At saturating substrate, the individual rate constants k_2 and k_{-2} can be calculated by combining the measured quantities k_{obs} and K_2. Semilogarithmic plots of these rate constants versus pressure yield the activation volumes ΔV_2^* and ΔV_{-2}^*. The viscosity dependence of the spin relaxation rate constants under conditions of saturating camphor was probed by using P-450 samples on the presence of 20%, 40%, 60%, and 90% glycerol (w/w) and 50% and 60% sucrose as a function of low temperatures. The results were then extrapolated to 0.1 MPa and at 4 °C.

RESULTS AND DISCUSSION

We have used the simple and reversible P-jump apparatus described here to follow relaxation kinetics of proteins versus pressure from 0.1 MPa to 400 MPa. The apparent relaxation rate constants, obtained by following the absorbance changes of P-450 between 417 and 404 nm after positive or negative P-jumps, were pressure dependent and camphor concentration dependent, as shown in Figure 12.1. All kinetic traces exhibited simple first-order kinetics. The relaxation rate constant k_{obs} decreased slowly as the pressure was raised from 200 to 100 MPa and increased as the camphor concentration was increased from 75 to 900 mM. In plots of k_{obs} at fixed pressure versus camphor concentration, the curves exhibited saturation behavior at high substrate concentrations (Figure 12.2), indicating that the binding of camphor to cytochrome P-450$_{cam}$ is a two-step process. Both steps are affected by pressure. When the pressure was increased from 20 MPa to 100 MPa, k_2 and k_{-2} decreased by factors of 1.5 and 1.2, respectively; this resulted in a decrease of the spin-state equilibrium constant K_2 by factor of 11.6. The diffusion step also decreased, with K_1^d increasing by a factor of 2.1.

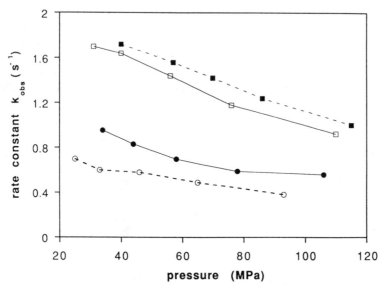

Figure 12.1. P-jump spin relaxation rate constants versus pressure at different concentrations of camphor: 75 μM (○), 150 μM (●), 600 μM (□), and 900 μM (■). Experimental conditions were 100 mM Tris-ethylene glycol 50% (v/v) buffer, 60 mM KCl, 12 μM P-450. The pH was 7.2 and the temperature was 0° C.

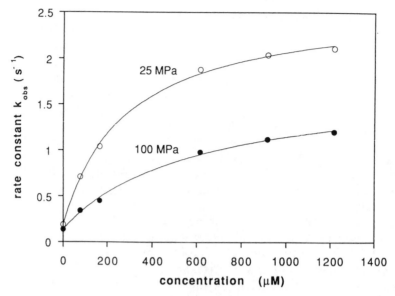

Figure 12.2. Observed spin relaxation rate constants versus camphor concentrations at two pressures: 25 MPa (○) and 100 MPa (●). Experimental conditions as described in Figure 12.1.

Under conditions of saturating substrate, 800 μM of camphor in Tris-ethylene glycol, it is possible to calculate the individual rate constants (k_2 and k_{-2}) of the second step from the measurements of k_{obs} and K_2 versus pressure as indicated above in Eqs. 1–6. Figure 12.3 shows the results from a series of postive and negative pressure jumps applied to the same sample. The semilogarithmic plots of the individual high-spin (k_2) and low-spin (k_{-2}) rate constants exhibited linear behavior. Extrapolation of the straight lines to 0.1 MPa gave the individual rate constants at atmospheric pressure. The corresponding activation volumes calculated from the slope of the lines gave $\Delta V_2^* = 37.4$ ml mol^{-1} and $\Delta V_{-2}^* = -6.7$ ml mol^{-1}.

The relaxation rate constants measured with samples in aqueous Tris buffer at the same temperature showed a similar dependence on pressure and degree of saturation with substrate. However, the kinetics were faster, indicating that the viscosity effect of ethylene glycol slows the reaction. Saturating levels of two substrates (fenchone and 5-bromocamphor, which differed in the position of two methyl groups and the substitution of a bromide atome) were used to probe the effects of steric and hydrophobic interactions at the active site on the spin relaxation kinetics of the second step. Figure 12.4 shows the dependence of the rate constants on the concentration of a specific monovalent cation, K$^+$. This cation is known to interact with a specific site at the end of the short B′ helix and to induce a conformational change in this helix which includes tyrosine 96. This leads to a change in the heme pocket and to formation of a hydrogen bond between the substrate camphor carbonyl and tyrosine 96, which expel the sixth water ligand out of the pocket. Saturation of the cation site shifts the spin state to high spin. All the high-spin (k_2) relaxation rate constants increase slightly. The presence of

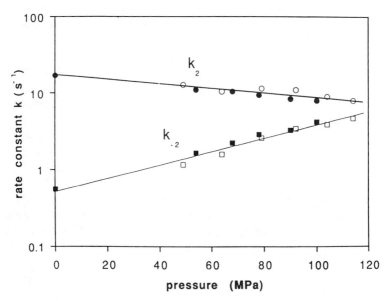

Figure 12.3. Semilogarithmic plot of the camphor-bound spin relaxation rate constants k_2 and k_{-2} versus pressure. Experimental conditions were Tris-ethylene glycol 50% (v/v) buffer, 16.9 μM P-450, 800 μM camphor, 60 mM KCl. The pH was 7.2 and the temperature was 0° C.

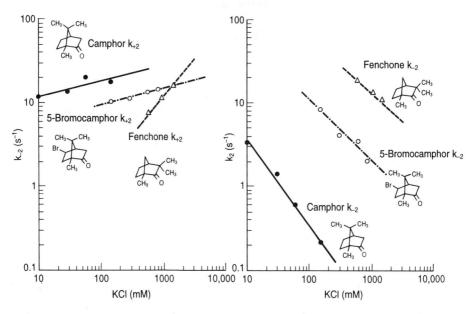

Figure 12.4. Extrapolated spin relaxation rate constants at 0.1 MPa versus potassium ion concentrations for camphor-, 5-bromocamphor-, and fenchone-bound cytochrome P-450$_{cam}$. The rate constants are for the low-spin to high-spin conversion (k_2) and high-spin to low-spin conversion (k_{-2}). The solvent was aqueous 100 mM Tris buffer at pH 7.2. The temperature was 0 °C.

a bromine atom at position 5 and the change of two methyl groups (from position 7 to 3) in the fenchone analogue led to a decrease in k_2. By contrast, the effect of K^+ concentrations on the low-spin rate constant (k_{-2}) was greater for the camphor complex. This is compatible with a potassium cation-induced increase of the equilibrium-binding constant for the substrate. The corresponding dissociation or low-spin relaxation rate constant (k_{-2}) with fenchone as the substrate was larger, but showed a similar dependence of K^+ concentration. The alteration of two methyl groups when going from camphor to the fenchone analogue changed specific interactions between the substrate and P-450, leading to lower binding affinity, resulting in an increase of k_{-2} and of the accessibility of bulk solvent into the active site. Similar changes were observed for the relaxation of the 5-bromocamphor-P-450 complex, but to a lesser extent. The introduction of bromine presumably had a steric effect on the dissociation rate of the analogue. In comparison with the camphor-bound complex, k_{-2} increased by a factor of 200 in the fenchone-bound complex and 100 in the bromocamphor-bound complex, respectively.

Spin-state reaction and activation volumes deduced from P-jump experiments were modulated by the binding of K^+ to camphor-bound cytochrome P-450 as shown in Figure 12.5. The reaction volume $(\Delta V° = V_{HS} - V_{LS})$ increased from 40 to 64 ml mol^{-1} with increasing concentration of KCl, as a saturating kinetic function. The difference between the partial molar volumes of the high- and low-spin states was positive at saturating K^+ concentrations. This means that starting at high potassium, in which the heme was pentacoordinated (95% HS), pressure shifts the heme to low spin, hexa-coordinated, by forcing water molecules into the active site. This suggests that water flux into the heme pocket makes a

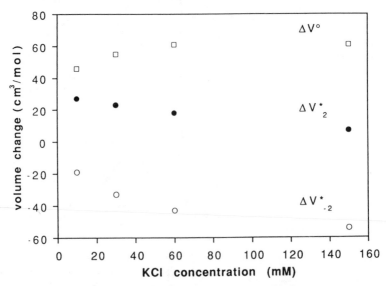

Figure 12.5. Experimental spin volume change versus potassium [ion] concentrations. Reaction volume $\Delta V°$ ($\square$), low- to high-spin activation volume ΔV^*_{-2} ($\bigcirc$). Same experimental conditions as in Figure 12.4.

major contribution to the reaction volume. By contrast, if the starting state of the camphor-bound complex was partially hydrated (mixed spin states) and had a low potassium concentration, the magnitude of the reaction volume change is smaller ($\Delta V^\circ = 40$ ml mol^{-1}). Similar results were obtained with the analogue-bound complexes: fenchone (46% HS), $\Delta V^\circ = 50$ ml mol^{-1}; bromocamphor (80% HS), $\Delta V^\circ = 46$ ml mol^{-1}. The activation volume for the forward rate ($\Delta V_2^* = V^* - V_{ls}$) decreased from 30 ml mol^{-1} to 7 ml mol^{-1} as the concentration of potassium increased, indicating that the partial molar volume of the transition state approaches that of the low-spin hexacoordinated state. The activation volume for the reverse rate ($\Delta V^*_{-2} = V^* - V_{HS}$) was negative and shifted from -10 to -54 ml mol^{-1} with increasing K$^+$ concentration, showing that the partial molar volume of the transition state was smaller than that of the high-spin penta-coordinated state. These results are compatible with the mechanism of hydration of the sixth ligand of heme by water molecules in the transition state.

Finally, P-jump relaxation kinetics were recorded with samples of camphor-bound cytochrome P-450 in different solvents and at different temperatures. A comparative study of pressure jumps from 105 MPa to 80 MPa with different samples at 4 °C gave half-times for the kinetic transition observed at 417 nm of 77 milliseconds for a sample in Tris buffer 7-glycerol 40% (w/w) and 112 milliseconds in Tris buffer pH 7-glycerol 60% (w/w). However, under certain extreme conditions of solvent (glycerol 80% w/w), low temperature (-20 °C), or high pressure (170 MPa), the viscosity of the solution can reach a value as high as 17 poise, increasing the relaxation time enough to allow full spectral measurements during the spin transition and the analysis of the effect of viscosity on the individual rate constants k_2 and k_{-2} (Marden & Hui Bon Hoa, 1982). Table 12.1 shows the experimental results from pressure jumps at different temperatures extrapolated to 0.1 MPa and 4 °C. Both rate constants were viscosity dependent, and the relaxation time of the spin transition was 1000 seconds in a solution of 1017 cP. The plot of log k versus log η was linear, with a slope near unity. The corrected rate constant, k'_{-2}, was obtained by multiplying k_{-2} by the ratio between K_2 in cosolvent and K_2 in water at 0.1 MPa. Since viscosity affects rotational and

Table 12.1. Experimental results of pressure-jump relaxation kinetics for camphour-bound cytochrome P-450 in different solvents at different temperatures extrapolated to 0.1 MPa and 4 °C

	log (η/cP)	log (k_2 s^{-1})	log (k'_{-2} s^{-1})
Glycerol (w/w)			
20%	+0.10	+0.28	+1.10
40%	+1.10	−0.80	+0.40
80%	+2.70	−2.50	
90%	+3.35	−3.90	
Sucrose (w/w)			
50%	+2.20	−1.30	−1.80
60%	+2.70	−1.10	−1.30

diffusional motion, the viscosity dependence of the spin transition in camphor-bound cytochrome P-450 suggests that the process may involve a conformational change whose motion is strongly coupled to the solvent. However, the composition of the solvent also affects the equilibrium properties of the reaction. Further studies will be required to characterize other solvent effects.

ACKNOWLEDGMENTS: This work was supported by grants from the Institut Nationalde la Santé et de la Recherche Médicale and the Institut National de la Recherche Agronomique.

REFERENCES

Atkins, W. M., & Sliger, S. G. (1989). Molecular recognition in cytochrome P450: alteration of regio-selective alkane hydroxylation via protein engineering. *J. Am. Chem. Soc.* **111**, 2715–2717.

Balny, C., Saldana, J. L., & Dahan, N. (1984). High-pressure stopped-flow spectrometry at low temperatures. *Anal. Biochem.* **139**, 178–189.

Eigen, M., & De Mayer (1962). Theoretical basis of realization methods. In *Techniques of Organic Chemistry*, ed. E. Weissberger, New York, Interscience, 8:901–1054.

Goldsack, D. E., Hurst, R. E., & Love, J. (1969). A pressure jump apparatus with optical detection. *Anal. Biochem.* **28**, 273–281.

Hui Bon Hoa, G., & Douzou, P. (1973). Stoped flow method at subzero temperatures. *Anal. Biochem.* **51**, 127–136.

Hui Bon Hoa, G., Douzou, P., Dahan, N., & Balny, C. (1982). High-pressure spectrometry at subzero temperatures. *Anal. Biochem.* **120**, 125–135.

Jaenicke, R., Bernhardt, G., Ludemann, H. D., & Stetter, K. O. (1988). Pressure induced alteration in the protein pattern of the thermophilic archaebacterium. *Methanococcus themolithothrophicus. Appl. Environ. Microbiol.* **54**, 2375–2380.

Kramers, H. A. (1940). Brownian motion in a field of force and the diffusion model of chemical reactions. *Physica* (Amsterdam) **7**, 284.

Lange, R., Hui Bon Hoa, G., Debey, P., & Gunsalus, I. C. (1979). Spin transition of camphor-bound cytochrome P-450: 2. Kinetics following rapid changes of the local paH at subzero temperatures. *Eur. J. Biochem.* **94**, 491–496.

Larson, R. S., & Kostin, M. D. (1980). Friction and velocity in Kramer's theory of chemical kinetics. *J. Chem. Phys.* **72**, 1392–1400.

Marden, M. C., & Hui Bon Hoa, G. (1982). Dynamics of the spin transition in camphor-bound ferric cytochrome P-450 versus temperature, pressure and viscosity. *Eur. J. Biochem.* **129**, 111–117.

Poulos, T. L., Finzel, B. C., & Howard, A. J. (1987). High-resolution crystal structure of cytochrome P-450$_{cam}$. *J. Mol. Biol.* **195**, 687–700.

Rigler, R., Rabl, C. R., & Jovin, T. M. (1974). A temperature-jump apparatus with optical detection. *Rev. Sci. Instrum.* **45**, 580–588.

Somero, G. (1940). Life at low volume change: hydrostatic pressure as a selective factor in the aquatic environment. *Am. Zool.* **30**, 123–35.

13

Pressure Effects on the Ligand-Binding Kinetics for Hemoproteins and Their Site-Directed Mutants

ISAO MORISHIMA

The effects of high pressure up to 1500 bar on the recombination kinetics of oxygen and carbon monoxide (CO) binding to human hemoglobin (intact and isolated chain forms), human myoglobin (and its mutants), and cytochrome P-450 were studied by the use of millisecond and nanosecond laser photolysis. The activation volumes for the binding of CO to the R- and T-quaternary states of hemoglobin (Hbs) were determined to be -9.0 and -31.7 ml, respectively. The characteristic pressure dependence of the activation volume was observed for the R-state Hb but not for the T-state Hb. More detailed studies were made with isolated α- and β-chains of human Hb. The kinetic data were analyzed on the basis of a simple three-species model, which assumes two elementary reaction processes of bond formation and steps of ligand migration. A pressure-dependent activation volume change from negative to positive values in the bimolecular CO association reaction was observed for both chains. This is attributed to a change of the rate-limiting step from the bond-formation step to the ligand-migration step. High-pressure ligand-binding kinetics were also examined for site-specific mutants of human myoglobin in which some amino acid residues at the heme distal sites, such as Leu 29, Lys 45, Ala 66, and Thr 67, are substituted by others. The pressure dependence of the CO binding rate for the L29 mutants was unusual: a positive value was obtained unexpectedly for overall CO binding. Corresponding to this anomaly was an unusual geometry of the iron-bound CO, which was determined by IR and NMR spectroscopies. The effects of camphor and camphor analogues as substrates on the CO-binding kinetics for P-450$_{cam}$ were also studied under pressure. The positive activation volumes for CO binding were obtained for substrate-free and norcamphor- and adamantane-bound P-450, whereas other substrate analogue-bound P-450 complexes exhibited the negative activation volumes. All of the present high-pressure results are discussed in relation to (1) the dynamic aspects of the protein conformation, and (2) the specific participation of amino acid residues in the heme distal site in each elementary step of the ligand-binding reaction process.

The ligand-binding dynamics of hemoproteins have been studied extensively in the last decade, since they serve as a typical model for protein dynamics. Transient absorption spectroscopic studies have been shown to be useful for delineating the mechanism of the ligand-binding reaction. The binding of a ligand to a heme iron from the solvent phase was observed on a millisecond time scale (Philo et al., 1988). In the nanosecond region, the geminate ligand binding (50–100 ns) was observed (Alpert et al., 1979; Friedman & Lyons, 1980; Olson et al., 1987). This nanosecond rebinding is a process in which the photodissociated ligand rebinds to the heme iron without having diffused into the surrounding solvent. From these millisecond and nanosecond kinetic studies, a three-state model has been proposed in which there are two elementary steps for the bimolecular association reaction. The first step is the entry of the ligand into the heme pocket; the second step is the subsequent ligand binding to the heme iron.

$$PCO \rightleftharpoons P \cdot CO \rightleftharpoons P + CO \tag{1}$$
$$ A B C$$

where PCO represents the bound state containing CO attached to the heme iron atom, $P \cdot CO$ is the geminate state formed after bond dissociation, and $P + CO$ is the state in which the CO molecule is in the solvent phase.

Despite these extensive studies on the ligand-binding kinetics for hemoproteins, several important questions about the ligand-binding mechanism, especially concerning the ligand-binding dynamics, have remained unanswered. One of the most challenging problems we encounter in studying the ligand-binding mechanism is directly characterizing the intermediate states in the ligand-binding dynamics in terms of protein structures. In this article, we describe the use of high-pressure laser photolysis to elucidate the dynamic aspects of ligand binding to hemoproteins.

The effects of pressure on the structures of some hemoproteins have been studied by the application of UV/vis absorption (Weber & Drickamer, 1983; Ogunmola et al., 1977; Alden et al., 1989), magnetic susceptibility (Messana et al., 1978), NMR (Morishima et al., 1979, 1980; Morishima & Hara, 1982, 1983a,b), and resonance Raman spectroscopy (Alden et al., 1989). The influence of pressure on the ligand-binding kinetics of some hemoproteins has also been studied (Hasinoff, 1974; Caldin & Hasinoff, 1975; Adachi & Morishima, 1989; Projahn et al., 1990; Frauenfelder et al., 1990). These pressure-dependent kinetic studies provide information on the volume profile of the ligand-binding process. The activation volume, the difference in partial molar volume between the activated complex and the reactants, is very easily visualized and is sensitive to the dynamics associated with the reaction process. In the present work, we have undertaken the study of CO and O_2 rebinding kinetics following laser photolysis for various hemoproteins such as the isolated α- and β-chains or intact R and T states of human Hb, cytochrome P-450, and site-directed mutants of human myoglobin under high pressure using laser photolysis.

Structural and kinetic differences between the α- and β-chains of Hb have been shown by many investigators. For the isolated chains of Hb, Olson et al. (1987) concluded that the protein structure of the β heme pocket is flexible and can accommodate ligands more readily than that of the α-chain. We have studied the effects of pressure on CO binding to the isolated α- and β-chains of Hb and have examined whether or not the volume profiles are different for the isolated chains (Unno et al., 1991).

One of the key issues to be addressed by the structural model of Hb allostery is the elucidation of differences in the mechanism of the R and T states of Hb. We have studied the effect of pressure on the bimolecular CO rebinding to intact Hb to characterize the intermediate states in the ligand-binding reaction for the R and T quarternary Hbs (Unno et al., 1990).

One of the interesting aspects of P-450$_{cam}$ is the substrate dependence of its reactivity and physical properties on the nature of the substrate. For example, the binding of substrate camphor lowers the ligand affinity of ferrous P-450$_{cam}$ for CO 10-fold, and this lower affinity is attributed mainly to a 100-fold decrease in the association rate (Peterson & Griffin, 1972). To gain further insight into the effects of substrate on the ligand-binding process, we have applied the high-pressure flash-photolysis technique in studies of the CO adducts of ferrous P-450$_{cam}$ in its camphor-bound, norcamphor-bound, and substrate-free forms [which we denote P450($+$cam), P450($+$nor), P450($-$cam), respectively]. We have also compared the effects of pressure on the CO-binding kinetics of P450($+$cam), in the presence and absence of putidaredoxin to obtain some new insight into the effects of putidaredoxin binding (Unno et al., 1994).

To date, studies of mutant myoglobins (Mbs) have examined the roles of the distal His 64(E7) and Val 68(E11) residues in the ligand-binding process, and discussions have focused mainly on the steric and polarity effects of these residues (Caldin & Hasinoff, 1975; Frauenfelder et al., 1979, 1988, 1990; Freidman & Lyons, 1980). A detailed analysis of the photolysis results for O_2 and CO adducts of several His 64(E7) mutants indicated that polarity is more important than size in its contribution to the ligand-binding barrier between the heme pocket and the outside of the protein. This suggests that the distal His has a role in blocking ligand migration from the heme pocket to the solvent and vice versa. Here we focus on the effects of the replacement of the highly conserved Leu 29(B10) residue in human Mb with Ile (L29I) or Ala (L29A) on the ligand-binding properties. Since Leu 29(B10) participates in a hydrophobic cluster with distal residues, which serves to restrict the movement of distal side chains, its modification is expected to affect the interactions of amino acids in the heme pocket and to lead to changes in the fractional distribution of each conformer (A_0–A_3) in MbCO. Overall rebinding rate constants of CO or O_2 for wild-type and Leu 29(B10) mutant human Mbs were determined at normal and elevated pressures. To elucidate the detailed mechanism of the ligand-binding process for Leu 29(B10) mutant Mbs, the kinetic results under normal and elevated pressures have been investigated and analyzed, especially in relation to the different fractional distributions of MbCO conformers (Adachi et al., 1992).

METHODOLOGY

The high-pressure photolysis experiments were performed with samples in a high-pressure cell with an inner capsule made of quartz. The sample's inner capsule consisted of an upper, cylindrical part and a lower, optical, square part with a path length of 5.0 mm. The details of the high-pressure cell and its inner capsule are described elsewhere (Hara & Morishima, 1988). The pressure was transmitted from an intensifier and was measured by using a Bourdon tube gauge. Laser flash photolysis experiments were carried out as described in our previous reports.

CO rebinding to the hemoprotein was measured in milliseconds, and was analyzed in terms of simple bimolecular association rate constants. The mono-phasic time courses of 436 nm absorption at all pressures were fitted to

$$\Delta A_t = \Delta A_o \exp(-k_{app} t) \qquad (2)$$

where ΔA_t is the absorbance change at any time t, ΔA_o is the total change, and k_{app} is the observed first-order rate constant. When the CO rebinding was biphasic, the time courses for all the pressures were fitted to

$$\Delta A_t = \Delta A_f \exp(-k_f t) + \Delta A_s \exp(-k_s t) \qquad (3)$$

where k_f and k_s are the observed fast and slow first-order rate constants, and ΔA_f and ΔA_s are amplitudes of the exponential expressions for the fast and slow phases, respectively.

On a nanosecond time scale, the rebinding time courses were fitted to

$$\Delta A_t = \Delta A_\infty + \Delta A_g \exp(-k_g t) \qquad (4)$$

where ΔA_t and ΔA_∞ are absorbance changes at time t and at the end of the geminate reaction, and k_g is the first-order geminate rate constant. ΔA_g is the extrapolated zero time absorbance change associated with the geminate recombination. The fractional recombination yield f (germinate yield) can be calculated from the parameters in Eq. 4.

$$f = \Delta A_g / (\Delta A_\infty + \Delta A_g) \qquad (5)$$

The rate constants for the elementary steps can be determined by the following

$$k_{BA} = k_g f \qquad (6)$$

$$k_{BS} = k_g(1 - f) \qquad (7)$$

$$k_{SB} = k_{on}/f \qquad (8)$$

where k_{BA} is the bond formation rate, k_{BS} is the rate for ligand escape, and k_{SB} is the second-order rate constant for the ligand migration to the geminate state. The activation volume is calculated as

$$\Delta V^{\ddagger} = -RT\left[\frac{\partial \ln(k_p/k_1)}{\partial P}\right]_T \tag{9}$$

RESULTS AND DISCUSSION

Pressure Effects on the Carbon Monoxide Rebinding to the Isolated α- and β-Chains of Human Hemoglobin

Pressure caused substantial changes in the bimolecular association rates for both isolated chains. The bimolecular association rate constants at various pressures were obtained for the isolated α- and the fast phase (major component) and for the isolated β-chain, they are plotted as $\ln[k_p/k_1]$ against pressure (Figure 13.1). The association rate constant for the isolated α-chain increased from 5.1×10^6 at 0.1 MPa to $6.9 \times 10^6\ M^{-1}\,s^{-1}$ as pressure increased to 100 MPa, and decreased to $6.2 \times 10^6\ M^{-1}\,s^{-1}$ at a higher pressure of 150 MPa. Elevating the pressure from 0.1 to ~110 MPa caused the association rate of the fast phase for the isolated β-chain to increase from 1.3×10^7 to $1.6 \times 10^7\ M^{-1}\,s^{-1}$ and it decreased to $1.6 \times 10^7\ M^{-1}\,s^{-1}$ at 150 MPa. High pressure also caused changes in the fraction of the fast and slow phases for the β-chain. The spectral amplitude of the slow phase decreased with rising pressure, and the recombination reaction appears to consist of the fast phase only at 150 MPa (results not shown). Since the signal change for the slow phase is small and the uncertainty in the parameters for the β-chain is large, we could not obtain the activation volumes for the slow phase of the β-chain.

On a nanosecond time scale, the geminate rebinding kinetics of CO to the isolated α- and β-chains were fitted to a single exponential (Eq. 2). The geminate

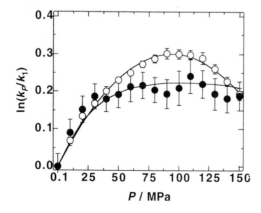

Figure 13.1. Logarithmic plots of the rate constants for the bimolecular CO association reaction versus pressure (○) for the isolated α-chain and (●) for the fast phase of the isolated β-chains of hemoglobin.

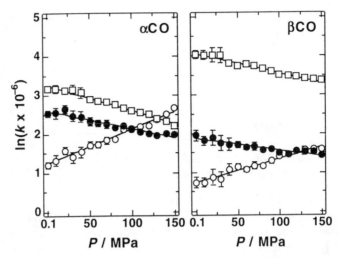

Figure 13.2. Logarithmic plots versus pressure of ($\bigcirc$, the bond formation rate, k_{BA}; ($\bullet$), the rate of ligand escape from the geminate state to the solvent, k_{BS}, and ($\square$), the rate of ligand migration from the solvent to the geminate state, k_{SB}. The straight lines were obtained by linear least-squares calculations.

rate constants at normal pressure are 16×10^6 and $9 \times 10^6 \, \text{s}^{-1}$ for the α- and β-chains, respectively, and are almost independent of pressure. The geminate yield for both chains is monotonously increased by pressure. Assuming a steady-state population of the geminate state, the rate constants for the elementary steps were calculated from Eqs. 6–8. To obtain the activation volumes for each elementary step in accordance with Eq. 9, these rate constants were plotted against pressure as semilog plots (Figure 13.2). For both chains, the slopes of the linear plots for the ligand migration process (k_{BS}, k_{SB}) are positive, whereas the slope for the bond formation process (k_{BA}) is negative. It is worth noting that pressurization reverses the relative magnitudes of the bond formation (k_{BA}) and ligand escape rates (k_{BS}) at about 100 MPa for both chains. This implies that the ligand migration process is rate determining under high pressure. The estimated activation volumes for the elementary steps at atmospheric pressure are collected in Table 13.1.

Table 13.1. Volume (ml mol^{-1}) activation for the bimolecular association reaction and the elementary steps of CO binding to the isolated α- and β-chains of hemoglobin[a]

Protein	$\Delta V_{BA}^{\ddagger}$	$\Delta V_{BS}^{\ddagger}$	$\Delta V_{SB}^{\ddagger}$	$\Delta V^{\ddagger a}$
α-chain	-21.8 ± 0.9	11.1 ± 0.8	16.0 ± 0.6	-18.4 ± 0.5
β-chain[b]	-15.4 ± 0.8	7.4 ± 0.7	12.1 ± 0.6	-22.4 ± 1.7

[a] The values are for the bimolecular association reaction.

[b] The values are for the fast phase of β-chains.

It has been reported that a negative activation volume is characteristic of the bond-formation reaction of neutral molecules (le Noble, 1965); yet the activation volume for the diffusion-controlled reaction of CO binding to protoheme is positive in both aqueous ethylene glycol and glycerol (Caldin & Hasinoff, 1975). A relationship between the activation volume and the rate-limiting steps was suggested by studies of pressure effects on ligand binding to Mb (Hasinoff, 1974; Adachi & Morishima, 1989; Projahn et al., 1990; Taube et al., 1990) and Hb (Hasinoff 1974; Unno et al., 1990, 1991). The activation volumes of the bimolecular association reaction for the isolated α-chain and those of the fast phase of the reaction for the isolated β-chain of Hb at atmospheric pressure are estimated here as -18.2 and $-22.4 \, \text{ml mol}^{-1}$, respectively. Such negative activation volumes suggest that the iron-ligand bond-formation step in the overall bimolecular association reaction predominates at ordinary pressure. In contrast to the situation at 0.1 MPa, activation volumes at higher pressure ($> \sim 100$ MPa) for bimolecular association with the isolated α- and β-chains are positive (Figure 13.1). Since the ligand-migration step is characterized by a positive activation volume, it is likely that the ligand-migration process becomes rate limiting at higher pressure. However, measurements involving overall association cannot rule out the possibility that the pressure-induced change in activation volume is caused by pressure-induced conformational changes. Since the nanosecond laser photolysis experiment offers information about the elementary steps, it supports a pressure-dependent alteration of the rate-limiting step. As shown in Figure 13.2, the bond-formation rate (k_{BA}) is accelerated by pressure, and the migration rates (k_{BS}, k_{SB}) are reduced for both isolated chains. The plots in Figure 13.2 are essentially linear for each elementary step, which suggests that pressure-induced conformational changes, if they occur, do not affect the elementary rate constants. In fact, proton NMR measurements of the CO complex with both subunits under pressure showed that the heme environmental structures are hardly affected by pressure (data not shown). Furthermore, k_{BS} is larger than k_{BA} at normal pressure, implying that the bond formation step is rate determining. At about 100 MPa, however, k_{BS} becomes greater than k_{BA}, indicating that the ligand-migration step becomes rate determining under high pressure.

A positive activation volume for the ligand-migration process was also observed for ligand binding to Mb (Adachi & Morishima, 1989; Taube et al., 1990). Taube et al. (1990) also observed a positive activation volume for the ligand-escape process and attributed it to conformational changes from the closed to open structure in Mb. Since the volume profile for the isolated Hb chains exhibits features similar to those of Mb (Adachi & Morishima, 1989; Taube et al., 1990), such conformational changes from the closed to open structure in the CO-migration process probably also describe the isolated α- and β-chains of Hb.

The observed activation volume for the bond-formation process (-15 to $-22 \, \text{ml mol}^{-1}$) is almost identical to the activation volume ($-19 \, \text{ml mol}^{-1}$) for the bimolecular addition of CO to model heme complexes (Taube et al., 1990). Therefore, it appears quite reasonable that the iron-ligand bond-making process contributes mainly to the observed activation volumes. This implies that the other conformational factors, such as motions of the surrounding protein, do not contribute much to the activation volume in the bond-formation process.

Effects of Pressure on the Carbon Monoxide Binding Reaction for R and T States of Hemoglobins

To delineate the pressure effects on the ligand-binding reaction rate for intact Hb, $\ln(k_P/k_1)$ was plotted against pressure to provide Figure 13.3. The activation volume was determined from the slope, and the resulting activation volumes were evaluated for both R- and T-state Hbs at atmospheric pressure. This analysis led to two important findings. First, the CO rebinding process for the T-state Hb undergoes a larger activation volume $(-32 \pm 2 \text{ ml mol}^{-1})$ than that for the R-state Hb $(-9 \pm 1 \text{ ml mol}^{-1})$. This difference between the R and T states implies that the CO rebinding mechanism in the two quaternary states has different features. Second, the activation volume for the T-state Hb is almost pressure independent, whereas that for the R-state Hb is reduced to nearly zero in going from 0.1 to 100 MPa.

Since the ligand-migration step is associated with a positive activation volume, it is likely that this step contributes to the rate-limiting process for CO rebinding to R-state Hb at high pressure; this implies that increased pressure reduces the ligand-migration rate and accelerates the bond-formation rate. Although we cannot rule out the possibility that the pressure-induced activation volume change is caused by the simple conformational change that does not affect the rate-limiting step, the observed pressure-induced increase of geminate rebinding also supports the reduction of the ligand-migration rate by pressurization. For the T-state Hb, however, pressure does not appear to affect the rate-limiting step as much as that for the R-state Hb. The free energy barrier between HbCO and Hb·CO is lowered, and the barrier between Hb·CO and Hb + CO is raised by

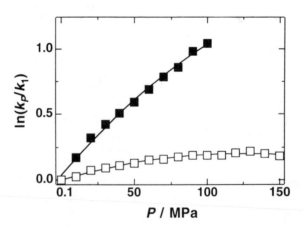

Figure 13.3. Logarithmic plots of the rate constants for the bimolecular CO association reaction versus pressure for (□) R- and (■) T-state hemoglobin. Experiments were performed using a millisecond laser photolysis apparatus; samples were in 0.05 M Tris-0.1 M Cl⁻, pH 7 at 293 K. The amplitude of the T-state Hb was markedly reduced by high pressure, so that the rate constants could not be determined at pressures above 100 MPa.

pressurization, which may explain the pressure-induced switching of the rate-limiting step from bond formation to migration for the R-state Hb. For the T-state Hb, however, no change of the rate-limiting step is caused by pressure because the difference in the free energy barrier between the two processes is too large to be affected. This explanation appears to be further supported by our finding that pressure induces an increase in the geminate yield, that is, the lowered free energy barrier for bond formation at high pressure leads to an increase in the geminate yield.

Effects of Pressure on Carbon Monoxide Binding Kinetics of Cytochrome P-450$_{cam}$

We performed laser photolysis experiments of CO complexes on P450(+cam), P450(+nor), and P450(−cam). The time courses were monophasic and fit well to a single exponential expression (Eq. 2). The resultant bimolecular association rate constant (k_{on}) at atmospheric pressure is $1.0 \times 10^5 \ M^{-1} s^{-1}$. The data are plotted versus pressure (Figures 13.4A, 4B, and 4C, respectively, for the three systems). The activation volumes were determined from slopes, and evaluated at 0.1 MPa (Table 13.2). Two features emerge from this analysis. First, the CO rebinding process for P450(+cam) exhibits a negative activation volume (-23 ml mol^{-1}), while those for P450(+nor) and P450(−cam) each exhibit a positive activation volume ($+3$ to $+6$ ml mol^{-1}). Second, the slopes of the plots for P450(+nor) and P450(−cam) are almost independent of pressure, whereas that for P450(+cam) is reduced by pressurization. As a result, the activation volume for P450(+cam) is reduced to -14 ml mol^{-1} at 200 MPa.

We have also investigated the effects of pressure on the CO recombination kinetics of P450(+cam) in the presence of putidaredoxin (the effects are ~fivefold greater than with P-450$_{cam}$ protein alone). Gunsalus and Sligar (1978) reported that the bimolecular association rate constant at ordinary pressure is almost unaffected by the presence of putidaredoxin. A similar lack of effect of putidaredoxin binding is demonstrated in Figure 13.4A (solid circles). The figure shows that the association rate constant is increased by pressurization in both the presence or absence of putidaredoxin. In fact, the activation volumes at 0.1 MPa in the absence or presence of putidaredoxin are nearly indistinguishable, although the association rate constant at elevated pressure is slightly increased by putidaredoxin binding.

It is suggested that the negative activation volume should be attributed to the iron-ligand bond-making step and the positive activation volume should be attributed to gate-like conformational changes of the protein matrix. Thus, it is reasonable to consider that the negative activation volume (-23 ml mol^{-1}) observed for CO binding to P450(+cam) implies that the bond-formation process is rate determining. In contrast, the positive activation volumes ($+3$ to $+6$ ml mol^{-1}) observed for the other species are unexpected. Further, the bimolecular association rate constants for P450(+nor) and P450(−cam) are ~100-fold larger than that of P450(+cam). These positive activation volumes and large association rate constants strongly suggest that the ligand-migration process is rate determining. In other words, binding of camphor causes a substantial increase in the barriers for iron-ligand bond formation, which leads to a negative

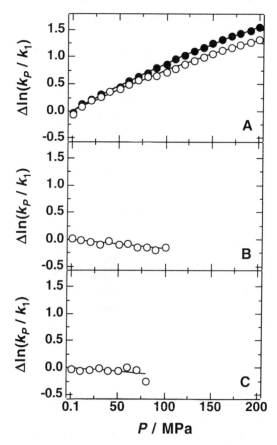

Figure 13.4. Logarithmic plots of the rate constants for the bimolecular CO association reaction versus pressure for (**A**) camphor-bound P-450$_{cam}$ [(●) + putidaredoxin; (○) − putidaredoxin]; (**B**) norcamphor-bound P-450$_{cam}$; (**C**) substrate-free P-450$_{cam}$.

activation volume and reduced association rate constant, whereas binding of norcamphor exhibits no effect. Poulos et al. (1985) have shown that camphor is buried in an internal pocket just above the heme distal surface adjacent to the CO binding site. Their further study of the CO adduct of P450(+cam) demonstrates that the camphor molecule moves about 0.8 Å to accommodate the CO molecule and that the CO appears to be bent from the heme normal due to steric interactions with the camphor molecule (Raag & Poulos, 1989b). Similar effects of camphor binding on structural changes in the heme active site have been suggested by several spectroscopic methods (that is, by infrared and resonance Raman studies of the CO or NO adducts of P-450$_{cam}$; O'Keefe et al., 1978; Uno et al., 1985; Jung & Marlow, 1987; Hu & Kincaid, 1991a, 1991b; and by a ^{15}N NMR study of CN$^-$ complexes by Shiro et al., 1989). The results of these reports indicate that the structural interpretation of the present observations is straightforward: the barrier to the bound camphor molecule interacts sterically with CO and leads to an increase in iron-ligand bond formation.

Table 13.2. Kinetic parameters for the bimolecular association reaction of CO with cytochrome P-450$_{cam}$

Substrate	$k_{on} \times 10^{-6}$ $(M^{-1} s^{-1})$	$\Delta V^{\ddagger}$ (ml mol^{-1})
+camphor[a]	0.10	−23.0
+camphor + putidaredoxin[a]	0.10[c]	−24.0
+norcamphor[a]	13.0	3.4
−camphor[b]	11.0	5.5

[a] 293 K.
[b] 278 K.
[c] Gunsalus and Sligar (1978).

One of the most striking results shown in Figure 13.4 and Table 13.2 is the lack of effect that norcamphor binding has on the CO recombination kinetics. This result contrasts markedly with the drastic effect of camphor binding just mentioned. However, this result is not entirely unexpected because the molecular size of norcamphor is smaller than that of camphor. Indeed, the lack of effect of norcamphor binding correlates well with the crystallographic data of Raag and Poulos (1989a). They show that since norcamphor lacks the 8-, 9, and 10-methyl groups of camphor, specific interactions between these groups and Phe-87, Val-247, and Val-295 are missing in the norcamphor-P-450$_{cam}$ complex. As a result, norcamphor binds about 0.9 Å farther from the CO binding site than does camphor, which may reduce the steric hindrance for iron-CO bond formation. Furthermore, a comparison of crystallographic temperature factors indicates that norcamphor is more loosely bound than camphor. These observations are consistent with recent resonance Raman studies of CO and NO adducts of P-450$_{cam}$ (Hu & Kincaid, 1991a,b). Therefore, it is reasonable to conclude that there is almost no steric effect of norcamphor on the iron–CO bond-formation process, which leads to the lower bond-formation barrier.

Effects of Pressure on the Ligand-Binding Kinetics for the Site-Directed Mutants of Human Myoglobin at Heme Distal Sites

Figure 13.5 shows the pressure dependence of binding rate constants of CO for wild-type and mutant Mbs (Adachi et al., 1992). The pressure dependence of the CO binding rate and the resultant positive activation volume for L29A Mb is unusual among various types of the CO form (Table 13.3). Two reasons for such an anomaly seem to be possible: (1) the rate-determining step of CO binding to L29A Mb is the ligand migration in the protein matrix, in contrast to wild-type Mb, or (2) the activation volumes for CO binding to the three conformers A_0–A_3 are different, with A_1 being associated with a large negative value, and A_3 being associated with a nearly zero or positive value, resulting in compensating effects in L29I and L29A Mb. Since the nanosecond laser photolysis study of mutant Mbs showed an absence of the CO geminate binding process, the former seems to be unlikely. Projahn and van Eldik (1990) suggested that the activation volume of about −10 ml mol^{-1} for CO binding to sperm whale Mb is actually too small

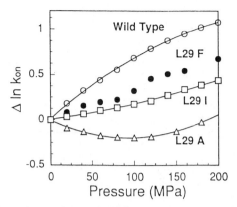

Figure 13.5. Pressure dependence of second-order rate constants for CO binding to myoglobin: (○) = wild type; (□) =)L29I; (△) = L29A; (●) = L29F.

(in terms of absolute magnitude) if the value arises only from the bond formation and the heme structural changes that accompany ligand binding. A positive value of the activation volume due to desolvation of CO upon entering into the protein matrix and possible conformational changes could also contribute to the negative activation volume for the A_3 conformational substate, resulting in apparent positive or negative activation volume, depending on the relative importance of these two opposing effects for each substate.

Note that Figure 13.5 also shows that the pressure dependence of k_{on}^{CO} for L29A Mb deviates from a linear relation, implying pressure dependence of the activation volume. The nonlinear pressure dependence of k_{on}^{CO} was also encountered for Hb A (Figure 13.1). The rate of CO binding to isolated α- and β-chains of Hb A first increases and then decreases. This was interpreted in terms of switching of the rate-limiting step from bond formation to ligand migration under pressure. However, it would not be the case for L29A MbCO, because the rate first decreases then increases with pressure and, as shown by the nanosecond laser photolysis of L29A MbCO, the CO migration step, which is responsible for positive activation

Table 13.3. Kinetic rate constants and activation volumes for CO and O₂ binding to the wild-type and mutant human myoglobins

Myoglobin	k_{on}^{CO} $(M^{-1}\,\mu s^{-1})$	$k_{on}^{O_2}$ $(M^{-1}\,\mu s^{-1})$	$\Delta V_{CO}^{\ddagger}$ $(ml\,mol^{-1})$	$\Delta V_{O_2}^{\ddagger}$ $(ml\,mol^{-1})$
Wild type	0.70	13.0	−21.0	3.6
L29I	0.19	4.1	−3.7	9.2
L29A	0.15	4.3	10.0	13.0
L72A	0.92	22.0	−21.0	−3.0
L104A	2.10	28.0	−23.0	0.11
K45R	1.41	18.5	−18.3	6.8
A66V	0.85	15.8	−19.6	3.3
T67V	1.02	16.7	−18.5	5.4
T67S	1.29	14.2	−16.2	7.2

volume, is not rate limiting at ambient pressure. It is more likely that pressurization affects the fractional distribution of conformers A_0–A_3 so that the faster rebinding conformers A_0 or A_1 are favored under pressure. To clarify this, we are planning to perform IR spectroscopy as a function of pressure.

REFERENCES

Adachi, S., & Morishima, I. (1989). The effects of pressure on oxygen and carbon monoxide binding kinetics for myoglobin: a high pressure laser flash photolysis study. *J. Biol. Chem.* **264**, 18896–18901.

Adachi, S., Sunohara, N., Ishimori, K., & Morishima, I. (1992). Structure and ligand binding properties of leucine 29(B10) mutants of human myoglobin. *J. Biol. Chem.* **267**, 12614–12621.

Alden, R. G., Satterlee, J. D., Mintorovitch, J., Constantindis, I., Ondrias, M. R., & Swanson, B. I. (1989). The effects of high pressure upon ligated and deoxyhemoglobins and myoglobin. *J. Biol. Chem.* **264**, 1933–1940.

Alpert, B., El Mohsni, S., Lindqvist, L., & Tfibel, F. (1979). Transient effects in the nanosecond laser photolysis of carboxyhemoglobin: "cage" recombination and spectral evolution of the protein. *Chem. Phys. Lett.* **64**, 11–16.

Caldin, E. F., & Hasinoff, B. B. (1975). Diffusion-controlled kinetics in the reaction of ferroprotoporphyrin IX with carbon monoxide: the reaction studied as a function of pressure, temperature and solvent by means of a laser flash-photolysis high-pressure apparatus. *Chem. Soc. Faraday* I **71**, 515–527.

Frauenfelder, H., Alberding, N. A., Ansari, A., Braunstein, D., Cowen, B. R., Hong, M. K., Iben, I. E. T., Johnson, J. B., Luck, S., Maredn, M. C., Mourant, J. R., Ormos, P., Reinisch, L., Scholl, R., Schulte, A., Shyamusnder, E., Sorensen, L. B., Steinbach, P. J., Xie, A., Young, R. D., & Yue, K. T. (1990). Proteins and pressure. *J. Phys. Chem.* **94**, 1024–1037.

Frauenfelder, H., Parak, E., & Young, R. D. (1988). Conformational substates in proteins. *Annu. Rev. Biophys. Biophys. Chem.* **17**, 451–479.

Frauenfelder, H., Petsko, G. A., & Tsernoglou, D. (1979). Temperature-dependent x-ray diffraction as a probe of protein structural dynamics. *Nature* **280**, 558–563.

Friedman, J. M., & Lyons, K. B. (1980). Transient Raman study of CO-haemoprotein photolysis: origin of the quantum yield. *Nature* (London) **284**, 570–572.

Geraci, G., Parkhurst, L. J., & Gibson, Q. H. (1969). The reaction of hemoglobin with some ligands. *J. Biol. Chem.* **244**, 4664–4667.

Gunsalus, I. C., & Sligar, S. G. (1978). Oxygen reduction by the P-450 monoxygenase systems. *Adv. Enzym.* **47**, 1–44.

Hara, K., & Morishima, I. (1988). High-pressure cell and its inner capsule for optical studies of liquids. *Rev. Sci. Instrum.* **59**, 2397–2398.

Hasinoff, B. B. (1974). Kinetic activation volumes of the binding of oxygen and carbon monoxide to hemoglobin and myoglobin studied on a high-pressure laser flash photolysis apparatus. *Biochemistry* **13**, 3111–3117.

Hu, S., & Kincaid, J. R. (1991a). Resonance Raman characterization of nitric oxide adducts of cytochrome P-450$_{cam}$: the effect of substrate structure on the iron-ligand vibrations. *J. Am. Chem. Soc.* **113**, 2843–2850.

Hu, S., & Kincaid, J. R. (1991b). Resonance Raman spectra of the nitric oxide adducts of ferrous cytochrome P-450$_{cam}$ in the presence of various substrates. *J. Am. Chem. Soc.* **113**, 9760–9766.

Jung, C., & Marlow, F. (1987). Dynamic behavior of the active-site structure in bacterial cytochrome P-450. *Stud. Biophys.* **120**, 241–251.

le Noble, W. J. (1965). Kinetic reactions in solutions under pressure. *Progr. Phys. Org. Chem.* **5**, 207–330.

Messana, C., Cerdania, M., Shenkin, P., Nable, R. W., Fermi, G., Perutz, R. N., & Perutz, M. F. (1978). Influence of quaternary structure of the globin on thermal spin equilibrium in different methemoglobin derivatives. *Biochemistry* **17**, 3652–3662.

Morishima, I. (1987). High-pressure NMR studies of hemoproteins. In *Current Perspectives in High Pressure Biology*, ed. H. W. Jannasch, R. E. Marquis, & A. M., Zimmerman. New York, Academic Press, pp. 315–332.

Morishima, I. (1994). Pressure effects on ligand binding kinetics for hemoproteins. In *Basic and Applied High Pressure Biology*, ed. P. B. Bennett & R. E. Marquis. New York, University of Rochester Press. pp. 131–140.

Morishima, I., & Hara, M. (1982). High-pressure NMR studies of hemoproteins: pressure-induced structural changes in the heme environments of cyanometmyoglobin. *J. Am. Chem. Soc.* **104**, 6833–6834.

Morishima, I., & Hara, M. (1983a). High-pressure NMR studies of hemoproteins: the effect of pressure on the tertiary and quarternary structures of human adult hemoglobin. *J. Biol. Chem.* **258**, 14428–14432.

Morishima, I., & Hara, M. (1983b). High-pressure nuclear magnetic resonance studies of hemoproteins: pressure-induced structural changes in the heme environments of ferric low-spin metmyoglobin complexes. *Biochemistry* **22**, 4102–4107.

Morishima, I., Ogawa, S., & Yamada, H. (1979). Nuclear magnetic resonance studies of the effects of pressure on the heme environmental structure of hemoproteins. *J. Am. Chem. Soc.* **101**, 7074–7076.

Morishima, I., Ogawa, S., & Yamada, H. (1980). High-pressure proton nuclear magnetic resonance studies of hemoproteins: pressure-induced structural changes in heme environments of myoglobin, hemoglobin and horseradish peroxidase. *Biochemistry* **19**, 1569–1575.

Ogunmola, G. B., Zipp, A., Chen, F., & Kauzmann, W. (1977). Effect of pressure on visible spectra of complexes of myoglobin, hemoglobin, cytochrome *c*, and horseradish peroxidase. *Proc. Natl. Acad. Sci. USA* **74**, 1–4.

O'Keefe, D. H., Ebel, R. E., Peterson, J. A., Maxwell, J. C., & Caughey, W. S. (1978). An infrared spectroscopic study of carbon monoxide binding to ferrous cytochrome P-450. *Biochemistry* **17**, 5845–5852.

Olson, J. S., Rohlfs, R. J., & Gibson, Q. H. (1987). Ligand recombination to the α and β subunits of human hemoglobin. *J. Biol. Chem.* **262**, 12930–12938.

Peterson, J. A., & Griffin, B. W. (1972). Carbon monoxide binding by *Pseudomonas putida* cytochrome P-450. *Arch. Biochem. Biophys.* **151**, 427–433.

Philo, J. S., Lary, J. W., & Schuster, T. M. (1988). Quarternary interactions in hemoglobin β subunit tetramers: kinetics of ligand binding and self-assembly. *J. Biol. Chem.* **263**, 682–689.

Poulos, T. L., Finzel, B. C., Gunsalus, I. C., Wagner, G. C., & Kraut, J. (1985). The 2.6-A crystal structure of *Pseudomonas putida* cytochrome P-450. *J. Biol. Chem.* **260**, 16122–16130.

Projahn, H.-D., & van Eldik, R. (1990). Effect of pressure on the formation and deoxygenation kinetics of oxymyoglobin. Mechanistic information from a volume profile analysis. *Inorg. Chem.* **30**, 3288–3293.

Projahn, H.-D., Dreher, C., & van Eldik, R. (1990). Volume profile analysis of the formation and dissociation of carboxymyglobin: comparison with the corresponding oxymyglobin system. *J. Am. Chem. Soc.* **112**, 17–22.

Raag, R., & Poulos, T. L. (1989a). The structural basis for substrate-induced changes in redox potential and spin equilibrium in cytochrome P-450. *Biochemistry* **28**, 917–922.

Raag, R., & Poulos, T. L. (1989b). Crystal structure of the carbon monoxide-substrate-cytochrome P-450$_{cam}$ ternary complex. *Biochemistry* **28**, 7586–7592.

Shiro, Y., Iizuka, T., Makino, R., Ishimura, Y., & Morishima, I. (1989). ^{15}N NMR study on cyanide($C^{15}N^-$) complex of cytochrome P-450$_{cam}$. Effects of *d*-camphor and putidaredoxin on the iron-ligand structure. *J. Am. Chem. Soc.* **111**, 7707–7711.

Taube, D. J., Projahn, H.-D., van Eldik, R., Magde, D., & Traylor, T. G. (1990). Mechanism of ligand binding of hemes and hemoproteins: a high-pressure study. *J. Am. Chem. Soc.* **112**, 6880–6886.

Unno, M., Ishimori, K., & Morishima, I. (1990). High pressure laser photolysis study of hemoproteins: effects of pressure on carbon monoxide binding dynamics for R- and T-state hemoglobins. *Biochemistry* **29**, 10199–10205.

Unno, M., Ishimori, K., & Morishima, I. (1991). Pressure effects on carbon monoxide rebinding to the isolated α and β chains of human hemoglobin. *Biochemistry* **30**, 10679–10685.

Unno, M., Ishimori, K., Ishimura, Y., & Morishima, I. (1994). High-pressure flash photolysis study of hemoprotein: effects of substrate analogues on the recombination of carbon monoxide to cytochrome P-450$_{cam}$. *Biochemistry* **33**, 9762–9768.

Uno, T., Nishimura, Y., Makino, R., Iizuka, T., Ishimura, Y., & Tsuboi, M. (1985). The resonance Raman frequencies of the Fe–CO stretching and bending modes in the CO complex of cytochrome P-450$_{cam}$.

Weber, G., & Drickamer, H. G. A. (1983). The effect of high pressure upon proteins and other biomolecules. *Q. Rev. Biophys.* **16**, 89–95.

14

Transient Enzyme Kinetics at High Pressure

CLAUDE BALNY

In a detailed study of an enzyme reaction pathway, a measured composite rate constant, for example, k_{cat}, can be interpreted in ways that lead to ambiguous conclusions. Two conditions must be met to solve this problem: (1) an elementary rate constant must be measured, and (2) a maximum number of physical-chemical parameters must be used to perturb the system under study. To gain access to elementary rate constants, cryobaroenzymology and/or transient methods, such as stopped-flow and flow-quench kinetics, can be used. Both perturbation and kinetics measurements performed under either high pressure or low temperatures can then be used to probe the thermodynamics of the interconversion of two successive intermediates to obtain parameters such as $\Delta G^{\ddagger}$, $\Delta S^{\ddagger}$, $\Delta H^{\ddagger}$, and $\Delta V^{\ddagger}$. The interdependence of the two major variables, namely temperature and pressure, is presented in this article, in which the role of organic cosolvents is considered as a third variable.

During catalytic reactions, enzymes undergo a number of conformational changes related to their dynamic structural flexibility. This appears as a succession of different steps. A complete study of such processes, which generally are very rapid, consists of the exploration of the properties of these steps, including thermodynamic features obtained by the action of temperature and pressure. As long ago as 1950, Laidler (1950) formulated the first theoretical basis for explaining the responses of enzymes to high hydrostatic pressures. Chemists used this parameter extensively, and in the early stages of high-pressure kinetics they attempted to analyze the observed results on the basis of collision theory (Asano, 1991) or transition-state theory (Evans & Polanyi, 1935). These theories are still used to describe pressure effects on enzyme reactions. It is postulated that between two successive intermediates there is a labile transition state which governs the energetics of the reaction (Glastone et al., 1941). But we must remember that this theory was first applied only to simple homogeneous reactions in gases. For solutions, the treatment can require the introduction of other parameters such as the viscosity.

If we now consider only the overall enzyme reaction pathway for the analysis of the pressure effect, the data obtained can lead to ambiguous conclusions due to the measurement of a composite rate constant, such as k_{cat}, or a composite binding constant, such as K_m. A convenient approach for a clear and detailed analysis of pressure effects on enzyme reactions is to study elementary reaction steps. When one is able to study an elementary step in isolation, for example, under subzero temperatures or by using an analytical technique specific for a given intermediate, it becomes possible to obtain thermodynamic values describing the interconversion of two successive intermediates (Balny et al., 1985). Moreover, the knowledge of the energetic level of transient states is important for the modulation of enzyme activity; reactions with negative activation volumes are accelerated by a pressure increase and vice versa (Groß & Jaenicke, 1994; Mozhaev et al., 1996). In addition to pressure and temperature, a third variable can be considered: the nature of the cryosolvent that allows experiments to be extended to subzero conditions in the first place, but which also acts as a perturbing agent on its own, thereby inducing controlled and reversible changes in equilibrium and rate processes. The interdependence of the two variables, namely temperature (T) and pressure (p), predicted by the general equation for the standard variation of free energy $\Delta G = f(T, p)$, is presented. By adjusting two of the three variables (T, p, solvent), it is possible either to amplify or cancel out the effect of the third.

To illustrate this approach, we present results with various enzyme-catalyzed reactions, including enzyme-substrate reactions related to a change in protein conformation (creatine kinase, horseradish peroxidase, butyrylcholinesterase), intra- and intermolecular protein-protein electron transfer reactions (hydroxylamine oxidoreductase–cytochrome c, methanol dehydrogenase–cytochrome cL), and binding of small ligand to proteins (CO binding to reduced hemoproteins). Results will be discussed in terms of changes in protein conformation or in solvation, where the application of high pressure can magnify phenomena that are scarcely detectable at atmospheric pressure.

GENERAL THERMODYNAMIC INTERPRETATIONS OF TRANSIENT ENZYME KINETICS

The application of the transition-state theory of Eyring to enzyme reaction has been described in several reviews (see, for example, Morild, 1981, Balny et al., 1989, Groß & Jaenicke, 1994, Mozhaev et al., 1996). It is postulated that between two successive complexes, for example, ES and E*S, there is a labile transition-state complex ES‡ (ES being the complex of collision, and E*S being associated with a conformational change). Any conversion of ES to E*S goes through a transition state ES‡ which has a higher energy than either ES or E*S according to the sequence

$$\text{ES} \underset{\xleftarrow{\hspace{1em}}}{\overset{k_+}{\rightleftharpoons}} \text{ES}^{\ddagger} \underset{\xrightarrow{\hspace{1em}}}{\overset{k_-}{\rightleftharpoons}} \text{E*S} \tag{1}$$

For the forward reaction, the conversion ES → E*S depends on the difference in the free energies of ES and ES‡ (the Gibbs free energy of activation $\Delta G^{\ddagger}$). By carrying out temperature dependency studies, the thermodynamic parameters associated with these constants are (example k_{+})

$$\Delta G^{\ddagger} = -RT \ln(k_{+} \cdot h/kT) = \Delta H^{\ddagger} - T \cdot \Delta S^{\ddagger} \qquad (2)$$

where k is the Boltzmann constant, h the Plank constant, T the absolute temperature, $\Delta G^{\ddagger}$ the transition-state activation free energy, $\Delta H^{\ddagger}$ the transition-state activation enthalpy, and $\Delta S^{\ddagger}$ the transition-state activation entropy. We can also vary another intensive parameter, the pressure (p):

$$\Delta V^{\ddagger} = (\delta \Delta G^{\ddagger}/\delta p)_{T} \qquad (3)$$

where $\Delta V^{\ddagger}$ is the transition-state activation volume, the variation in the molar volume between ES and ES‡, at constant temperature T. If the activation volume itself does not depend on pressure, the dependence of the logarithm of the reaction rate (k_{obs}) on pressure will be a linear function, with the activation volume as a proportionality factor.

$$k_{obs} = A \cdot \exp(-p\Delta V^{\ddagger}/RT) \qquad (4)$$

where $R = 82$ atm $\cdot$ K$^{-1} \cdot$ ml mol^{-1} (1 atm $= 1.01 \cdot 10^{5}$ Pa). Deviations from linearity may reflect either medium compression or viscosity variation. However, Eq. 4 is an over-simplification since the preexponential term A is itself viscosity dependent and suggests that the pressure could affect it. One must bear in mind that pressure increases the viscosity of a system significantly, and if the rate constant is strongly viscosity dependent, the data should be evaluated by using the Kramers equations (Frauenfelder & Wolynes, 1985). The quantities $\Delta G^{\ddagger}$, $\Delta H^{\ddagger}$, $\Delta S^{\ddagger}$, and $\Delta V^{\ddagger}$ for a given reaction as a function of T and p, respectively, are related according to the Maxwell relationships

$$-(\delta \Delta V/\delta T)_{p} = (\delta \Delta S/\delta P)_{T}; \quad (\delta H/T)_{p} = \Delta V - T(\delta \Delta V/\delta T)_{p}. \qquad (5)$$

For many examples, the overall enzyme reactions can be described by the Michaelis-Menten mechanism. Under some experimental conditions (see the following discussion of the experiments carried out at subzero temperatures in hydroorganic solvent), the binding of substrate to enzyme can be analyzed using the induced-fit theory, which implies that the binding of the substrate S to the enzyme E is a two-step process: (1) formation of a collision complex ES (that is, rapid equilibrium related to the association constant K_{1}), and (2) an isomerization step described by the rate constants k_{2} and k_{-2} (Koshland, 1958). The simplest expression is then

$$\text{E} + \text{S} \underset{}{\overset{K_{1}}{\rightleftharpoons}} \text{ES} \underset{k_{-2}}{\overset{k_{2}}{\rightleftharpoons}} \text{E*S} \qquad \qquad \text{Scheme 1}$$

The observed rate constant of the reaction is $k_{obs} = k_{-2} + k_{2}[\text{S}]/(K_{1} + [\text{S}])$. When the concentration of S is increased, a plateau for k_{obs} must be reached.

EXPERIMENTAL CONSIDERATIONS

Cryoenzymology

Cryoenzymology is used to slow down enzyme-catalyzed reactions according to the Arrhenius expression so as to record transients or to stabilize intermediates for further study (Douzou, 1977). However, cryoenzymology could also be used for other purposes. First, the cryosolvent which maintains the fluidity of the medium at subzero temperatures itself acts as a perturbant (Barman et al., 1986). Second, in combination with pressure, cryoenzymology can be a tool to study the thermodynamic properties of the individual rate constants controlling the interconversions of reaction intermediates (Balny et al., 1985).

Stopped-Flow Kinetics

If the system under study can be characterized via optical detection (absorbancy, fluorescence), the stopped-flow method is easily used. The relatively long deadtime of high-pressure techniques using spectroscopic detection is the first important limitation to exploring kinetics. The second limitation deals with efficient control of temperature, which must be able to compensate for the heat of compression. To resolve these problems, devices were designed to mix samples under high pressure and at controlled temperatures. The earliest apparatus was described more than 20 years ago by Grieger and Eckert (1970) and was later modified by Sasaki et al. (1976). Both systems were thermostated and used the break of a foil diaphragm to permit mixing of two components. The mixing was satisfactory within 5 seconds, and experiments were reported at pressures up to 2 kbar. In 1979, Sasaki et al. improved their system, and a high-pressure stopped-flow apparatus was constructed which could be operative at hydrostatic pressures up to 3 kbar with a deadtime of a few milliseconds. Nearly the same year, the Heremans' group described a stopped-flow apparatus designed for spectroscopic detection of fast reactions at pressures up to 1.2 kbar by immersing a stopped-flow unit in a high-pressure cell (Heremans et al., 1980). The highest pseudo-first-order reaction rate constant that could be measured using this device was 35 s^{-1} (that is, a deadtime of about 20 milliseconds). Since then, other high-pressure stopped-flow devices have been described (Smith et al., 1982; Taniguchi & Iguchi, 1983; Ducommun et al., 1984), and a commercial apparatus is now proposed by High-Tech Scientific, Salisbury, UK.

To reduce the deadtime with respect to the reactions studied, in our laboratory we have developed an apparatus that combines low temperatures, which decrease reaction velocities according to the Arrhenius equation, with stopped-flow, which provides rapid mixing of two solutions. The design of our instrument incorporates certain features of previous stopped-flow systems described for cryoenzymological studies (Markley et al., 1981) and for investigations under high pressure (Heremans et al., 1980). The general design, already published (Balny et al., 1984, 1987), consists of a syringe driving mechanism, two vertical drive syringes, a mixing chamber, an observation chamber, and a waste syringe

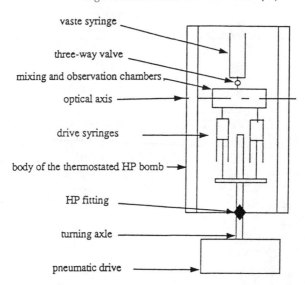

vaste syringe

three-way valve

mixing and observation chambers

optical axis

drive syringes

body of the thermostated HP bomb

HP fitting

turning axle

pneumatic drive

Figure 14.1. Schematic diagram of the stopped-flow apparatus operating at high pressure and subzero temperatures (HPSF).

(Figure 14.1). Both temperature and pressure homogeneities are maintained by housing the entire apparatus in a thermostated high-pressure cell. The stopped-flow apparatus can operate in absorbance or fluorescence mode over a wide range of temperatures ($+40$ to $-35\,°C$) and pressure (1 to 3000 bar). The system is mounted either on an Aminco DW2 spectrophotometer or on a spectrofluorometer specially designed in the laboratory. The deadtime, nearly independent of pressure, is less than 5 milliseconds for aqueous solutions at room temperature. The actual high-pressure stopped-flow apparatus (HPSF) described here differs in certain important respects from that described previously. In particular, the initial electric drive motor has been replaced by a more powerful pneumatic system (see Figure 14.1).

Flow-Quench Kinetics at High Pressure

When no optical signal is present, or when the assignment of an optical signal is difficult, the rapid-flow-quench method must be used. The general principle of this method is as follows: after mixing enzyme and substrate, the reaction mixture is allowed to age and is then stopped (quenched) by the addition of a suitable quencher (for example, acid). The quenched-reaction mixture is then assayed at leisure by any suitable chemical, physical, or enzymatic method. This method is more generally applicable than stopped-flow. On the other hand, it is a point-by-point method which is laborious and reagent consuming (Barman & Travers, 1985).

At atmospheric pressure, the chemical sampling and quenching of reaction mixtures aged 10 seconds or more can be done by hand. The situation is different for high-pressure measurements, and a reactor permitting injection and sampling for steady-state studies of enzymatic reactions at high pressure has been described by Hui Bon Hoa et al. (1990). It is the only existing apparatus operating at pressures up to 4 kbar. It allows injection, stirring, and sampling without release of the pressure. The deadtime of sampling is 10–15 seconds, which allows reactions with pseudo-first-order rate constants smaller than about 1 minute^{-1} to be recorded.

Unfortunately, the rapid-flow-quench method under high-pressure conditions has not yet been extended to studies of reaction with half times shorter than 10 seconds; several technical problems have proved difficult to solve (nature of materials, metal-ion contamination, leaks of samples at the connection levels, and so on).

Relaxation Kinetics

For reversible reactions, relaxation methods have been introduced, such as flash photolysis (Hasinoff, 1974), temperature-jump (Nakatani, 1984), and pressure-jump apparatus in which pressure jumps can be achieved at various pressures so that activation volumes can be obtained. The first device was developed in 1968 for pressure-jump perturbations of 30–50 bar. It was fit for use at any pressure up to 2 kbar and could be applied to spectroscopic relaxation studies of chemical reactions (Brower, 1968, Knoche, 1986). To circumvent the limitation of the small perturbations induced, other systems were improved; they increased the range of pressure jump, increased the higher or lower final pressure, and reduced mechanical relaxation time. For biochemical studies, various improvements were also described (Davis & Gutfreund, 1976).

ENZYME-SUBSTRATE REACTIONS RELATED TO CHANGES IN PROTEIN CONFORMATION

The first experiments using cryobaroenzymatic studies performed in this laboratory were carried out on the reaction of formation of a transition-state analogue complex of creatine kinase: enzyme-ADP-nitrate-creatine, where nitrate mimics the transferable phosphate of ATP. Formation of the analogue complex is accompanied by a conformational change that manifests itself by tryptophan perturbation, allowing studies using the high-pressure stopped-flow method in absorbancy mode (Balny et al., 1985). This work, combining temperature, pressure, and solvent effects, has provided results interpreted within a thermodynamic framework, but it has also demonstrated the limitations of classical transition-state theory. In 40% ethylene glycol, it appeared that the activation volumes, related to the rate-limiting step of the modification of conformation (k), are independent of temperature in the range of 1 to $-15\,°C$ (see Figure 14.2). The absence of temperature sensitivity on $\Delta V^{\ddagger}$ found in this system differs from the situations observed in studies of certain other enzymes in which there are large temperature effects (see the HAO reaction). The values obtained for $\Delta G^{\ddagger}$, $\Delta H^{\ddagger}$, $\Delta S^{\ddagger}$, and $\Delta V^{\ddagger}$ give rise to an accurate fit to the Maxwell's relationships, indicating that the results are internally consistent.

Using a different approach, namely the classical steady-state parameters of

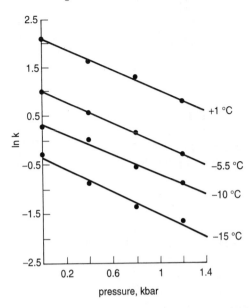

Figure 14.2. Dependence of k on pressure in 40% ethylene glycol at different temperatures. Kinetics were recorded using the HPSF apparatus in absorbance mode at 296.5 nm. One syringe contained creatine kinase (50 μM); the other syringe contained KNO_3 (40 μM). The solution in both syringes was 0.15 mM ADP, 0.65 mM magnesium acetate, 0.1 M sodium acetate, 20 mM creatine, 50 mM Tris acetate buffer, pH 8 (from Balny et al., 1985 with permission).

an enzymic reaction, (Butz et al., 1988) the group of Ludwig analyzed volume changes during fumarase catalysis with hypotheses of enzyme pulsation. This is an excellent example of the analyzing results involving the action of pressure on steady-state parameters. A nonlinearity in plots of $\ln k_{cat}$ versus p has been observed and this deviation interpreted in terms of different sensitivity of individual constants, modulated by the ionic strength. By analyzing the different ΔV values obtained by studying the reaction in two directions (forward and reverse), they concluded that both direct and opposite reactions have a common activated state that governs pulsations of fumarase during catalysis. The volume changes during this pulsation are very small compared with the changes in total volume of the enzyme.

An example of substrate-induced conformational change involves the two-step formation of horseradish peroxidase (HRP) compound I in which the effects of temperature, pressure, and solvent on the reaction of hydrogen peroxide or ethyl peroxide with enzyme were studied. The kinetics of formation of compound I (k_{obs}) were followed at 403 nm using HPSF apparatus. Under cryoenzymic conditions, by measuring k_{obs} as a function of peroxide concentration, we were able to show that the binding kinetics followed a hyperbolic relationship, suggesting a two-step process which involved a modification of the protein conformation (Balny et al., 1987). Moreover, we have also shown that solution conditions, as well as temperature and pressure effects, modulate the energetic barrier of the transition-state level. These observations are very important since it is known that the structure of proteins fluctuates and that this fluctuation is the driving factor for the catalytic process. Similar conclusions were given concerning the kinetics of the reduction of the quinoprotein glucose dehydrogenase by its substrate, xylose (Frank et al., 1991).

Substrate-induced conformational changes and changes in water structure have been studied in various laboratories. In collaboration with P. Masson (CRSSA, La Tronche-Grenoble), we have investigated the effects of high pressure on the single-turnover kinetics of the carbamylation of butyrylcholinesterase. Experiments were carried out using the high-pressure stopped-flow apparatus designed for measuring fluorescence changes at pressures up to 1 kbar (formation of the fluorescent ion N-methyl-7-hydroxyquinolinium). The large positive activation volume measured (119 ml mol^{-1}) reflected extended structural and hydration changes (Masson & Balny, 1988). More recently, this conformational plasticity of butyrylcholinesterase has been confirmed using different solvents in high-pressure experiments (Masson & Balny, 1990).

ELECTRON-TRANSFER REACTIONS

Pressure has great potential in the study of electron-transfer reactions, as shown in the case of reactions of the multiheme protein, hydroxylamine oxidoreductase (HAO), in which an intramolecular electron transfer takes place between one specific cytochrome P-460-containing site and a series of c-hemes. This enzyme from *Nitrosomonas europaea* catalyzes the oxidation of NH_2OH. The kinetics of reduction (k) of the enzyme by NH_2OH were studied using HPSF in various media, in a temperature range of $+20$ to $-15\,°C$, and at pressures up to 1 kbar. Changing the solvent from water to 40% ethylene glycol resulted in an increase in activation volume from -3.6 to 57 ml mol^{-1} and changes in other thermodynamic parameters. The Arrhenius plot shows a downward inflection at about $0\,°C$, which is enhanced at high pressures (see Figure 14.3). For example, at 800 bar, the activation enthalpy value, $\Delta H^{\ddagger}$, was approximately 51 kJ·ml mol^{-1} at subzero temperatures, whereas this value was -18 kJ·ml mol^{-1} above $0\,°C$. This dependency was interpreted as a conformational change of HAO. Apparently, the physical orientation of the electron-transfer centers in the protein was affected by temperature so that the ease of electron transfer was modified. In all cases, the application of high pressure magnified the phenomena recorded at the elementary rate constant. The interaction of the redox centers involved is particularly susceptible to a change in protein structure, as well as in the nature of the solvent. It is an example in which the adjustment of two variables, solvent (40% ethylene glycol to maintain the medium fluid at subzero temperatures) and pressure, makes it possible to amplify the effect of the third variable: temperature. An unusual consequence was observed; at high pressures, a decrease in temperature induced an increase in the velocity of the reaction (Balny & Hooper, 1988).

In biochemistry, there are a limited number of studies of the pressure dependence of intermolecular protein-protein electron transfer. However, recent work devoted to the pressure dependence of inorganic electron-transfer reactions in solution has demonstrated the potential of this parameter (van Eldik, 1991).

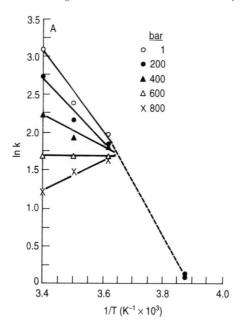

Figure 14.3. Arrhenius plot of k (kinetics of reduction of HAO) at different pressures. Absorbance changes were recorded against time at 552 nm. One syringe of the HPSF contained NH_2OH (40 μM); the other syringe contained HAO (0.9 μM). The solutions in both syringes were 0.2 M Tris-HCl aqueous buffer (pH 7.5) containing 40% ethylene glycol (from Balny & Hooper, 1988 with permission).

Technical difficulties in recording reactions that are generally very fast are one reason such studies are limited. Using the HPSF method and thermodynamic approach (in addition to the classical ionic strength variable), we have investigated two systems: hydroxylamine oxidoreductase (HAO)—mammalian cytochrome c as a model reaction; and methanol dehydrogenase (MDH)—cytochrome cL, a physiological reaction in which cytochrome cL is the natural substrate of MDH. The two enzyme systems exhibit very different behavior. In the case of the reaction of reduced HAO with ferric cytochrome c, the changes in the activation volume can be interpreted in terms of solvational changes. The activation volume appears to be independent of temperature but strongly dependent on ionic strength (Heiber-Langer et al., 1992a). In contrast to the reduction reaction of HAO with its substrate, hydroxylamine, presented here, no evidence for a conformational change was obtained. The situation differs in the other system (Heiber-Langer et al., 1992b) in which the activation volume of the reaction of reduction of cytochrome cL with MDH changes considerably with temperature. Break in the Arrhenius plot leads to the assumption that different enzyme conformations exist. Previous experiments with model reactions help us interpret the data in terms of cytochrome cL conformational changes. However, recent results using structural determinations (infrared and 4th derivative spectroscopy under high pressure) seem not to confirm this conclusion, showing that the modification of conformation could occur at the MDH level. Experiments are in progress to clarify these results.

BINDING OF CARBON MONOXIDE TO REDUCED HEMOPROTEINS

There are two differents methods for studying the properties of the elementary rate constants describing the binding of carbon monoxide to hemoproteins: the classical flash-photolysis method or the stopped-flow mixing method. The latter approach was used to avoid any cage effect which can be observed when the flash-photolysis method is used. The kinetics of the reactions was followed easily by recording absorbance change in the Soret band as a function of time. We observed a linear relationship between k_{obs} and the CO concentration for all experiments, and we interpreted the kinetic data in terms of the one-step binding model. However, the induced-fit theory implies that binding the CO ligand to the protein is a two-step process:

$$E + CO \underset{}{\overset{K_1}{\rightleftharpoons}} E - CO \underset{k_{-2}}{\overset{k_2}{\rightleftharpoons}} E^* - CO \qquad \text{Scheme 2}$$

This formulation is similar to Scheme 1. The linearity of k_{obs} as a function of CO concentration implies that K_1 and k_2 remain too high for the maximum allowed concentration of CO (0.5 mM). The reaction equation is then reduced to

$$E + CO \underset{k_-}{\overset{k_+}{\rightleftharpoons}} E^* - CO$$

$$k_{obs} = k_+[CO] + k_-$$

$$K_1 = k_+/k_-$$

Scheme 3

This dependency was observed for all the systems studied.

Detailed studies of the CO binding to reduced horseradish peroxidase under various conditions have shown that the medium composition (presence of organic cosolvents: 40% ethylene glycol, 50% methanol), as well as the temperature and the pressure effects, modulate the energetic barrier of the transition-state level. The large variations observed in the activation volume, $\Delta V^{\ddagger}$, under different conditions suggest that solvent reorganization is the predominant factor, and this drives the response of the system (see Figure 14.4). The activation volume observed is the sum of several components; the binding of CO, a subsequent conformational change occurring in the protein, and the associated reorganization of the solvation shell.

$$\Delta V^{\ddagger}_{observed} = \Delta V^{\ddagger}_{binding} + \Delta V^{\ddagger}_{conformational} + \Delta V^{\ddagger}_{solvation}$$

These events are interconnected and thus difficult to analyze independently. However, as in the case of the reaction of HAO with its substrate, a detailed interpretation of data showed that by adjusting two variables (chosen among pressure, temperature, or solvent) one can minimize the effect of the third (Balny & Travers, 1989).

This investigation has been extended to the CO binding to other reduced hemoproteins by stopped-flow rapid mixing as a function of pressure (up to 2 kbar)

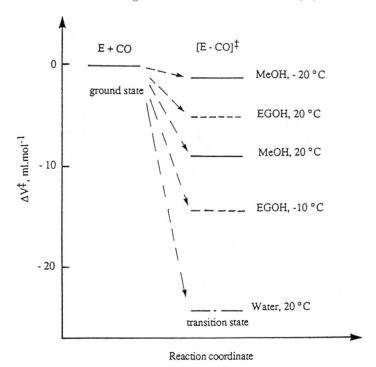

Figure 14.4. Energy diagram in $\Delta V^{\ddagger}$ of the reaction HRP + CO under various experimental conditions. $\Delta V^{\ddagger}$ were calculated from reaction rate constants obtained in recording absorbance change against time at 423 nm minus 500 nm (dual mode of the Aminco DW2 spectrophotometer). One syringe of the HPSF contained HRP (6 μM); the second syringe, CO (1 mM). The solutions in both syringes were 0.2 M Tris buffer, pH 7.5, containing either 40% ethylene glycol (EGOH) or 50% methanol (MeOH) (adapted from Balny & Travers, 1989).

and temperature (ranging from 4 to 35 °C) in aqueous buffer. We studied several varieties of cytochrome P-450, as well as chloroperoxidase and lactoperoxidase, and compared the results to the data reported for other hemoproteins (Lange et al., 1994). The CO binding was then used as a model for the interaction of cytochrome with oxygen, the oxygen activation being the key reaction for substrate hydroxylation in the case of cytochromes P-450. For these reactions, the dependence on the proximal axial heme ligand is a determining factor. Two classes of hemoproteins were examined: those with a thiolate as the fifth heme iron ligand (various isoenzymes of cytochrome P-450) and those in which an imidazole ligand is involved (peroxidases). Activation parameters of the CO binding to hemoproteins are given in Table 14-1. The pressure effect on the binding was different for the enzymes of these two classes. The values are positive ($\Delta V^{\ddagger}$ of few ml mol^{-1}) for P-450 isoenzymes and negative ($\Delta V^{\ddagger}$ of about -10, to 40 ml mol^{-1}) for the other enzymes, respectively. These data suggested the involvement of protein conformational changes in the N-ligand hemoproteins (lactoperoxidase, HRP, myoglobin, hemoglobin, cytochrome P-460, cytochrome P-420), whereas practically

Table 14.1. Activation volumes of the CO binding to various hemoproteins

Enzyme	Ligand	k_1 (25 °C, 0.1 MPa)	$\Delta V^{\ddagger}$
		$M^{-1} s^{-1}$	$ml\ mol^{-1}$
P-450$_{scc}$	S^- (cys)	2.3×10^5	2 ± 2
P-450LM2	S^- (cys)	3.1×10^6	3 ± 2
P-450LM3	S^- (cys)	2.2×10^6	6 ± 5
P-450$7\alpha$	S^- (cys)	2.8×10^4	2 ± 2
CPO	S^- (cys)	1.4×10^5	1 ± 2
LP	N (his)	8.0×10^2	-10 ± 3
HRP	N (his)	2.8×10^3	-24
Myoglobin	N (his)	3.8×10^5	-9
			-18.8
Hemoglobin	N (his)	2.8×10^6	-3
		6.0×10^4	-21
P-460	N (his)	2.8×10^3	-36
P-420	Unknown	$66 \times 10^{3\dagger}$	-25 ± 5
		$12 \times 10^{3\dagger}$	-11 ± 6
		$0.96 \times 10^{3\dagger}$	-42 ± 8

From Lange et al. (1994).

no structural changes were observed with the S-ligand hemoproteins (in the case of isoenzymes of cytochrome P-450). It can be hypothesized that for enzymes with an axial S-ligand, the transition state of the CO binding reaction has an electronic structure similar to the ferrous ground state.

Other studies of enzyme transition states focus on the preexponential factor, A, of the Arrhenius equation, indicating that factor A depends on the viscosity of the solvent and then relates to the interaction of the enzyme with the solvent (Marden & Hui Bon Hoa, 1982). Though Dunker et al. (1980) recommended the investigation of the prefactor A when a multiplicity of protein conformations is suspected, our results indicated that a study of the preexponential factor alone is not informative in a comparative study of different enzymes. In contrast, a kinetic investigation combining pressure and temperature effects appears to correlate with structural features and thus permits a more general understanding of the nature of the enzyme reactional transition state.

CONCLUSION

It is clear from the present results that by collecting as many data as possible and by observing their interdependence, one might eventually be able to formulate general concepts concerning the dynamics of the reaction under study. The action of pressure is one of the few avenues available for studying the problem of protein dynamics in solution. However, one suspects that the use of the transition-state theory is imperfect for enzyme reactions because it is too simplistic. We have shown that the Kramers formulation is difficult to use for many studies. Basic work and theoretical progress are necessary to clarify these points.

In spite of these deficiencies, the pressure parameter has many applications in biotechnology and food science (Balny et al., 1992; Balny & Masson, 1993; Mozhaev et al., 1994). For example, the stereospecific bioconversion of phenyl-pyruvate to L-phenylalanine is catalyzed under high hydrogen pressure by an integrated, immobilized system including bacterial cells and the enzyme alanine dehydrogenase. Further developments of this technology look promising.

ACKNOWLEDGMENTS: The author thanks Drs. F. Travers, I. Heiber-Langer, T. Barman, A. B. Hooper, J. Frank, P. Masson, and R. Lange, Mme. Bec, and Mr. J-L. Saldana for help and fruitful discussions. I acknowledge particularly Dr. Vadim V. Mozhaev for his stimulating discussions and critical reading of the manuscript.

REFERENCES

Asano, T. (1991). Basic principles and mechanisms. In *Organic Synthesis at High Pressures*, ed. K. Matsumoto & R. M. Acheson. New York, Wiley, pp. 7–75.

Balny, C., & Hooper, A. B. (1988). Effect of solvent, pressure and temperature on reaction rates of the multiheme hydroxylamine oxidoreductase: evidence for conformational change. *Eur. J. Biochem.* **176**, 273–279.

Balny, C., & Masson, P. (1993). Effects of high pressure on proteins. *Food Rev. Inter.* **9**, 611–628.

Balny, C., & Travers, F. (1989). Activation thermodynamics of the binding of carbon monoxide to horseradish peroxidase: role of pressure, temperature and solvent. *Biophys. Chem.* **33**, 237–244.

Balny, C., Hayashi, R., Heremans, K., & Masson, P. (eds.) (1992). *High Pressure and Biotechnology*, vol. 224. London, Colloque INSERM-John Libbey Eurotext.

Balny, C., Masson, P., & Travers, F. (1989). Some recent aspects of the use of high pressure for protein investigations in solution. *High Press. Res.* **2**, 1–28.

Balny, C., Saldana, J-L., & Dahan, N. (1984). High pressure stopped-flow spectrometry at low temperatures. *Anal. Biochem.* **139**, 178–189.

Balny, C., Saldana, J-L., & Dahan, N. (1987). High pressure stopped-flow fluorometry at subzero temperatures: application to kinetics of the binding of NADH to liver alcohol dehydrogenase. *Anal. Biochem.* **163**, 309–315.

Balny, C., Travers, F., Barman, T., & Douzou, P. (1985). Cryobaroenzymic studies as a tool for investigating activated complexes: creatine kinase ADP·Mg·nitrate·creatine as a model. *Proc. Natl. Acad. Sci. USA* **82**, 7495–7499.

Balny, C., Travers, F., Barman, T., & Douzou, P. (1987). Thermodynamics of the two-step formation of horseradish peroxidase compound I. *Eur. Biophys. J.* **14**, 375–383.

Barman, T., & Travers, F. (1985). The rapid-flow-quench method in the study of fast reactions in biochemistry: extention to subzero conditions. *Meth. Biochem. Anal.* **31**, 1–59.

Barman, T., Travers, F., Balny, C., Hui Bon Hoa, G., & Douzou, P. (1986). New trends in cryoenzymology: probing the functional role of protein dynamics by single-step kinetics. *Biochimie* **68**, 1041–1051.

Brower, K. R. (1968). A method for measuring the activation volumes of fast reversible reactions: the ferric thiocyanate complex. *J. Am. Chem. Soc.* **90**, 5401–5406.

Butz, P., Greulich, K. O., & Ludwig, H. (1988). Volume changes during enzyme reactions:

indications of enzyme pulsation during fumarase catalysis. *Biochemistry* **27**, 1556–1563.

Davis, J. S., & Gutfreund, H. (1976). The scope of moderate pressure changes for kinetic and equilibrium studies of biochemical systems. *FEBS Lett.* **72**, 199–207.

Douzou, P. (1977). *Cryobiochemistry: An Introduction*. London, Academic Press.

Ducommun, Y., Nichols, P. J., Helm, L., Elding, L. I., & Merbach, A. E. (1984). Variable pressure oxygen-17 FTNMR and stopped-flow kinetic study of water exchange and DMSO substitution on square-planar tetraaqua-palladium (II) and -platinum (II). *J. Phys.* **45**, C8-221–224.

Dunker, A. M., Lusk, J. E., & Gibbs, J. H. (1980). Enzyme reaction rates and the stochastic theory of kinetics. *Biophys. Chem.* **11**, 9–16.

Evans, M. G., & Polanyi, M. (1935). Kinetic effects of pressure. *Trans. Faraday Soc.* **31**, 875–880.

Frank, J., Duine, J. A., & Balny, C. (1991). Preliminary studies on quinoprotein glucose dehydrogenase under extreme conditions of temperature and pressure. *Biochimie* **73**, 611–613.

Frauenfelder, H., & Wolynes, P. G. (1985). Rate theories and puzzles of hemeprotein kinetics. *Science* **229**, 337–345.

Glastone, S., Laidler, K., & Eyring, H. (1941). *The Theory of Rate Processes*. New York, McGraw-Hill.

Grieger, R. A., & Eckert, C. A. (1970). A new technique for chemical kinetics at high pressure. *AIChE J.* **16**, 766–770.

Groβ, M., & Jaenicke, R. (1994). Proteins under pressure: the influence of high hydrostatic pressure on structure, function and assembly of proteins and protein complexes. *Eur. J. Biochem.* **221**, 617–630.

Hasinoff, B. B. (1974). Kinetic activation volumes of the binding of oxygen and carbon monoxide to hemoglobin and myoglobin studied on a high-pressure flash photolysis apparatus. *Biochemistry* **13**, 3111–3117.

Heiber-Langer, I., Hooper, A. B., & Balny, C. (1992a). Pressure modulation of cytochrome-to-cytochrome electron transfer: models and enzyme reactions. *Biophys. Chem.* **43**, 265–277.

Heiber-Langer, I. Clery, C., Frank, J., Masson, P., & Balny, C. (1992b). Interactions of cytochrome *cL* with methanol dehydrogenase from *Methylphaga marina* 42: thermodyanimic arguments for conformational change. *Eur. Biophys. J.* **21**, 241–250.

Heremans, K., Snauwaert, J., & Rijkenberg, J. (1980). Stopped-flow apparatus for the study of fast reactions in solution under high pressure. *Rev. Sci. Instrum.* **51**, 806–808.

Hui Bon Hoa, G., Hamel, G., Else, A., Weill, G., & Hervé, G. (1990). A reactor permitting injection and sampling for steady state studies of enzymatic reactions at high pressure: tests with aspartate transcarbamylase. *Anal. Biochem.* **187**, 258–261.

Knoche, W. (1986). Pressure-jump methods. In *Investigations of Rates and Mechanisms of Reactions*. ed. C. Bernasconi. New York, Wiley, **2:** 191–218.

Koshland, D. E. (1958). Application of a theory of enzyme specificity to protein synthesis. *Proc. Natl. Acad. Sci. USA* **44**, 98–101.

Laidler, K. J. (1950). The influence of pressure on the rate of biological reactions. *Arch. Biochem. Biophys.* **30**, 226–236.

Lange, R., Heiber-Langer, I., Bonfils, C., Fabre, I., Negishi, M., & Balny, C. (1994). Activation volume and energetic properties of the binding of CO to hemoproteins. *Biophys. J.* **66**, 89–98.

Marden, M. C., & Hui Bon Hoa, G. (1982). Dynamics of the spin transition in camphor-bound ferric cytochrome P-450 versus temperature, pressure and viscosity. *Eur. J. Biochem.* **129**, 111–117.

Markley, J. L., Travers, F., & Balny, C. (1981). Lack of evidence for a tetrahedral intermediate in the hydrolysis of nitroanilide substrates by serine proteinases. *Eur. J. Biochem.* **120**, 477–485.

Masson, P., & Balny, C. (1988). Effects of high pressure on the single-turnover kinetics of the carbamylation of cholinesterase. *Biochim. Biophys. Acta* **954**, 208–215.

Masson, P., & Balny, C. (1990). Conformational plasticity of butyrylcholinesterase as revealed by high pressure experiments. *Biochim. Biophys. Acta* **1041**, 223–231.

Morild, E. (1981). The theory of pressure effects on enzymes. *Adv. Protein Chem.* **34**, 93–166.

Mozhaev, V. V., Heremans, K., Frank, J., Masson, P., & Balny, C. (1994). Exploiting the effects of high hydrostatic pressure in biotechnological applications. *TIBTECH* **12**, 403–501.

Mozhaev, V. V., Heremans, K., Frank, J., Masson, P., & Balny, C. (1996). High pressure effects on protein structure and function. *Proteins: Structure, Function and Genetics* **24**, 81–91.

Nakatani, H., & Hiromi, K. (1984). Kinetic study of β-cyclodextrin dye system by high pressure temperature jump method. *J. Biochem.* (Tokyo) **96**, 69–72.

Sasaki, M., Amita, F., & Osugi, J. (1979). High pressure stopped-flow apparatus up to 3 kbar. *Rev. Sci. Instrum.* **50**, 1104–1107.

Sasaki, M., Okamoto, M., Tsuzuki, H., & Osugi, J. (1976). High pressure cells for the study of moderately fast reaction in solution. *Chem. Lett.* 1289–1292.

Smith, P., Beile, E., & Berger, R. (1982). An observation chamber for a high pressure stopped-flow apparatus. *J. Biochem. Biophys. Meth.* **6**, 173–177.

Taniguchi, Y., & Iguchi, A. (1983). Effect of pressure on the rate of alkaline fading of triphenylmethane dyes in cationic micelles. *J. Am. Chem. Soc.* **105**, 6782–6786.

van Eldik, R. (1991). Pressure dependence of inorganic electron transfer in solution. *High Press. Res.* **6**, 251–259.

15

Steady-State Enzyme Kinetics at High Pressure

DEXTER B. NORTHROP

Pressure effects on enzyme-catalyzed reactions were traditionally interpreted within a simplistic kinetic mechanism in which the transition state presented the highest energy barrier, and the chemical transformation of substrate to product was considered to be a singular, rate-limiting step. This was also true of isotope effects on enzyme-catalyzed reactions, but extensive isotopic studies have led to the conclusion that this transition state is rarely the highest barrier. Rather, the release of products (or the conformational change preceding product release) is usually the slowest step, often accompanied by several other partially rate-limiting steps. Thus, our interpretations of pressure effects must be shifted accordingly. Values attributed to $\Delta V^{\ddagger}$ have been determined for more than 50 enzymes, more of them with a positive sign than negative, and most in the range of 20 to 40 ml mol^{-1} but these may or may not have anything to do with the activation volume associated with the transition state. Volume changes specifically associated with the binding of ligands to enzymes have been reported as well, including some very large values, as high as $\Delta V = 85$ ml mol^{-1}. Kinetically, these equilibrium pressure effects also originate in conformational changes because water is not very compressible; hence, rates of diffusion to and from enzymes are virtually unaffected by pressure. Much larger changes, as high as $\Delta V = -391$ ml mol^{-1}, have been observed during disaggregation and denaturation of enzymes. Thus, while it is possible for pressure effects to be expressed on every step of an enzymatic reaction, and to cause denaturation as well, making kinetic data from pressure effects hopelessly complex and uninterpretable, it appears likely that the most significant pressure effects will be expressed on conformational changes associated with product dissociations, without much kinetic complexity. This makes sense from another point of view—that the largest volume changes are probably on solvation equilibria during ligand binding and protein folding. Pressure effects on isotope effects have the potential of specifically identifying whether or not a volume change occurs

211

upon attaining the transition state. With the exception of hydrogen tunneling, intrinsic isotope effects are independent of pressure. When substrates labeled in a transferable position are used, isotope effects generally arise solely from one transition state. It therefore follows that if pressure decreases the rate of an enzyme-catalyzed reaction because of a positive volume change in the same transition state, then the expression of the isotope effect will *increase* because the step from which it arises will become more rate limiting. In contrast, if pressure decreases the reaction rate by an effect on the rate of some other step such as product release, then the expression of an isotope effect will *decrease*. Given the likelihood of the latter as a general rule, pressure effects and isotope effects should complement each other. Pressure effects may turn out to be the more useful tool because they can produce much larger effects (a $\Delta V^{\ddagger}$ of -40 ml mol^{-1} will generate a 963-fold increase in a rate of 4 kbar at 25 °C, for example). Also, pressure effects can be studied as a continuous function, which makes it possible to sort out some forms of kinetic complexity.

HISTORICAL BACKGROUND

Throughout most of the 80-year history of the kinetics of enzyme-calculated reactions, the dominant paradigm was a simplistic kinetic mechanism in which the transition state presented the highest energy barrier, and the chemical transformation of substrate to product was considered to be a singular, rate-limiting step. This traditional paradigm is illustrated in the reaction coordinate energy diagram shown in Figure 15.1. For pressure effects, this paradigm suggests that in the familiar kinetic equation from transition-state theory (Glastone et al., 1941); in which k_0 is the rate constant under atmospheric conditions, p stands for pressure, R is the gas constant, and T is the temperature.)

$$k/k_0 = e^{-\Delta V^{\ddagger} p/RT} \tag{1}$$

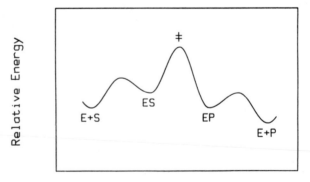

Figure 15.1. Reaction coordinate versus energy diagram of the traditional minimal mechanism of an enzyme-catalyzed reaction.

This is often written in logarithmic form

$$\ln(k/k_0) = -\frac{\Delta V^{\ddagger} p}{RT} \tag{2}$$

The volume change indicated by $\Delta V^{\ddagger}$ represents the difference between volume of the enzyme-substrate complex and the volume in the transition state, and as such is correctly signified by a double dagger superscript. EP has a lower energy level than ES (Figure 15.1), indicating that the reaction is essentially irreversible. The small barrier for substrate binding compared with the large barrier for the chemical transformation of substrate to product is consistent with the rapid equilibrium binding hypothesis employed by Michaelis and Menten (1913) in the original formulation of their rate equation. Despite the demonstration by Briggs and Haldane (1925) that this restriction was unnecessary and that their alternative steady-state hypothesis could equally well account for the kinetic behavior of enzymes, this paradigm has persisted, particularly in introductory textbooks and in the minds of many who work with enzymes but do not call themselves enzymologists.

During the past two decades, a kind of renaissance has occurred in the use of isotope effects to determine both chemical and kinetic mechanisms of enzymes, triggered primarily by the sixth Steenbock Symposium, Isotope Effects on Enzyme-Catalyzed Reactions, held in 1976 (Cleland et al., 1977). The first round of discoveries of isotope effects on enzyme-catalyzed reactions was made primarily by researchers who identified either with nuclear chemistry and had access to the early isotopes or with organic chemistry and had an interest in transition-state theory. These researchers also employed the paradigm in Figure 15.1, which for isotope effects implies that the intrinsic deuterium isotope on a bond-breaking step will be fully expressed on reaction rates such as the maximal vecloity (V), as in Eq. 3:

$$V_{\mathrm{H}}/V_{\mathrm{D}} = k_{\mathrm{H}}/k_{\mathrm{D}} \tag{3}$$

The second round of discoveries was made by researchers who identified specifically with enzyme kinetics. One of the conclusions of the second analysis is that the transition stateady state is rarely the highest barrier, and isotope effects are rarely expressed fully. Instead, the release of products (or more specifically, the conformational change preceding product release) is usually the slowest step, often accompanied by several other partially rate-limiting steps. This newer paradigm is represented in the energy diagram shown in Figure 15.2. Its features relevant to the current discussion are multiple steps, high barriers for conformational changes (that is, for ES $\rightarrow$ E*S and E*P $\rightarrow$ EP), more than one step representing the chemical transformation of substrate to product (that is, for E*S $\rightarrow$ E*X and E*X $\rightarrow$ E*P), relatively low barriers of these chemical steps, and equilibria associated with the chemical steps at values near $K_{eq} = 1$. This last feature means that even for highly exothermic reactions, the chemical steps can be readily reversible (Nageswara Rao & Cohn, 1979). Dueterium and ^{18}O isotope effects expressed during fumarase catalysis, for example, are not kinetic isotope

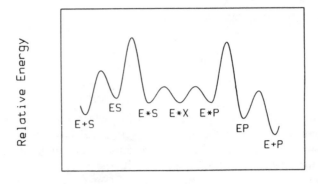

Figure 15.2. Reaction coordinate versus energy diagram of a realistic mechanism of an enzyme-catalyzed reaction, incorporating the kinetic complexities consistent with experimental observations.

effects at all, but rather are equilibrium isotope effects because of the rapid equilibrium of fumarate and malate bound to the enzyme as a result of the slow release of products (Blanchard & Cleland, 1980).

Obviously, our paradigm for pressure effects on enzyme-catalyzed reactions must be shifted from Figure 15.1 to Figure 15.2, and our interpretations of pressure effects must be based on rate equations similar to those developed for the expression of isotope effects (Northrop, 1981a). Moreover, those equations show that a perturbation of a transition state is expressed differently in the maximal velocity, V, than in V/K (the fundamental kinetic parameters in which K is the Michaelis-Menten constant), which is another way of saying that isotope effects may vary with changes in the concentration of substrate. Except for the singular analysis of fumarase by Butz et al. (1988) discussed later in this chapter, the paradigm shift has not yet occurred with respect to pressure. It is the opinion of this observer that the shift is inevitable and that this 1994 Steenbock symposium will be the trigger for another renaissance in physical techniques for investigating enzyme kinetics.

As an illustration of the strength of the old paradigm, the excellent review of pressure effects on enzymes by Morild (1981) lists values of $\Delta V^{\ddagger}$ for more than 50 enzymes. These are single values specifically attributed the transition state activation volume; many of them were obtained from assays at a single, saturating concentration of substrate; more of them have a positive sign than a negative one; and most are in the range of 20–40 ml mol^{-1}. The point is, considering the possibilities illustrated in Figure 15.2, we now know that these values may or may not have anything to do with volume changes associated with transition state chemistry. Moreover, volume changes specifically associated with the binding of ligands to enzymes and other proteins have been reported as well, including some very large values, as high as $\Delta V = 85$ ml mol^{-1} (Kornblat et al., 1988). Much larger changes, as high as $\Delta V = -391$ ml mol^{-1}, have been observed during disaggregation and denaturation of enzymes (Ikkai & Ooi, 1975). Because isotopic studies have shown that chemical transformations are rarely rate limiting, it follows that many of the observed pressure effects probably have their origins in something other than the transition states of bond-breaking chemical steps, and many are likely to be composite volumes from multiple steps. Indeed, unlike isotope effects, it is possible for pressure effects to be expressed on every step of an enzymatic reaction, which would make kinetic data from pressure effects hopelessly complex and uninterpretable. However, it appears more likely that the most significant pressure effects originate in conformational changes without much kinetic complexity. This makes sense from another point of view, namely that the largest volume changes are probably on solvation equilibria during ligand binding and protein folding associated with conformational changes.

To be sure, Morild and a few others have addressed substrate dependence, but mostly in terms of pressure effects on the Michaelis-Menten constant, which in the reaction coordinate diagram of Figure 15.1 is rendered equal to the dissociation constant. Some have also considered the possibility of multiple steps and derived rate equations to address the added complexity. Laidler and Bunting (1973), for example, considered the mechanism shown in Scheme I:

$$E + S \underset{k_{-1}}{\overset{k_1}{\rightleftharpoons}} ES \overset{k_2}{\rightarrow} EP \overset{k_3}{\rightarrow} E + P \qquad \text{Scheme I}$$

The maximal velocity for this mechanism depends on two steps, the chemical step and the product release step, governed by Eq. 4:

$$V = \frac{k_2 k_3}{k_2 + k_3} \tag{4}$$

Laidler and Bunting then derived an equation to define $\Delta V_c^{\ddagger}$, which they termed the weighted mean of $\Delta V_2^{\ddagger}$ and $\Delta V_3^{\ddagger}$

$$\Delta V_c^{\ddagger} = \frac{k_2 \Delta V_3^{\ddagger} + k_3 \Delta V_2^{\ddagger}}{k_2 + k_3} \tag{5}$$

This weighted mean volume change allowed them to formulate an expression analogous to Eq. 2 for the effect of pressure on a maximal velocity:

$$\ln(V/V_0) = -\frac{\Delta V_c^{\ddagger} p}{RT} \tag{6}$$

The derivation was done by allusion and was too abstract for this reader to grasp. Consequently, to obtain a better understanding of exactly what $\Delta V_c^{\ddagger}$ represents, Eq. 6 was simulated on computer, using $k_2 = k_3 = 1$, $\Delta V_2^{\ddagger} = 10 \text{ ml mol}^{-1}$, and $\Delta V_3^{\ddagger} = 90 \text{ ml mol}^{-1}$. The result, shown in Figure 15.3, is a curve with a tangent to zero pressure whose slop corresponds to a volume change of 50 ml mol^{-1}, the weighted mean volume change in this numerical exercise. Hence, it appears that Laidler and Bunting's equation is the differential at $p = 0$. As such, it is useful for

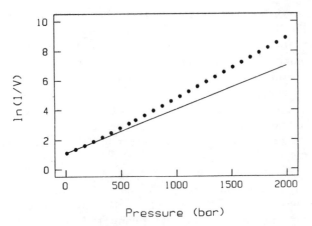

Figure 15.3. Simulated pressure effects on the maximal velocity of the reaction mechanism of Scheme I, assuming rate constants of equal value at atmospheric pressure and volume changes of 10 and 90 ml mol^{-1}. The solid line is the tangent to the curve at zero pressure with a slope of 50 ml RT mol^{-1}.

addressing low-pressure effects in the real world, but it ignores the mechanistic information that is expressed in the curve at artificial but useful high pressures.

Laidler and Bunting (1973) also explored the effects of pressure on V/K within the mechanism of Scheme I:

$$\frac{V}{K} = \frac{k_1 k_2}{k_{-1} + k_2} \tag{7}$$

They considered two possibilities, the extremes of $k_2 \gg k_{-1}$ and $k_2 \ll k_{-1}$. In the former, $V/K = k_1$ and $\Delta V_{V/K}^{\ddagger} = \Delta V_1^{\ddagger}$. For the latter,

$$\frac{V}{K} = \frac{k_1 k_2}{k_{-1}} \tag{8}$$

Hence, the apparent volume change expressed in V/K as defined by Eq. 8 was described as

$$\Delta V_0^{\ddagger} = \Delta V_1^{\ddagger} + \Delta V_2^{\ddagger} - \Delta V_{-1}^{\ddagger} \tag{9}$$

The operative pressure equation analogous to Eq. 3 is therefore

$$\ln\left(\frac{V/K}{(V/K)_0}\right) = -\frac{\Delta V_0^{\ddagger} p}{RT} \tag{10}$$

Equation 10 resembles Eq. 6 but differs in that the volume change is neither a weighted mean nor applicable only at low pressures. A numerical simulation gives a straight line—which illustrates how the math works in pressure kinetics: when rate constants are multiplied or divided by each other, the pressure effects is a linear function with a slope dependent on a single *composite* volume change, which in turn consists of sums and differences of the individual volume changes. Only when rate constants are added within a rate equation does the algebraic form generate curves. Ludwig and Greulich (1978) and Butz et al. (1988) extended the analysis of V/K to the full expression of Eq. 7 and obtained

$$\Delta V_0^{\ddagger} = \Delta V_1^{\ddagger} + \frac{\Delta V_2^{\ddagger} + \Delta V_1^{\ddagger}}{1 + \left[\dfrac{k_2}{k_{-1}}\right] e^{(-(\Delta V_2^{\ddagger} - \Delta V_{-1}^{\ddagger})p/RT)}} \tag{11}$$

By simple substitution in Eq. 11, an analogous equation for maximal velocities can be formulated (compare Eqs. 4 and 7)

$$\Delta V_c^{\ddagger} = \Delta V_2^{\ddagger} + \frac{\Delta V_3^{\ddagger} + \Delta V_2^{\ddagger}}{1 + \left[\dfrac{k_3}{k_2}\right] e^{(-(\Delta V_3^{\ddagger} - \Delta V_2^{\ddagger})p/RT)}} \tag{12}$$

It appears to this observer that Laidler and Bunting's analysis had a negative effect on the development of pressure kinetics of enzyme-catalyzed reactions because it casts the analysis in terms of single composite volume changes, perpetuating the paradigm in Figure 15.1 long after most enzymologists had embraced the implications of Figure 15.2 in other areas of enzyme kinetics. Even Ludwig's group, which extended graphical analysis to curves, used this cast—which may account for why their work failed to get the recognition it deserves; it simply does not look that different from Laidler and Bunting's approach, through in fact Ludwig's analysis was two steps more advanced. Moreover, this form of equation is extremely difficult for traditional kineticists to relate to intuitively; enzyme kineticists are used to thinking in terms of slopes and intercepts of reciprocal plots, not composite rate constants and their associated composite volume changes.

PROPOSED CHANGES IN THE ANALYSIS OF PRESSURE EFFECTS

Enzyme kinetics, like all other kinetics and most analytical science, has its origins in analytical geometry; algebraic functions were routinely transformed to linear forms that could be approached by graphical analysis. With the advent of computers, however, analysis of curves become possible, and most areas of enzyme kinetics have replaced graphics with nonlinear regression as a means of analysis, while still retaining the linear transformations as the preferred means of display (Cleland, 1967, 1979). A search of the literature reveals that such a change has not happened with respect to pressure effects. Therefore, as an exercise, the data in Figure 15.3 were fitted to Eq. 4 (in the form of Eq. 13, see next paragraph) to determine if pressure effects were amenable to this method of analysis. The result was positive, with the regression quickly converging on the original values. However, depending on the initial estimates used, either $\Delta V_2^{\ddagger}$ or $\Delta V_3^{\ddagger}$ regressed to a value of 10 while the other regressed to 90. Obviously, Eq. 4 (and Eq. 13) is symmetrical with respect to the two reactive steps, and to establish which volume change is associated with which step, it becomes necessary to study pressure effects as a function of something else. That, too, is a typical approach within other areas of enzyme kinetics, but it is rare in pressure kinetics (again, the important exception of Butz, Greulich, & Ludwig (1988) should be noted, see below and chapter 14 on transient enzyme kinetics by Claude Balny, who stresses this same point).

As an alternative to the form of pressure equations shown in the opening section, which employ composite volume changes, it is here proposed that pressure effects be expressed within traditional rate equations of steady-state enzyme kinetics, with each volume change represented individually with its associated rate constant. To illustrate, Eqs. 4–6 and 12 can be represented by the single expression

$$V = \frac{k_2 e^{-(\Delta V_2^\ddagger p/RT)} k_3 e^{-(\Delta V_3^\ddagger p/RT)}}{k_2 e^{-(\Delta V_2^\ddagger p/RT)} + k_3 e^{-(\Delta V_3^\ddagger p/RT)}} \tag{13}$$

which can also be written

$$V = \frac{1}{\dfrac{1}{k_2 e^{-(\Delta V_2^\ddagger p/RT)}} + \dfrac{1}{k_3 e^{-(\Delta V_3^\ddagger p/RT)}}} \tag{14}$$

If the reaction in Scheme I were reversible, the pressure effect may appear to be identical, but the definition of the first volume change would be different. Because V equals the reciprocal of the sum of the reciprocals of the net rate constants (Cleland, 1975), the rate equations can be written

$$V = \frac{1}{1/k_2' + 1/k_3'} \tag{15}$$

where the primes represent net rate constants, $k_2' = k_2 k_3/(k_{-2} k_3)$ and $k_3' = k_3$. The pressure effect on a reversible two-step reaction then becomes

$$V = \frac{1}{1/k_2' e^{-\Delta V_2^\ddagger p/RT} + 1/k_3 e^{-\Delta V_3^\ddagger p/RT}} \tag{16}$$

Note that $\Delta V_2'$ is the volume change for the net rate constant for the second step of the mechanism. If $k_{-2} \ll k_3$ then $\Delta V_2' = \Delta V_2^\ddagger$, but if $k_2 \gg k_3$, then $\Delta V_2' = \Delta V_2^\ddagger + \Delta V_3^\ddagger - \Delta V_{-2}^\ddagger$, or to put it another way, $\Delta V_2' = \Delta V_{Keq(2)} + \Delta V_3^\ddagger$, where $\Delta V_{Keq(2)}$ is the volume change on the equilibrium constant of step 2. Thus, the individual volume changes on a net rate constant need not factor out and may appear as a single composite colume change. Such composite volume changes may or may not refer to a transition state but may instead refer to an equilibrium constant between ground states. When $k_{-2} \approx k_3$, then the full rate equation applies

$$V = \frac{k_2 e^{-\Delta V_2^\ddagger p/RT} k_3 e^{-\Delta V_3^\ddagger p/RT}}{k_2 e^{-\Delta V_2^\ddagger p/RT} + k_3 e^{-\Delta V_3^\ddagger p/RT} + k_{-2} e^{-\Delta V_{-2}^\ddagger p/RT}} \tag{17}$$

A problem with Eq. 17 is that it involves three rate constants and three volume changes, and at the present time it appears doubtful that nonlinear regression will converge with that number of parameters because of the high degree of covariance—which gives us an additional reason for studying pressure effects as a function of something else to factor out parts of a kinetic mechanism. On the surface, adding yet another variable to a kinetic system appears to be adding to the complexity of the system, but often it works the other way. The extra variable often makes it possible to tease apart the kinetic functions and parameters.

V/K may be described similarly in terms of volume changes on the individual rate constants. Eq. 7 yields

$$V/K = \frac{k_1 e^{-\Delta V_1^\ddagger p/RT} k_2 e^{-\Delta V_2^\ddagger p/RT}}{k_{-1} e^{-\Delta V_{-1}^\ddagger p/RT} + k_2 e^{-\Delta V_2^\ddagger p/RT}} \tag{18}$$

which can also be written

$$V/K = \cfrac{1}{\cfrac{1}{\left[\dfrac{k_1 k_2}{k_{-1}}\right] e^{-\Delta V_{(1.2/-1)} p/RT}} + \cfrac{1}{k_1 e^{-\Delta V_1^\ddagger p/RT}}} \tag{19}$$

A complex V/K may thus appear as a function of two volume changes, one composite ($\Delta V_{(1.2/-1)}$) and the other kinetic ($\Delta V_1^\ddagger$). If one assumes that diffusion-controlled rate constants are affected negligibly by pressure because water is not very compressible (see later in this chapter), then only k_2 will be affected by pressure:

$$V/K = \frac{k_1 k_2 e^{-\Delta V_2^\ddagger p/RT}}{k_{-1} + k_2 e^{-\Delta V_2^\ddagger p/RT}} \tag{20}$$

Nevertheless, Eq. 20 may generate complex results. If $k_{-1} < k_2$, then V/K will appear pressure independent because k_2 and its volume change will cancel out from the numerator and the denominator—unless $\Delta V_2^\ddagger$ is positive and very large, in which case k_2 will approach k_{-1} at high pressures. Such an example was simulated in Figure 15.4A. Alternatively, if $k_- > k_2$, then V/K will appear linearly dependent at low pressures, but may become pressure independent if $\Delta V_2^\ddagger$ is negative and very large, in which case k_2 will again ultimately approach and surpass k_{-1}. That possibility is simulated in Figure 15.4B.

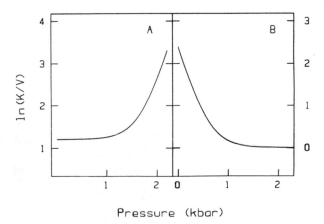

Figure 15.4. Simulated pressure effects on V/K of the reaction mechanism of Scheme I using Eq. 20. (**A**) $k_1 = 0.3$, $k_{-1} = 1$, $k_2 = 1000$, and $\Delta V_2^{\ddagger} = 100$ ml mol^{-1} (**B**) $k_1 = 1$, $k_{-1} = 10$, $k_2 = 1$, and $\Delta V_2^{\ddagger} = -100$ ml mol^{-1}.

Because enzyme kineticists have a tradition of plotting reciprocals of velocities, I propose here that graphical analysis of pressure data also be done in reciprocal form. For example, the reciprocal form of Eq. 6 is (using V as the volume parameter subscript instead of Laidler's c)

$$\ln(1/V) = \frac{\Delta V_V^{\ddagger} p}{RT} + \ln(1/V_0) \tag{21}$$

Plots of $\ln(1/V)$ versus p have a slope of $\Delta V_V^{\ddagger}/RT$, an intercept of $\ln(1/V_0)$, and resemble the intercept replots in the kinetics of dead-end inhibition. Similarly, Eq. 10 becomes (using subscript V/K)

$$\ln(K/V) = \frac{\Delta V_{V/K}^{\ddagger}}{RT} + \ln(K/V)_0 \tag{22}$$

Eq. 22 is analogous to slope replots from dead-end inhibition patterns. Casting the pressure functions in the form of Eqs. 13–22 can be done by simple inspection, using existing rate equations, which means that pressure kinetics can be applied to a variety of complex kinetic mechanisms without recourse to new derivations. In contrast, the single composite volume approach to Eqs. 4–12 required a difficult new derivation for each change in mechanism because the volume change itself had to be desribed as a changing function of pressure. That difficulty may also have held back the development of pressure kinetics of enzyme-catalyzed reactions.

ANALYSIS OF PUBLISHED RESULTS

There are very few examples in the literature in which pressure was varied systematically as a function of substrate concentrations or something else, and most of those come from the laboratory of Eddie Morild at the Norwegian Underwater Institute in Bergen, Norway. Two enzymes studied by Morild, yeast and liver alcohol dehydrogenase, plus fumarase studied at Ludwig's laboratory in Heidelberg, Germany, will be discussed as a means of illustrating where the field is and where it can go.

Yeast Alcohol Dehydrogenase

This enzyme obeys a rapid equilibrium random kinetic mechanism during alcohol oxidation, but the mechanism of aldehyde reduction is ordered (Brändén et al., 1975) outlined in Scheme II,

$$
\begin{array}{c}
\text{EA} \\
A \overset{}{\underset{K_{ia}}{\diagup}} \quad \underset{K_b}{\overset{B}{\diagdown}} \\
E \qquad\qquad \text{EAB} \overset{k_{cat}}{\to} E + P + Q \qquad\qquad \text{Scheme II} \\
B \underset{\diagup}{\overset{K_{ib}}{\diagdown}} \quad \underset{\diagdown A}{\overset{K_a}{}} \\
\text{EB}
\end{array}
$$

in which A represents NAD^+, B represents ethanol, P represents acetaldehyde, and Q represents NADH. Because catalysis is rate limiting, Michaelis constants at high and low concentration of cosubstrates are simple dissociation constants (for example, K_a and K_{ia} at high and low [B], respectively). Morild varied the concentration of NAD^+, at both low and high concentrations of ethanol, as function of pressure up to 1500 bar. Reciprocal plots of velocities versus pressure were linear. A large volume change was associated with the binding of NAD^+ to free enzyme, but a negligible volume change was associated with binding NAD^+ to the enzyme-ethanol binary complex, as indicated by the values of ΔV_{ia} and ΔV_a, respectively, in Table 15.1. Morild also varied the concentration of ethanol, at both low and high concentrations of NAD^+, and obtained very similar results with respect to binding ethanol to free enzyme and to the binary complex EB. A small but significant volume change was associated with the chemical segment, represented by ΔV_{cat} (Table 15.1).

Table 15.1. Molal volume changes of yeast alcohol dehydrogenase kinetics[a]

	Reactive step	(ml mol^{-1})
ΔV_{ia}	EA + A → EA	-75 ± 23
ΔV_{ib}	E + B → EB	-69 ± 25
ΔV_a	EB + A → EAB	-15 ± 10
ΔV_b	EA + B → EAB	-9 ± 10
ΔV_{cat}	EAB → EPQ	-20 ± 5

[a] Data from Morild (1977).

These results, plus similar volume data for invertase, dextranase, and dextran-sucrase (Ludwig & Greulich, 1978), invite speculation about the general effects of pressure on enzyme kinetics. The lack of a significant pressure effect on binding of either NAD^+ or ethanol to binary complexes suggests that diffusion-controlled reactions to and from the enzyme are not very sensitive to pressure, which makes sense because water is not very complressible.[1] Therefore, the large volume changes upon binding to free enzyme, suggest that something else is happening. Because of relatively slow ($\sim 10^3$ s^{-1}) changes in the spectrum of NADH upon binding, it has long been known that binding of the nucleotide coenzyme to alcohol dehydrogenase (Czerlinski, 1962) and many other dehydrogenases (Hammes & Schimmel, 1970) is a two-step process that includes a significant conformational change. What was not known before Morild's pressure data were published was that a similar (or perhaps identical?) conformational change occurs when ethanol binds to free enzyme.[2] The large values for ΔV_{ia} and ΔV_{ib} (Table 15.1) are also consistent with direct measurements of ligand binding in other proteins, as mentioned. The modest volume change on traversing the chemical segment of the reaction is consistent with the popular view of enzymatic catalysis: within the transition state, strained and tightly bound substrates are converted to products with a concerted movement of electrons and little or no movement of atoms. In the extreme, this view predicts that the chemical step of an enzyme-catalyzed reaction reaction should not undergo any change in volume—which suggests a means of testing the theory. Thus, while it is theoretically possible for pressure effects to be expressed during every step of an enzymatic reaction, which would make pressure kinetics hopelessly complex and uninterpretable, Morild's results on yeast alcohol dehydrogenase suggest that substrate binding and product release steps per se will be independent of pressure and that the most significant pressure effects will be expressed on conformation changes. This means that pressure effects on steady-state enzyme kinetics may be relatively free of kinetic complexity and may, at least sometimes, yield clean mechanistic interpretations. Moreover, what makes pressure effects particularly interesting is that the steps most likely to have the largest effects are precisely those that are most often rate limiting—changes in the conformations of protein.

Fumarase

Fumarase is one of the most studied of all enzymes, and its kinetics as a function of high pressure were examined in careful detail by Butz et al. (1988). Some of their data are shown in Figure 15.5. Note that initial velocities of malate

---

1. Diffusion-controlled reactions will be subject to a small pressure effect because of changes in viscosity. For example, at 30 °C, the viscosity of water increases by 12.7% at 2 kbar (Bridgeman, 1949) which translates into an apparent $\Delta V_{\ddagger}$ of -1.39 ml mol^{-1}. Interpolating Bridgeman's data yields $\Delta V_{\ddagger} = 0$ below 2 kbar at 19.1 °C, which is therefore the recommended temperature for kinetic studies.

2. This conclusion applies only to the random kinetic mechanism. Most dehydrogenases have ordered binding, with coenzyme binding first, and this order is thought to be driven by the nucleotide-dependent conformational change.

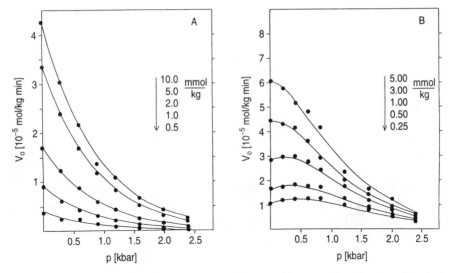

Figure 15.5. Pressure dependence of the initial velocities of fumarase for (**A**) malate dehydration and (**B**) fumarate dehydration as a function of substrate concentration. Data were reproduced from Butz et al. (1988).

dehydration decrease uniformly with increasing pressure, but that the initial velocities of fumarate hydration first increase slightly at low pressure and then decrease uniformly at high pressure. Something is present in the latter that is not present in the former, which is difficult to reconcile intuitively with a simple single-substrate single-product enzymatic reaction. Butz et al. (1988) noted that a plot of $\ln(k/k_0)$ as a function of pressure according to Eq. 2 is normally a straight line through the origin, but their fumarate hydration data generated curves. Furthermore, they stated that curvature in such plots is usually addressed within the context of compresssibility (that is, staying within the old paradigm, the single composite volume itself might be considered subject to a pressure effect), but they went on to say that "these are of the order of less than a few percent per kilobar." Consequently, they undertook the difficult task of addressing multiple reactive steps using the three-step mechanism of Scheme I and obtained the composite volume in Eqs. 12 and 13 in the previous section—they made a partial break from the old paradigm.

Several years after their work was published, other kinetic data were collected that are inconsistent with the three-step mechanism in Eq. 4. Dead-end inhibition (Rebholz & Northrop, 1994a), progress curve analysis, and Britton's isotopic tracer perturbation (Rebholz & Northrop, in preparation), plus isotopic exchange (Hanson et al., 1969; Rose et al., 1992) all require a fourth step in the form of a kinetically signifcant isomerization of free fumarase.

$$E + S \underset{k_2}{\overset{k_1}{\rightleftharpoons}} ES \underset{k_4}{\overset{k_3}{\rightleftharpoons}} FP \underset{k_6}{\overset{k_5}{\rightleftharpoons}} P + F \underset{k_8}{\overset{k_7}{\rightleftharpoons}} E \qquad \text{Scheme III}$$

This type of a kinetic mechanism has been termed an *isomechanism*. It is described by Eq. 23 (Cleland, 1963; Fromm 1975; Rebholz, 1993; Rebholz & Northrop, 1993):

$$V = \frac{V_f V_r \left([A] - \dfrac{[P]}{K_{eq}} \right)}{K_a V_r + V_r[A] + \dfrac{V_f[P]}{K_{eq}} + \dfrac{V_r[A][P]}{K_{iip}}} \tag{23}$$

Because of the K_{iip} term, Eq. 23 predicts a noncompetitive product inhibition pattern, whereas the pattern for the mechanism in Scheme I would be competitive. For a single-substrate single-product, reversible enzymatic reaction, the best way to determine the form of product inhibition is by an analysis of progress curves (Rebholz & Northrop, 1994b). Foster-Neiman plots are linear with competitive product inhibition but develop marked curvature with a noncompetitive component in the inhibition mechanism, which is turn can occur with an isosegment that is only partially rate limiting in a catalytic turnover. Figure 15.6 shows that

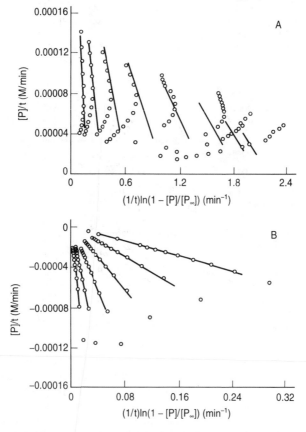

Figure 15.6. Foster-Neimann plots of progress curves for fumarase during (**A**) malate dehydration and (**B**) fumarate dehydration as a function of substrate concentration. Data were reproduced from Rebholz (1993).

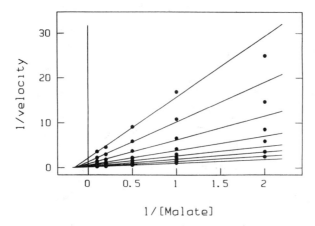

Figure 15.7. Double reciprocal plot of pressure kinetics on malate dehydration. Data were extracted from Figure 15.5A.

Foster-Neiman plots of fumarate hydration are distinctly curved, in contrast to plots of malate dehydration which are nearly linear. When comparing Figure 15.5 and 15.6, it is obvious that some kinetic function is expressed during fumarate hydration that is not expressed during malate dehydration. Because of the number and range of kinetic data supporting an isomechanism for fumarase,[3] it seems reasonable to consider that the missing kinetic function in the pressure kinetics is also the isosegment.

To address that question, Ludwig supplied the initial velocity data that were used to construct Figure 15.5, plus similar data of different ionic strength that had not been published, and these were graphed in double reciprocal plots familiar to enzyme kineticists. Figure 15.7 shows the effect of increasing pressure on malate dehydration; the linear plots resemble a noncompetitive inhibitory pattern of a dead-end inhibitor. The general equation for mixed noncompetitive inhibition (written in reciprocal form for the purpose of illustrating the analytical geometry) is

$$\frac{1}{v} = \frac{K}{V[S]}\left(1 + \frac{[I]}{K_{is}}\right) + \frac{1}{V}\left(1 + \frac{[I]}{K_{ii}}\right) \tag{24}$$

Therefore, an analogous reciprocal equation was constructed to account for a mixed noncompetitive inhibitory pressure effect:

$$\frac{1}{v} = \frac{K}{V[S]}\, e^{-\Delta V_{is}\,p/RT} + \frac{1}{V}\, e^{-\Delta V_{ii}\,p/RT} \tag{25}$$

---

3. In addition to the pressure data, many other anomalies in the extensive kinetics of fumarase have been reported, all of which may be resolved by reinterpretation within an isomechanism. These include biphasic Ahrrenius plots (Massey, 1953), unexplained shifts in the values of pK_a for the two active site carboxyl groups (Blanchard & Cleland, 1980), and unusual effects of viscosity on maximal velocities (Sweet & Blanchard, 1990).

The normal initial velocity data used to construct Figure 15.7 were therefore fitted to Eq. 25 using the nonlinear regression computer program of Duggleby (1984). The slope and intercept volume changes were $\Delta V_{is} = 25.5 \pm 1.9$ ml mol^{-1} and $\Delta V_{ii} = 27.6 \pm 1.6$ ml mol^{-1}, respectively. These are not significantly different, and when the same, the lines have a common intersection point on the horizontal axis at $1/[S] = -1/K_m$. For dead-end inhibition, such a pattern is called classical non-competitive inhibition, and it obeys the equation

$$\frac{1}{v} = \left(\frac{K}{V[S]} + \frac{1}{V} \right)\left(1 + \frac{[I]}{K_l} \right) \tag{26}$$

Consequently, the analogous classical noncompetitive pressure effect equation would be

$$\frac{1}{v} = \left(\frac{K}{V[S]} + \frac{1}{V} \right)e^{-\Delta V_i p/RT} \tag{27}$$

Fitting the normal data of Figure 15.7 to Eq. 27 gives $\Delta V_i = 26.6 \pm 0.5$ ml mol^{-1}. This is an excellent fit, with a precision (1.9%) that is unheard of in kinetics of dead-end inhibition. The reason for the better precision is that with dead-end inhibitors it is often impossible or undesirable to raise inhibitor concentrations much beyond two or three times K_l because of secondary solute effects, such as changes in ionic strength and so forth. Pressure, on the other hand, is noninvasive, so very high pressures can be applied to yield much larger changes in reaction rates, without necessarily encountering secondary effects. (One does have to worry about protein denaturation and subunit dissociation, however; see the section on pressure effects on isotope effects). Hence, the larger range of measurements account for the increased precision.

For slope and intercept to be equally affected—that is, V and V/K are equally inhibited—it is necesssary for some step common to both kinetic functions to be rate limiting in both. Because (a) the binding of substrate is not expressed in V, (b) the chemical step is at isotopic equilibrium (Blanchard & Cleland, 1980), (c) diffusion-controlled processes are probably pressure independent, and (d) malate dehydration is endothermic, the most likely candidate for the step expressing the volume change during malate dehydration is the conformational change preceding the release of fumarate. The kinetic arguments supporting that conclusion are difficult to describe because the algebraic definitions of rate-limiting steps are different for V and V/K (Northrop, 1981b; Ray, 1983). The simplest way of making the point is illustrated in the energy diagram of a five-step reaction sequence shown in Figure 15.8, and comes down to this: for a step to be equally rate-limiting to V and V/K it must come late in a reaction sequence with an "uphill" reaction coordinate diagram.

Initial velocity data for fumarate hydration were analyzed similarly and are shown in reciprocal form in Figure 15.9. The individual plots are again linear, but the pattern lacks a common intersection point; also, some of the lines cross to the right of the vertical axis, which is unusual in more familiar kinetic patterns. The

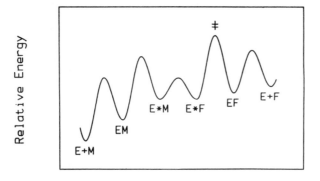

Reaction Coordinate

Figure 15.8. Hypothetical reaction coordinate versus energy diagram for fumarase during the endothermic dehydration of malate.

usual approach to understanding complex inhibition patterns is to construct replots of slopes and intercepts as a function of inhibitor concentrations. By analogy, and in keeping with Eqs. 21 and 22, replots of $\ln(K/V)$ and $\ln(1/V)$ versus pressure were constructed and are shown in Figure 15.10. For reference, lines corresponding to replots from Figure 15.7 are shown as well. A biphasic function is obvious in the data points of the slope replot and is suggested in the intercept replot by a comparison with the reference line. At extremely high pressure, the same transition state should become rate limiting in both forward and reverse reactions; hence, both forward and reverse intercept replots should approach similar asymptotes. The asymptotes will not be identical, but will differ by the difference in volumes of substrates and products. For the fumarase reaction, $\Delta V_{malate} - \Delta V_{fumarate} = 6.8$ ml mol^{-1} (Butz et al., 1988).

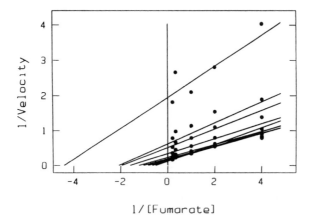

Figure 15.9. Double reciprocal plot of pressure kinetics on fumarate hydration. Data were extracted from Figure 15.5B.

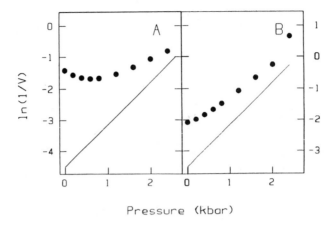

Pressure (kbar)

Figure 15.10. (A) Slope and (B) intercept replots of data extracted from Figure 15.9. The solid lines are reference replots constructed from data in Figure 15.7.

To understand slope (K/V) and intercept $(1/V)$ replots of an isomechanism, the mechanism in Scheme III must be written in a slightly different form:

$$\overset{k_1'}{\overbrace{F \rightleftharpoons E + S \rightleftharpoons \underset{k_{chem}}{\underbrace{ES \rightleftharpoons FP}} \rightarrow P + \underset{k_{iso}}{\underbrace{F \rightarrow E}}}} \qquad \text{Scheme IV}$$

Because V/K represents reaction velocities at $[S] \rightarrow 0$, isomerizations of free enzyme are at equilibrium (represented by $F \rightleftharpoons E$ on the left of Scheme IV, and by $K_{eq(F/E)}$ in Eq. 28 and 29). The combination of enzyme and substrate is determined by the net rate constant, k_1', which extends from the binding of substrate through the release of the first product

$$V/K = \frac{k_1'}{1 + K_{eq(F/E)}} \qquad (28)$$

Consequently, the slope data were fitted to the pressure equation

$$V/K = \frac{k_1' e^{-\Delta V_1^{\ddagger} p/RT}}{1 + K_{eq(F/E)} e^{-\Delta V_{eq(F/E)} p/RT}} \qquad (29)$$

V/K at atmospheric pressure was $2.2 \times 10^6 \ M^{-1} s^{-1}$, and the fitted parameters were: $k_1' = 6.18 \pm 0.012 \times 10^6 \ M^{-1} s^{-1}$, $K_{eq(F/E)} = 1.85 \pm 0.05$, $\Delta V_1 = 16.1 \pm 0.2$ ml mol^{-1}, and $\Delta V_K = 55.5 \pm 1.0$ ml mol^{-1}. The fitted curve is shown in Figure 15.11A, along with lines representing the pressure-dependent values of k_1' and $K_{eq(F/E)}$. There is excellent agreement between point and line, and again the precision of the fitted parameters is remarkable.

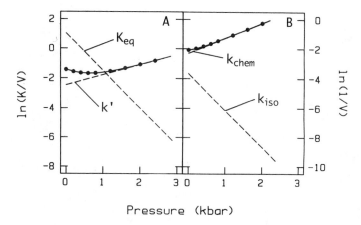

Figure 15.11. Results of nonlinear regression on data from Figure 15.10. (**A**) Apparent values of V/K were fit to Eq. 29. (**B**) Apparent values of V were fit to Eq. 30.

Similarly, because V represents reaction velocities at $[S] \rightarrow \infty$, isomerizations of free enzyme are irreversible (represented by $F \rightarrow E$ on the right of Scheme IV). The maximal velocity is determined by two apparent rate constants, k_{chem} and k_{iso}, which represent the chemical segment and isomerization segment, respectively. Analogous to Eq. 14, the pressure equation may be written as

$$V_{max} = \frac{1}{1/k_{chem}^{-\Delta V_{chem}^{\ddagger} p/RT} + 1/k_{iso}^{-\Delta V_{iso}^{\ddagger} p/RT}} \qquad (30)$$

Because of the symmetry of Eq. 30, k_{chem} and k_{iso} cannot be identified without some additional information. Luckily, Ludwig's group provided that in the form of pressure data at different concentrations of sodium chloride, and Hanson et al. (1969) showed that chloride inhibits isotopic exchange between ^{14}C-malate and ^{12}C-fumarate, the chemical segment of Scheme IV. Figure 15.12 shows intercept replots of pressure data at 0 and 0.3 M NaCl. Note that the data at high salt are linear, with a slope parallel to an apparent high-pressure asymptote of the nonlinear data at low salt. This identifies k_{chem} as being associated with the high-pressure asymptote. Using the slope of this asymptote multiplied by RT as an initial estimate for ΔV_{chem}, the intercept data were fit to Eq. 27 with V_0 at 80 s^{-1} to obtain the following results: $k_{chem} = 100.4 \pm 0.1$ s^{-1}, $k_{iso} = 390 \pm 10$ s^{-1}, $\Delta V_{chem} = 24.0 \pm 0.1$ ml mol^{-1}, and $\Delta V_{iso} = -58.4 \pm 4.6$ ml mol^{-1}. The fitted curve is shown in Figure 11B, along with lines representing the pressure-dependent values of k_{chem} and k_{iso}. Again, there is excellent agreement between point and line, and again the precision of the fitted parameters is remarkable. The tight fits reflect favorably on the technical skills of Peter Butz. These are very good data.

The agreement between ΔV_1 of malate dehydration and ΔV_{chem} of fumarate dehydration suggests that the similar steps are rate limiting in both directions, but the reaction coordinate diagrams in Figure 15.8 will not allow them to be the

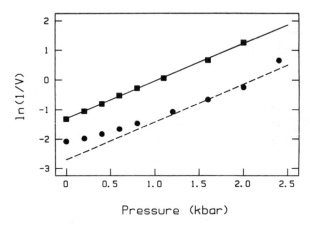

Figure 15.12. Effect of NaCl on volume changes expressed during fumarate hydration: (●) = intercept replot from Figure 15.9 of the effect of pressure in the absence of added soldium chloride; (■) = intercept replot of the effect of pressure in the presence of 0.3 M NaCl (H. Ludwig, personal communication). The dashed line was drawn parallel to the high salt line.

same step (see Northrop, 1981b, Case 5b & Figure 2). If the conformation change represented by E*F → EF is rate-limiting to V during malate dehydration, perhaps E*M → EM is rate-limiting to k_{chem} during fumarate hydration. The agreement between the kinetic ΔV_{iso} and thermodynamic ΔV_K suggests a more complex isomerization than outlined in Scheme III, perhaps one consisting of multiple steps, with the pressure-sensitive step in rapid equilibrium. Certainly, these volume changes will prove useful to characterizing the nature of the isomerization. It should be pointed out, however, that the agreements and goodness of fits do not prove that fumarase obeys the mechanism of Scheme III; Butz et al. (1988) also had a good fit to their three-step mechanism. To distinguish between the two mechanisms, specific experiments must be designed to address their differences, such as an analysis of progress curves as a function of pressure; if an isomechanism is present, then K_{iip} in Eq. 23 ought to be pressure dependent. Nevertheless, this analysis illustrates some of the possibilities of pressure kinetics. As a case in point, in addition to volume changes, the pressure kinetics of isomechanisms provide precise estimates of k_{chem}, k_{iso}, and $K_{eq(F/E)}$, unlike other kinetic approaches to isomechanisms which only offer combinations of these. For example dead-end inhibition kinetic patterns offer the relationship (Rebholz & Northrop, 1994a)

$$\frac{K_{is}}{K_{ii}V_r} = \frac{1}{k_{fiso}} + \frac{1}{k_{riso}} \qquad (31)$$

but cannot isolate the individual rate constants. Hence, pressure kinetics offers several exciting new tools with which to probe enzymatic mechanisms.

Liver Alcohol Dehydrogenase

This enzyme follows a Theorell-Chance kinetic mechanism (Theorell & Chance, 1951; Brändén et al., 1975) which may be diagrammed as follows:

$$
\begin{array}{ccccc}
A & B & P & Q & \\
\downarrow & \searrow \nearrow & & \uparrow & \\
\hline
E & EA & EQ & E &
\end{array}
\qquad \text{Scheme V}
$$

where A is again NAD^+, B is ethanol, P is acetaldehyde, and Q is NADH. The chemical segment is so fast as to be kinetically insignificant, and the rate-limiting segment is the release of NADH. Morild (1977) varied both NAD^+ and ethanol against against each other, in effect generating a complete initial velocity pattern for each of five different pressures, and obtained the complex curves reproduced in Figure 15.13. What is most obvious is that the rate of the reaction first increased substantially with pressure up to about 1500 bar, then decreased dramatically at even higher pressures, at all substrate concentrations. A closer look reveals that the maximum activity occurs at the same pressure when NAD^+ was varied, but at different pressures when ethanol was varied. Moreover, the curves cross when ethanol was varied, but not when NAD^+ was varied. Regarding these data, Morild wrote the following: "The results from the studies of liver alcohol dehydrogenase

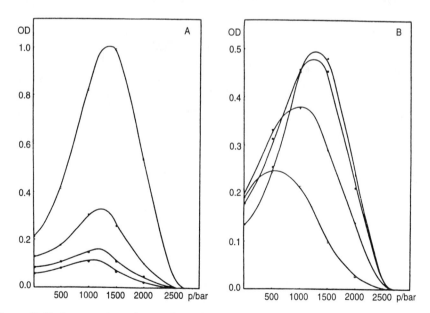

Figure 15.13. Pressure dependence of initial velocities of liver alcohol dehydrogenase, (**A**) as a function of varied NAD^+ and (**B**) as a function of varied ethanol. NAD^+ concentrations were, top-to-bottom, 0.767, 0.1, 0.04, and 0.02 nM. Ethanol concentrations were 200, 100, 50, and 10 mM. Data were extracted from Morild (1977).

are largely qualitative, and are at present under refinement. From the special role played by the ethanol concentration it is believed that these studies will reveal important features concerning the inhibition mechanism. The strange phenomenon of a pressure dependent activation volume indicates that two counteracting effects are present, each dominating different parts of the pressure range" (p. 361). Morild's first statement shows that he just didn't know what to do with these data; despite his considerable technical skills and curiosity, he was overwhelmed by the theoretical complexity. His second statement shows that his intuition was excellent, as will be demonstrated here later. His third statement documents the mental trap he and others of the time were in—he thought only in terms of the single composite volume change of the old paradigm. Data inconsistent with that mental image, his "strange phenomenon," could be understood only as a secondary pressure effect on the primary pressure effect, which was beyond his capacity to analyze. If we approach these data within the traditions of the kinetics of inhibition of enzyme-catalyzed reactions as outlined by Eqs. 13–22 and begin with the construction of double reciprocal plots, the kinetic complexity becomes manageable, and some interexting details emerge.

Figure 15.14 shows four plots of reciprocal velocities versus reciprocal concentrations of ethanol. Panel A is at atmospheric pressure. Instead of the normal linear function associated with the Lineweaver-Burk (1934) transformation, the data curve upward at high concentrations of ethanol (low 1/[S]) in an

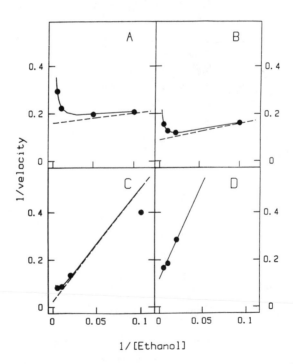

Figure 15.14. Double reciprocal plots of data extracted from Figure 15.13B showing pressure-dependent substrate inhibition by high concentrations of ethanol. Pressures were (**A**) 1 bar, (**B**) 500 bar, (**C**) 1500 bar, and (**D**) 2000 bar.

expression of substrate inhibition. In panels B and C, one can see that increasing hydrostatic pressure reduces the substrate inhibition, until at 2000 bar in panel D substrate inhibition is no longer detectable. Pressure-dependent substrate inhibition makes perfect mechanistic sense with what we already know about this enzyme from other kinetic studies. Because the release of NADH is rate limiting, it follows that the rate enhancement caused by increasing pressure must be an effect on the rate of dissociation of NADH from the enzyme.[4] A more rapid dissociation of NADH reduces the steady-state concentration of EQ during turnovers. The substrate inhibition of liver alcohol dehydrogenase by ethanol has long been thought to be due to the formation of an abortive complex, EQB, as outlined in the following paragraphs (Dalziel & Dickinson, 1966):

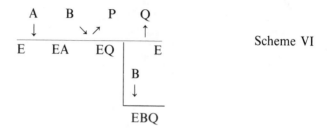

Scheme VI

Hence, by decreasing the steady-state concentration of EQ, pressure reduces the formation the EQB complex and thereby abolishes the substrate inhibition by ethanol. Morild's data confirm the mechanism in Scheme VI in a direct manner which had been deduced only indirectly before, the primary evidence being the uncompetitive substrate inhibition pattern versus NAD. However, that is not a unique pattern; uncompetitive substrate inhibition can also arise by secondary binding of the inhibitory substrate to a central complex (Northrop & Cleland, 1974).

Initial velocities versus [NAD] were fitted to the Michaelis-Mentern equation at different pressures:

$$v = \frac{V[A]}{K_a + [A]} \tag{32}$$

Similarly, initial velocities versus [ethanol] were fitted to the general equation for substrate inhibition (Cleland, 1970):

$$v = \frac{V[B]}{K_b + [B] + [B]^2/K_1} \tag{33}$$

Despite the paucity of data points representing the uninhibited velocities in the latter, reasonably good values of V and V/K were obtained. Intercepts $(1/V)$ from

---

4. Actually, the slow step is believed to be the conformational change preceding the release of the product nucleotide, E*Q → EQ. Coates et al. (1977) have obtained direct evidence from pressure-relation studies that pressure weakens the binding of nucleotides to liver alcohol dehydrogenase.

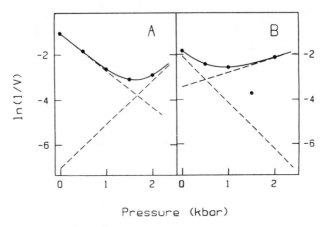

Figure 15.15. Intercept replots of data extracted from Figure 15.13, derived from (**A**) varied NAD$^+$ and (**B**) varied ethanol.

both sets of replots were fitted to Eq. 16 for two pressure-dependent rate constants. Reciprocal plots of these fittings are shown in Figure 15.15. The low-pressure volume changes were -38 ± 1 and -49 ± 22 ml mol^{-1}, and the high-pressure volume changes were 45 ± 3 and 15 ± 9 ml mol^{-1}, respectively, for varied NAD$^+$ and ethanol. These pairs ought to be identical, but the precision of the ethanol intercepts is lowered because of the longer extrapolation associated with substrate inhibition. Because of the parallel effect on substrate inhibition as described in Schme VI (another example of "something else"), the negative volume changes can be attributed to the conformational change preceding the release of NADH. The origin of the positive volume change, which leads to loss of activity at high pressure, is uncertain. It could be a transition state of the chemical segment, or it could be inactivation of the enzyme.

Slopes (K/V) from both sets of replots were similarly fit to Eq. 20, and the reciprocal pressure plot is shown in Figure 15.16. The NAD slopes shown in Figure 15.16A lack precision and enough points at high pressure to overcome the covariance and resolve all the kinetic parameters during nonlinear regression. Nevertheless, the shape of the curve is consistent with the kinetic mechanism. In an ordered addition of substrates, saturation with the second substrate (B) renders V/K_a irreversible and equal to k_1 in the absence of a conformational change (Cleland, 1963), or alternatively, if a conformational change is present, equal to Eq. 7 where k_2 refers to the conformational change. The flat portion of the line in Figure 15.16A at low pressure is consistent with $V/K_a \approx k_1$, which, being a diffusion-controlled rate constant, ought to be merely pressure independent. However, a conformational change has been observed spectroscopically upon binding NAD, as discussed here; hence, at low pressure k_2 must greatly exceed k_{-1} for V/K to approach the diffusion-controlled rate constant. The upturn at high pressure is consistent with k_2 being diminished to the point that V/K becomes dependent on all rate constants in Eq. 7 and identifies a pressure-dependent step located between the binding of NAD and ethanol with $\Delta V_2^\ddagger > 60$ ml mol^{-1}. The ethanol slopes shown in Figure 15.16B similarly lacked precision and enough

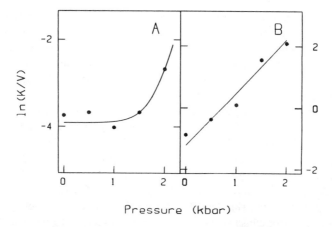

Figure 15.16. Slope replots of data extracted from Figure 15.13, derived from (**A**) varied NAD$^+$ and (**B**) varied ethanol.

points at low pressure to resolve all kinetic parameters, but the regression did gravitate toward a $\Delta V_2^{\ddagger}$ of 48 ± 43 ml mol^{-1}. Therefore, kinetic complexity was ignored, and the data were fitted to Eq. 1, which greatly improved the precision of this number to 43 ± 4 ml mol^{-1}. The lack of a plateau at low pressures suggests $k_2 < k_{-1}$ for ethanol binding (referring again to Eq. 7 but applying it to Scheme I and ignoring saturating NAD$^+$), and this is consistent with the higher K_m for ethanol than for acetaldehyde (Wratten & Cleland, 1963) and the large isotope effect on V/K for the oxidation of deuterated ethanol (Cook & Cleland, 1981).

PRESSURE EFFECTS ON ISOTOPE EFFECTS

The data from Morild's and Ludwig's laboratories show that the location of pressure-dependent steps with respect to substrate addition and product release can be determined if the pressure effects are measured as a function of "something else," such as a salt or a cosubstrate. Further resolution of different steps within the chemical segment may be possible if the "something else" is an isotope effect, because isotope effects are associated almost exclusively with a single chemical step in enzymatic catalysis (Northrop, 1981a). No pressure effect on an intrinsic isotope effect is predicted because of the Born-Oppenheimer approximation, which assumes that the isotopic and nonisotopic potential energy surfaces are the same (Maggiora & Christoffersen, 1978). However, if the chemical step is not fully rate limiting and pressure affects another partially rate-limiting step, then the size of an apparent isotope effect, expressed either on V or V/K, must change. On the one hand, if the other step has a negative volume change, then an increase in pressure will make that step less rate limiting, the reaction rate will increase, and the apparent isotope effect will also increase; if the volume change is postive, then the rate and isotope effect will both decrease. On the other hand, if the isotopically sensitive step itself has a negative volume change, then the increase in reaction rate will be accompanied by decrease in the isotope effect; if the volume change is positive, decrease in reaction rate will be accompanied by an increase in the isotope effect. Finally, if both steps undergo changes in volume, they may still be differential if only one of them is isotopically sensitive. Combining pressure effects with isotope effects has the potential of identifying whether or not a volume change occurs specifically in a transitiom state of a chemical step within a chemical segment, apart from the conformational changes that often precede the product release at the end of a chemical segment.

The Born-Oppenheimer approximation does not always hold with regard to isotopes of hydrogen, however. If hydrogen tunneling occurs, the transition states of deuterium and tritium will differ from protium, and a volume difference if possible (Isaacs, 1984). For example, hydride transfer during the oxidation of leuco-crystal violet is believed to proceed with tunneling, and $\Delta V_H^{\ddagger} = -25$ ml mol^{-1} as opposed to $\Delta V_D^{\ddagger} = -35$ ml mol^{-1} (Issacs et al., 1978). Because tunneling recently has been proposed for hydride transfer during the oxidation of benzyl alcohol by yeast alcohol dehydrogenase (Cha et al., 1989), we decided to test this hypothesis by looking for a pressure effect on the isotope effect. Benzyl alcohol is a much slower substrate than ethanol, and hydride transfer is believed to be fully rate limiting for benzyl alcohol oxidation (Klinman, 1976). Assays were run at concentrations of benzyl alcohol well below its K_m, so the reaction rates approximate V/K values. Figure 15.17A shows that the reaction rates for both normal and dideutero-benzyl alcohol first increased with pressures up to 1.5 kbar and then decreased in a similar way. Figure 15.17B shows that the deuterium isotope effect decreased as a continuous, approximately linear, function.

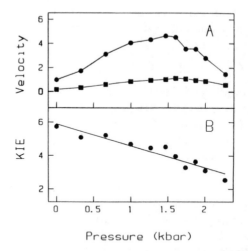

Figure 15.17. Effect of pressure on the oxidation of normal and dideutero-benzyl alcohol by liver alcohol dehydrogenase: (**A**) The initial velocities as a function of pressure; (**B**) the kinetic isotope effect, v_H/v_D, calculated from data in panel A. Assays contained 48 U/ml of Sigma horse liver alcohol dehydrogenase, 1.65 mM benzyl alcohol ($K_m = 3$ mM), 10.5 mM NAD$^+$ ($K_m = 1$ mM) in 80 mM Tris-HCl pH 8.5 at 25 °C. Absorbancy changes at 340 nm were measured using an OLIS spectrophotometric data acquisition system and a Gilford 240 monochrometer attached to an SLM high-pressure bomb. Pressures were generated with an APCS-1 automated pressure-control system from Advanced Pressure Products.

This change in the isotope effect provides us with an assurance that the decrease in activity is due to modulation of a catalytic rate constant and not to inactivation of the enzyme. Inactivation simply removes the catalyst from the system but will not alter the isotope effect because the isotope only reports properties of the active catalyst that remains behind. This is an important new idea because we know that high pressure causes subunit dissociations (for example, <3 kbar, Weber & Drickamer, 1983) and protein unfolding (for example, >4 kbar, Royer, 1994), so some means is needed to distinguish modulation from inactivation if we are to interpret positive volume changes in enzymatic activity. The principle behind the idea is not new, however, Parmentier et al. (1992) used isotope effects to address a similar question regarding the allosteric regulation of aspartate transcarbamylase. They wanted to know if the changes in activity by allosteric activators and inhibitors followed the Monod model, in which an equilibrium is perturbed between active and inactive enzyme, as opposed to changes in a mix of active enzyme forms with different levels of activity. They found that the ^{15}N-isotope effects on the V/K of aspartate were identical for the holo-enzyme, the catalytic subunit stripped of the regulatory sununit, and for holo-enzyme in the presence of activating ATP or inhibiting CTP. Their results are consistent with the Monod model and contradict the pressure experiment shown in Figure 15.17B.

The data in Figure 15.17B were fit to an equation similar to Eq. 14, for two irreversible steps

$$v = \frac{1}{1/k_1 e^{-\Delta V_1^{\ddagger} p/RT} + 1/k_2 e^{-\Delta V_2^{\ddagger} p/RT}} \tag{34}$$

where k_1 represents the apparent rate constant for the conversion of the enzyme-NAD complex to enzyme-NADH and k_2 is the apparent rate constant for the conversion of the latter to free enzyme and free NADH. Both processes are rendered irreversible under initial velocity conditions, where the concentrations of acetylaldehyde and NADH are near zero. The results are plotted in Figure 15.18 are tabulated in Table 15.2. The two rate constants differ by two orders of magnitude, with a deuterium isotope effect clearly associated with the smaller rate constant, k_1. The larger rate constant may have an isotope effect as well but, if so, it is lost in the standard error. A moderately large negative volume change $(-34.4 \pm 5.6$ ml mol$^{-1})$ is associated with k_1, and this volume change is very interesting because it is the first that can be specifically attributed to a transition state of a chemical step—*because of the isotope effect*—which means that this is the first structural information about a transition state in an enzyme-catalyzed reaction! The volume change with the dideutero-benzyl alcohol $(-34.4 \pm 4.7$ ml mol$^{-1})$ is identical, which argues against the hydrogen tunneling hypothesis.[5]

---

[5] Note added in proof. This preliminary conclusion was based on a derivative analysis of an isotope effect on a pressure effect (that is, regressions were performed separately on protium and deuterium data sets, and the results were compared). Later studies showed that such comparisons can be insensitive to model differences. A global analysis combining protium and deuterium data sets in a single regression supports the model containing the hydrogen tunneling hypothesis.

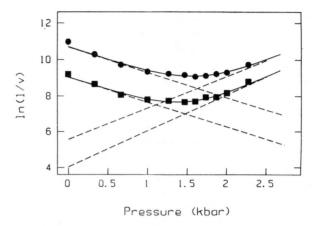

Figure 15.18. Results of nonlinear regression on data from Figure 15.17A. Initial velocities using (●) normal benzyl alcohol and (■) dideutero-benzyl alcohol were fitted to Eq. 34 and plotted as reciprocal logarithms.

A larger positive volume change is associated with k_2, which may be another chemical step, conformational change, or dissociation of nucleotide product. Morild (1977) observed large negative volume changes for the binding of substrates to free enzyme, so this positive change would be consistent with the opposite process: the dissociation of the nucleotide product.

CONCLUSION

Pressure effects on the steady-state kinetics of enzyme-catalyzed reactions promise to reveal many new things about enzymes, if approached with the sophistication that is common to other fields of enzyme kinetics. The most obvious procedure to include in pressure kinetics is a variation of the concentration of substrates to separate effects on V and V/K. The results may be displayed on graphs, but the analysis should employ some fitting procedure such as nonlinear regression to extract kinetic constants and standard errors. Most important, volume changes should be determined as a function of "something else" to assign the volumes to specific steps of a kinetic mechanism. Changes resulting from such things as ionic strength, pH, alternative substrates, and isotopes are a few of the more obvious

Table 15.2. Analysis of the pressure effect on the isotope effect in Figure 15.18b

	H	D	k_H/k_D	$\Delta V_H - \Delta V_D$
k_1 (m$_{-1}$)	0.360 ± 0.054	0.066 ± 0.012	5.2 ± 1.0	
ΔV_1 (ml mol^{-1})	-34.4 ± 5.6	-34.0 ± 4.7		0.03 ± 10.3
k_2 (m^{-1})	56.4 ± 44.4	12 ± 9	4.8 ± 5.3	
ΔV_2 (ml mol^{-1})	49.8 ± 9.4	43.3 ± 8.7		6.5 ± 18

variables that should be explored. Finally, when interpreting pressure data, it is important to keep in mind that not all kinetic volume changes are necessarily of the double dagger variety. Without an isotope effect or other reporter of bond breaking, it is always possible that one is dealing with a composite volume change dominated by a volume difference between intermediates along the reaction pathway. Given the likelihood that pressure effects and isotope effects will originate predominantly from different steps, then these two kinetic techniques should complement each other. Pressure effects may turn out to be the more useful tool because they can produce much larger effects (a $\Delta V^{\ddagger}$ of -40 ml mol^{-1} will generate a 963-fold increase in a rate at 4 kbar and 25 °C, for example, whereas k_H/k_D greater than 5 are uncommon). Moreover, pressure effects can be studied as a continuous function, which enables sorting out some forms of kinetic complexity that remain insoluble to isotopic approaches.

ACKNOWLEDGMENTS: The author is indebted to Karen L. Rebholz for determining the progress curves for fumarase, to Yong-Kweon Cho for determining the pressure effect on the isotope effect of yeast alcohol dehydrogenase, to Horst Ludwig for providing ûnpublished pressure effects on fumarase and for careful reading of the manuscript, to Catherine A. Royer for many helpful discussions, to the National Science Foundation for a grant to study pressure effects on enzymes, and to the National Institutes of Health for a grant to study isomechanisms of enzyme-catalyzed reactions.

REFERENCES

Blanchard, J. S., & Cleland, W. W. (1980). Use of isotope effects to deduce the chemical mechanism of fumarase. *Biochemistry* **19**, 4505–4513.

Brändén, C-I., Jörnvall, H., Eklund, H., & Furugren, B. (1975). Alcohol dehydrogenase. In *The Enzymes*, 3rd ed., ed. P. Boyer, New York, Academic Press, **11**, 103–190.

Bridgeman, P. W. (1949). *The Physics of High Pressure*, 2nd ed. London, G. Bell & Sons, p. 346.

Briggs, G. E., & Haldane, J. B. S. (1925). A note on the kinetics of enzyme action. *Biochem. J.* **19**, 338–339.

Butz, P., Greulich, K. O., & Ludwig, H. (1988). Volume changes during enzyme reactions: indications of enzyme pulsation during fumarase catalysis. *Biochemistry* **27**, 1556–1563.

Cha, Y., Murray, C. J., & Klinman, J. P. (1989). Hydrogen tunneling in enzyme reactions. *Science* **243**, 1325–1330.

Cleland, W. W. (1963). The kinetics of enzyme-catalyzed reactions with two or more substrate or products: I. Nomenclature and rate equations. *Biochem. Biophys. Acta* **67**, 104–137.

Cleland, W. W. (1967). The statistical analysis of enzyme kinetic data. *Adv. Enzymol.* **29**, 1–29.

Cleland, W. W. (1970). Steady state kinetics. In *The Enzymes*, 3rd ed., ed. P. Boyer, New York, Academic Press, **2**, 1–65.

Cleland, W. W. (1975). What limits the rate of an enzyme-catalyzed reaction. *Acc. Chem. Res.* **8**, 145–151.

Cleland, W. W. (1979). Statistical analysis of enzyme kinetic data. *Mech. Enzymol.* **63**, 103–138.

Cleland, W. W., O'Leary, M. H., & Northrop, D. B. (1977). *Isotope Effects on Enzyme-Catalyzed Reactions.* Baltimore, University Park Press.

Coates, J. H., Hardman, M. J., Shore, J. D., & Gutfreund, G. (1977). Pressure relaxation studies of isomerisations of horse liver alcohol dehydrogenase linked to NAD^+ binding. *FEBS Lett.* **84**, 25–28.

Cook, P. F. & Cleland, W. W. (1981). Mechanistic deductions from isotope effects in multireactant enzyme kinetic mechanisms. *Biochemistry* **20**, 1790–1796.

Czerlinski, G. (1962). Two ternary complexes of liver alcohol dehydrogenase with reduced diphosphopyridine nucleotide and the inhibitor imidazole. *Biophys. Biochim. Acta* **64**, 199–201.

Dalziel, K., & Dickinson, F. M. (1966). The kinetics and mechanism of liver alcohol dehydrogenase with primary and secondary alcohols as substrates. *Biochen. J.* **100**, 34–46.

Duggleby, R. G. (1984). Regression analysis of nonlinear arrhenius plots: an empirical model and a computer program. *Comput. Biol. Med.* **14**, 447–455.

Fromm H. J. (1975). *Initial Rate Enzyme Kinetics.* New York, Springer-Verlag, chap. 7.

Glastone, S., Laidler, K. J., & Eyring, H. (1941). *The Theory of Rate Processes.* New York, McGraw-Hill.

Hammes, G. G., & Schimmel, P. R. (1970). Rapid reactions and transient states. In *The Enzymes*, 3rd ed., ed. P. Boyer, New York, Academic Press, **2**, 67–114.

Hanson, J. N., Dinovo, E. C., & Boyer, P. D. (1969). Initial and equilibrium ^{18}O, ^{14}C, ^3H, and ^2H exchange rates as probes of the fumarase reaction mechanism. *J. Biol. Chem.* **244**, 6270–6279.

Ikkai, T., & Ooi, T. (1975). Actin: volume changes on transformation of G-form to F-form. *Science* **152**, 1756–1757.

Isaacs, N. S. (1984). The effect of pressure on kinetic isotope effects. In *Isotopes in Organic Chenistry*, ed. E. Buncel, &. C. C. Lee, New York, Elsevier, **6**, 67–105.

Isaacs, N. S., Javaid, K., & Rannala, E. (1978). Reactions at high pressure: Part 5. The effect of pressure on some primary kinetic isotope effects. *J. Chem. Soc., Perkin II*, 709–711.

Klinman, J. P. (1976). Isotope effects and structure-reactivity correlations in the yeast alcohol dehydrogenase reaction: a study of the enzyme-catalyzed oxidation of aromatic alcohols. *Biochemistry* **2**, 2018–2026.

Kornblat, J. A., Hui Bon Hoa, G., & Heremans, K. (1988). Pressure-induced effects on cytochrome oxidase: the aerobic steady state. *Biochemistry* **27**, 5122–5128.

Laidler, K. J., & Bunding, P. S. (1973). *The Chemical Kinetics of Enzyme Action*, 2nd ed. Oxford, Clarendon, pp. 220–232.

Lineweaver, H., & Burk, D. (1934). The determination of enzyme dissociation constants. *J. Am. Chem. Soc.* **56**, 658–666.

Ludwig, H., & Greulich, K. O. (1978). Volume changes during enzyme reactions: the influence of pressure on the action of invertase, dextranase, and dextransucrase. *Biophys. Chem.* **8**, 163–169.

Maggiora, G. M., & Christoffersen, R. E. (1978). Quantum-mechanical approaches to the study of enzymic transition states and reaction paths. In *Transition States of Biological Processes*, ed. R. D. Gandour & R. L. Schowen. New York, Plenum, pp. 119–163.

Massey, V. (1953). Studies on fumarase 3: the effect of temperature. *Biochem. J.* **53**, 72–79.

Michaelis, L., & Menten, M. L. (1913). Die Kinetic der Invertinwirkung. Biochem. Z. **49**, 333–369.

Morild, E. (1977). Pressure variation of enzymatic reaction rates: yeast and liver alcohol dehydrogenase. *Biophys. Chem.* **6**, 351–362.

Morild, E. (1981). The theory of pressure effects on enzymes. *Adv. Protein Chem.* **34**, 93–166.

Nageswara Rao, B. D., & Cohn, M. (1979). ^{31}P NMR of enzyme-bound substrates of rabbit muscle creatine kinase: equilibrium constants, interconversion rates, and NMR parameters of enzyme-bound complexes. *J. Biol. Chem.* **254**, 2689–2696.

Northrop, D. B. (1981a). The expression of isotope effects on enzyme-catalyzed reactions. *Ann. Rev.* **50**, 103–131.

Northrop, D. B. (1981b). Minimal kinetic mechanism and general equation for deuterium isotope effects on enzymic reactions: uncertainty in detecting a rate-limiting step. *Biochemistry* **20**, 4056–4061.

Northrop, D. B., & Cleland, W. W. (1974). The kinetics of pig heart TPN-isocitrate dehydrogenase II: dead end and multiple inhibition studies. *J. Biol. Chem.* **249**, 2928–2931.

Parmentier, L. E., O'Leary, M. H., Schackman, H. K., & Cleland, W. W. (1992). ^{13}C isotope effects as a probe of the kinetic mechanism and allosteric properties of *Escherichia coli* aspartate transcarbamylase. *Biochemistry* **31**, 6570–6576.

Ray, W. J., Jr. (1983). Rate-limiting step: a quantitative definition. Application to steady-state enzymic reactions. *Biochemistry* **22**, 4625–4637.

Rebholz, K. L. (1993). Enzymatic iso-mechanisms: alanine racemase, fumarase, and aspartic proteinases. Ph.D. thesis, University of Wisconsin, Madison, BC., p. 124.

Rebholz, K. L., & Northrop, D. B. (1993). Kinetics of enzymes with iso-mechanisms: analysis of product inhibition. *Biochem. J.* **296**, 355–360.

Rebholz, K. L., & Northrop, D. B. (1994a). Kinetics of enzymes with iso-mechanisms: analysis and display of progress curves. *Anal. Biochem.* **216**, 285–290.

Rebholz, K. L., & Northrop, D. B. (1994b). Kinetics of enzymes with iso-mechanisms: dead-end inhibition of fumarase and carbonic anhydrase II. *Arch. Biochem. Biophys.* **312**, 227–233.

Rose, I. A., Warms, J. V. B., & Kuo, D. J. (1992). Proton transfer in catalysis by fumarase. *Biochemistry* **31**, 9993–9999.

Royer, C. A. (1994). The application of pressure to biochemical systems: What can we learn from the other thermodynamic variable? *Meth. Enzymol.* In press.

Sweet, W. L., & Blanchard, J. S. (1990). Fumarase: viscosity dependence of the kinetic parameters. *Arch. Biochem. Biophys.* **277**, 196–202.

Theorell, H., & Chance, B. (1951). Studies on liver alcohol dehydrogenase II: the kinetics of the compound of horse liver alcohol dehydrogenase and reduced diphospho-pyridine nucleotide. *Acta Chem. Scand.* **5**, 1127–1144.

Weber, G., & Drickamer, H. G. (1983). The effect of high-pressure on yeast and liver alcohol dehydrogenase. *Q. Rev. Biophys.* **16**, 89–112.

Wratten, C. C., & Cleland, W. W. (1963). Product inhibition studies on yeast and liver alcohol dehydrogenase. *Biochemistry* **2**, 935–941.

16

Effects of High Pressure on the Allosteric Properties of Phosphofructokinase from *Escherichia coli*

JASON L. JOHNSON and GREGORY D. REINHART

Phosphofructokinase (PFK) from *Escherichia coli* is subject to allosteric regulation by phosphoenolpyruvate (PEP) and MgADP. These ligands inhibit and activate, respectively, by binding to a single allosteric binding domain and thereby altering the affinity the enzyme displays for its substrate, fructose-6-phosphate (Fru-6-P). The effect of hydrostatic pressure of up to 1.4 kbar on the binding of each of these ligands to PFK has been evaluated. This pressure range is insufficient to cause significant dissociation of the PFK tetramers. However, the logarithm of the equilibrium constant for each ligand binding to free enzyme decreases in a linear manner, and to virtually the same extent, when pressure is increased from 1 to 700 bar. Consequently, the ΔV associated with the binding of the inhibitor ligand PEP is virtually identical to the ΔV for the binding of the activator ligand MgADP or the substrate Fru-6-P, which falls within the range of 40–45 ml mol^{-1}. The apparent ΔV for Fru-6-P binding decreases with increasing concentration of PEP until it is equal to $+18$ ml mol when PEP is fully saturating. Similarly, ΔV for Fru-6-P binding decreases to $+26$ ml mol^{-1} when MgADP is fully saturating. These data are interpreted as implying that both PEP and MgADP improve the "fit" of Fru-6-P to its binding domain despite the fact that the ligands have opposing effects on Fru-6-P binding affinity.

Phosphofructokinase (PFK) from *E. coli* is a prototypical allosteric enzyme which was one of the first to be studied in depth (Blangy & Buc, 1967; Blangy et al., 1968) after Monod et al. (1965) published their famous proposal that allosteric behavior results from the concerted transition between discrete functional states of a protein (the MWC two-state model). PFK from *E. coli* is a tetrameric enzyme with a single allosteric binding domain that can bind either the activator

MgADP or the inhibitor PEP. Under many circumstances the substrate, fructose 6-phosphate (Fru-6-P), binds to the enzyme with positive cooperativity, and the affinity and cooperativity that Fru-6-P exhibits is modulated by the allosteric ligands in classic K-type fashion. In recent years, primarily through the efforts of Philip Evans and coworkers in Cambridge, three-dimensional structures of *E. coli* PFK have been determined from x-ray crystallography (Rypniewski & Evans, 1989; Shirakihara & Evans, 1988). These structures, along with companion structures of the related isozyme from *B. stearothermophilus* (Evans & Hudson, 1979; Evans et al., 1981), have revealed various structural perturbations that result from the binding of different ligands, and attempts have been made to reconcile the regulatory properties of the enzyme to these evident perturbations (Evans et al., 1986; Schirmer & Evans, 1990). These attempts have generally been inspired by the MWC two-state model used to interpret the enzyme's regulatory properties.

Recently we (Johnson & Reinhart, 1992; 1994a,b) and others (Lau & Fersht, 1987; Kundrot & Evans, 1991; Deville-Bonne et al., 1991; Zheng & Kemp, 1992) have documented discrepancies in the ability of the two-state model to adequately explain the functional behavior of *E. coli* PFK. We have found it much more satisfactory and informative to describe instead the actions of the allosteric ligands using the principles of thermodynamic linkage (Johnson & Reinhart, 1992, 1994a,b). This approach entails the determination of coupling parameters between various ligands capable of binding simultaneously to the enzyme that quantitatively describe both the nature and the magnitude of a K-type allosteric effect without any need to invoke a priori assumptions regarding particular conformational states or functional forms that the enzyme may or may not be able to adopt.

The coupling free energy between a substrate, A, and an allosteric ligand, X, (ΔG_{ax}) is defined simply as the difference between the binding free energy of substrate to free enzyme compared with that for the binding of substrate to the enzyme while the allosteric ligand remains bound. The principle of thermodynamic linkage mandates that the same difference should be obtained when the binding of the allosteric ligand to free enzyme is compared with that obtained when substrate is bound (Weber, 1972, 1975; Reinhart, 1983, 1988). Formally, this definition can be stated as

$$\Delta G_{ax} = \Delta G^o_{x/a} - \Delta G^o_x = \Delta G^o_{a/x} - \Delta G^o_a \tag{1}$$

where ΔG^o_a and ΔG^o_a equal the standard free energy of binding X and A, respectively, to free enzyme, and $\Delta G^o_{x/a}$ and $\Delta G^o_{a/x}$ equal the standard free energy of binding X and A, respectively, to enzyme already saturated with A and X, respectively. It can be readily seen from this definition that if X is an activator, (that is, if the presence of X causes the affinity for A to increase), then $\Delta G_{ax} < 0$, and conversely, if X is an inhibitor, (that is, A binds with lower affinity after X has bound), then $\Delta G_{ax} > 0$. Hence, the sign of the coupling free energy denotes the nature of the allosteric effect, while the absolute value of ΔG_{ax} conveys the magnitude, (that is, the maximum extent) of the activation or inhibition.

An interesting feature of the coupling free energy is that, since it is the free energy difference between two binding free energies, it can be considered the standard free energy of the equilibrium that summarizes the difference in those equilibria, which results in the following disproportionation equilibrium

$$XE + EA \rightleftharpoons E + XEA \qquad (2)$$

where E, XE, EA, and XEA refer to free enzyme, enzyme with allosteric ligand bound, enzyme with substrate bound, and enzyme with both allosteric ligand and substrate bound, respectively. Therefore, it is the poise of this equilibrium that actually establishes both the nature and the magnitude of an allosteric effect. Of particular significance for the following discussion is the notable absence of free ligand as a contributor to this equilibrium and hence to the coupling free energy. Moreover, the contribution of empty binding sites and bound ligands of each type is balanced in that each is given equal representation on both sides of the equilibrium.

In an effort to more fully characterize the actions of the allosteric ligands of *E. coli* PFK, we examined the variation of this equilibrium with pressure to ascertain the apparent change in volume associated with various pairs of ligands exhibiting activation and inhibition. We were assisted greatly in this effort by the fact that the intrinsic fluorescence of *E. coli* PFK is very responsive to the binding of ligands at either the active site or the allosteric site (Johnson & Reinhart, 1992). By measuring changes in intrinsic fluorescence intensity and/or the intrinsic fluorescence polarization at different ligand concentrations, we were able to measure directly the binding of the substrate Fru-6-P and allosteric ligands MgADP and PEP.

METHODOLOGY

Materials

All reagents used in buffers, PFK purification, fluorescence, and enzymatic assays were of analytical grade, purchased from either Sigma, Fisher, or Aldrich. Creatine phosphate, creatine kinase, and the potassium salts of ADP and Fru-6-P were obtained from Sigma; the coupling enzymes, aldolase, triosephosphate isomerase, and glycerol-3-phosphate dehydrogenase in ammonium sulfate suspensions were purchased from Boehringer Mannheim. Coupling enzymes were dialyzed extensively against a buffer consisting of 50 mM MOPS-KOH, 100 mM KCl, 5 mM $MgCl_2$, and 100 mM EDTA at pH 7.0. Punctilious grade absolute ethanol, obtained from Quantum Chemicals, was used as the pressurizing medium. Deionized distilled water was used throughout.

PFK Purification

PFK, overexpressed in DF1020 cells (an *E. coli* strain in which both PFK genes have been deleted) from a plasmid kindly provided by Dr. Robert Kemp (Zheng

& Kemp, 1992), was purified via a modification of the method of Kotlarz and Buc (1982) as described previously (Johnson & Reinhart, 1992).

Enzyme Activity Determination

Measurements of maximal activity were carried out as described previously (Johnson & Reinhart, 1992) in 1.0 ml of an EPPS buffer adjusted to pH 8.0 and containing 50 mM EPPS, 10 mM MgCl$_2$, 10 mM NH$_4$Cl, 0.1 mM EDTA, 2 mM DTT, 0.2 mM NADH, 250 µg of aldolase, 50 µg of glycerol-3-phosphate dehydrogenase, 5 µg of triosephosphate isomerase, 1 mM creatine phosphate, 10 µg/ml creatine kinase, and saturating substrate concentrations ([Fru-6-P] = 4 mM, [MgATP] = 2 mM).

Protein Determination

Protein determinations were accomplished using the BCA Protein Assay Reagent (Pierce) and/or absorbance readings (ε_{278} = 0.6 cm^2 mg^{-1}; Kotlarz & Buc, 1977).

Steady-State Fluorescence Measurements under High Hydrostatic Pressure

Intrinsic PFK fluorescence was monitored on a spectrofluorometer equipped with a 450W xeonon-arc lamp, optical module monochromators, and the controller electronics of an SLM 4800, coupled with ISS PX01 photon counting electronics. Samples were subjected to hydrostatic pressure in an SLM-Aminco high-pressure spectroscopy cell built in the design of Paladini and Weber (1981). The ~1.5 ml sample was contained within a bottle cuvette sealed by flexible plastic tubing that served as a diaphragm to separate the sample from the pressurizing medium (ethanol) while transmitting the pressure to the sample. Pressure was generated and maintained by an Advanced Pressure Products, computer-controlled, automated, screw-drive pump with a feedback pressure sensor. ISS Spectral Software served in data collection and analyses.

In both steady-state intensity and anisotropy/polarization measurements, emission subsequent to excitation at 300 nm (to avoid tyrosine excitation and/or energy transfer) was collected through a 2-mm thick Schott WG-345 cut-on filter and a Corning 7-54 band-pass filter. Relative changes in fluorescence intensity were obtained by first monitoring the change concomitant to the addition of a particular titrant concentration outside the pressure bomb, and subsequently by setting this value as the initial relative intensity at 1 bar, to which all pressure-induced changes were referenced. Correction factors for the pressure-induced change in the birefringence of the pressure cell's quartz windows were independently determined from a fluorescein solution in glycerol at -15 °C (Paladini & Weber, 1981) and routinely used to provide the anisotropy values in ligand binding profiles. Polarization artifacts associated with the emission monochromator were avoided by using excitation and emission polarizers oriented at angles of 55° and 0°, respectively. All experiments were performed in 50 mM EPPS-KOH (pH = 8.0), 10 mM MgCl$_2$, 10 mM NH$_4$Cl, and 0.2 mM EDTA. PFK subunit concentration was equal to 2.6 µM, unless otherwise noted. Fluorescence intensity and anisotropy associated with pressurized samples were fully reversible, and enzyme activity after pressurization was checked periodically and found to be more than 96% recoverable, provided pressure never exceeded 1.2 kbar.

RESULTS

Since *E. coli* PFK is a tetramer, the first issue that must be considered is the influence pressure has on the oligomeric structure of the enzyme. Deville-Bonne and Else (1991) have reported that the enzyme is most subject to pressure dissociation in the absence of bound ligands when pressures exceed 0.8 kbar. We have confirmed this general observation in several ways. First, the tryptophan intensity of the free enzyme increases with pressure in a continuous manner until just over 1.5 kbar in pressure, above which one observes a precipitous decrease in intensity (Figure 16.1). The center of mass of the emission, expressed in wavenumbers, shows a continuous gradual decline through this pressure region, suggesting to us that the decline in intensity is due to the onset of precipitation that occurs after an appreciable concentration of dissociated subunits has built up and which is clearly evident upon inspection of the sample after pressure is subsequently released. A more sensitive probe of the effects of pressure on the enzyme can be seen from the inclusion of increasing concentrations of I^-, which can quench the native enzyme's tryptophan with low efficiency at atmospheric

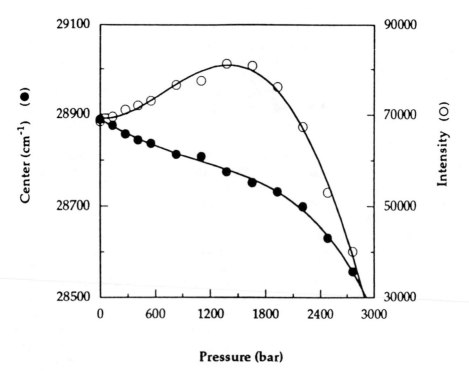

Figure 16.1. Influence of hydrostatic pressure on the relative intrinsic tryptophan fluorescence and center of mass of PFK from *E. coli*.

pressure (data not shown). The ability of I⁻ to quench remains relatively constant until the pressure exceeds approximately 0.7 kbar, above which a slight increase in quenching efficiency can be noticed. A substantial increase in quenching is evident above 1 kbar. Because we are interested in the effects of pressure on the binding of ligands to the active tetramer, we focused our attention on pressures less than 800 bar.

The binding of the substrate Fru-6-P can be followed by the decrease in the intrinsic fluorescence it produces upon binding to *E. coli* PFK. In Figure 16.2 we see the results of a titration of the enzyme with Fru-6-P from 2 μM to 1 mM as a function of pressure. It is apparent that Fru-6-P binds with lower affinity at higher pressures, giving rise to the unusual upward curvature of the pressure

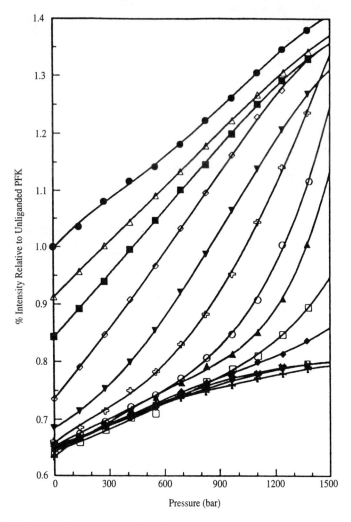

Figure 16.2. Influence of pressure on the relative intrinsic tryptophan fluorescence at various concentrations of Fru-6-P which increment two-fold from 2 μM (△) to 1 mM (#). The top curve (●) represents the behavior in the absence of Fru-6-P.

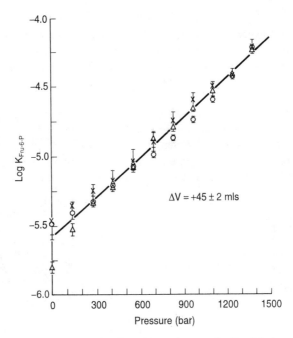

Figure 16.3. Influence of pressure on the dissociation constant for Fru-6-P determined from data of the type shown in Figure 16.2 determined at PFK concentrations of 0.025 (×), 0.1 (○), and 0.4 (△) mg/ml. Slope of the line yields the indicated value of ΔV for Fru-6-P binding according to Eq. 3.

dependence at intermediate Fru-6-P concentrations. From such data, the dissociation constant for Fru-6-P as a function of pressure can be determined, and is presented in Figure 16.3. The data exhibit a linear dependence on pressure through 1 kbar, with some curvature apparent at higher pressures. From these data the change in volume produced by the binding of Fru-6-P to free enzyme, ΔV_{F6P}, can be determined according to the following relationship (Torgerson et al., 1979),

$$\Delta V_{F6P} = RT\left[\frac{\partial(\ln K_{F6P})}{\partial(P)}\right] \tag{3}$$

where K_{F6P} is the *dissociation* constant of Fru-6-P and ΔV_{F6P} refers to the change in volume of the *binding* reaction. To confirm that this value was not being influenced significantly by dissociation of the enzyme, these experiments were repeated at enzyme concentrations fourfold lower and fourfold higher than that which we adopted as standard (0.1 mg/ml). It is evident from Figure 16.3 that changing enzyme concentration over this 16-fold range did not significantly influence the pressure dependence, with an average value equal to $+45$ ml mol^{-1} indicated by these data.

To assess the effect of pressure on the coupling between Fru-6-P and PEP, which *defines* the inhibition that characterizes PEP's action, these experiments were repeated at several different fixed concentrations of PEP. The variation of

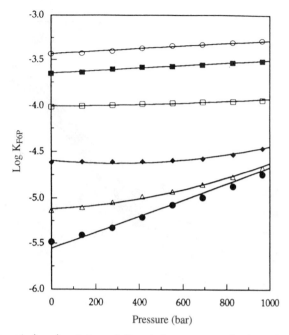

Figure 16.4. Pressure-induced variation of dissociation constants for Fru-6-P at concentrations of 0 (●), 0.25 (△), 1 (◆), 4 (□), 15 (■), and 60 (○) m*M* PEP.

the apparent dissociation constant for Fru-6-P with pressure at several concentrations of PEP is shown in Figure 16.4. The data at the extremes of PEP concentrations are reasonably linear over several hundred bar, with intermediate concentrations of PEP generating curved dependences. It is clear that the binding of PEP causes the ΔV associated with Fru-6-P binding to decrease. By replotting these same data so that the K_d for Fru-6-P is shown as a function of PEP concentration at various pressures, it is also evident that pressure increases the K_d for PEP (Figure 16.5). The data at each pressure were analyzed by the following simple monomeric linkage expression (Reinhart, 1983, 1985)

$$K_a = K_a^o \left[\frac{K_{ix}^o + [X]}{K_{ix}^o + Q_{ax}[X]} \right] \tag{4}$$

where K_a = the apparent dissociation constant for Fru-6-P at a particular concentration of PEP (X), K_a^o = the dissociation constant for Fru-6-P when [PEP] = 0, K_{ix}^o = the dissociation constant for PEP when [Fru-6-P] = 0, and $Q_{ax} = \exp(-\Delta G_{ax}/RT)$. Variation of K_a^o, K_{ix}^o, and Q_{ax} with pressure then yields the corresponding values for ΔV according to Eq. 3. These values are given in Tables 16.1 and 16.2.

Similar analyses were performed on the ability of pressure to influence the binding of MgADP and its coupling to Fru-6-P binding. MgADP binding to the allosteric site can be followed easily by the decrease it causes in the intrinsic fluorescence intensity of PFK from *E. coli* (Johnson & Reinhart, 1992). Moreover,

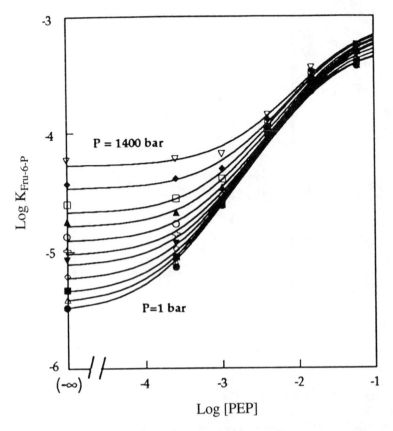

Figure 16.5. Variation in the dissociation constant for Fru-6-P induced by PEP at various pressures. Pressure ranged from 1 bar to 1400 bar as indicated.

since the maximal change in fluorescence intensity induced by MgADP binding is less than that observed by the binding of Fru-6-P, binding of Fru-6-P to the enzyme saturated with MgADP can still be followed, and the extent to which MgADP perturbs Fru-6-P's binding can be ascertained as a function of pressure. The ΔV associated with both MgADP binding and the coupling between MgADP and Fru-6-P deduced from these data are also presented in Tables 16.1 and 16.2.

Table 16.1. Volume changes of ligand binding equilibria

Ligand	Other saturating ligands	$\Delta V(\text{ml mol}^{-1})$
Fru-6-P	none	$+45 \pm 2$
"	PEP	$+18 \pm 8$
"	MgADP	$+26 \pm 5$
PEP	none	$+40 \pm 11$
"	Fru-6-P	$+13 \pm 12$
MgADP	none	$+44 \pm 3$
"	Fru-6-P	$+25 \pm 6$

Table 16.2. Volume changes of disproportionation
equilibria

Coupled pair	ΔV(ml mol^{-1})
F6P-PEP	-27 ± 8
F6P-Mg/ADP	-19 ± 5

DISCUSSION

It is clear that pressure has an effect on E. coli phosphofructokinase. For pressures below 1 kbar, which are insufficient to cause the active tetramers to dissociate, noticeable changes in the intrinsic tryptophan fluorescence are nonetheless very pronounced, indicating that the environment around the single tryptophan is changing and likely becoming more exposed to the solvent. Because the try-potophan is partially exposed to the surface in the native enzyme and because its emission is easily perturbed by the binding of ligands at sites approximately 20 Å away, it is reasonable to conclude that the conformation of the enzyme is responding to pressure and that the ensuing increase in exposure of the tryptophan to the solvent could be evident without the separation of subunits. This conclusion is also supported by the observation that the dissociation constant for Fru-6-P, while influenced by pressure over the first kbar, was not affected significantly by changing enzyme concentration over 16-fold (Figure 16.1).

Table 16.1 summarizes the values of ΔV determined for binding the substrate Fru-6-P and the allosteric effector ligands PEP and MgADP, individually, to PFK from E. coli. For each of these ligands, ΔV is positive, indicating that increasing pressure causes dissociation of the bound ligand. An obvious issue to consider when interpreting these results is the likely contribution that electrostriction mechanisms will make toward the observed values of ΔV. Since each of the unbound ligands carries multiple charges which are effectively neutralized to some degree upon binding, it is probable that the bulk water density is higher in the vicinity of the free ligand because the charge-dipole interactions between the ligand and water molecules would be stronger, and hence shorter, than the otherwise prevailing dipole-dipole interactions between water molecules in bulk solution. This effect alone would suggest a positive sign for ΔV consistent with our observations. Indeed, even the magnitude of the volume changes observed can be rationalized by this mechanism, since the neutralization of dibasic phosphate by ethylamine in aqueous solution can be calculated to have a ΔV greater than 30 ml mol^{-1} (Heremans, 1993).

$$\Delta V = +30.4 \text{ ml mol}^{-1}$$

$$HPO_4^{--} + EtNH_3^+ \rightleftharpoons H_2PO_4^- + EtNH_2 \tag{5}$$

The comparable neutralization of a carboxylic acid by ethylamine occurs with a ΔV greater than 17 ml mol^{-1}. Consequently, a substantial fraction, if not the entire value, of the ΔV for binding Fru-6-P. PEP, and MgADP might be rationalized on this basis, and one might conclude from these data that it is impossible to reach any explicit interpretation of the possible contributions protein conformational changes resulting from ligand binding might make to the observed ΔV for binding. It is of interest to note, however, how similar the value of ΔV is for each ligand, given the different charges they carry at pH 8.

The role played by electrostriction, and solvent in general, can be expected to be greatly minimized when considering the influence of pressure on the coupling constants $Q_{F6P\text{-}PEP}$ and $Q_{F6P\text{-}MgADP}$. Since these parameters are the equilibrium constants to the corresponding disproportionation equilibria designated in Eq. 2, it is evident that free ligand in solution, regardless of charge, makes no contribution to their respective values. This conclusion derives from the fact that when any of the ligands binds *after* another ligand has already bound, the volume change is greatly diminished relative to the binding of that ligand to free enzyme, as seen by the data summarized in Table 16.1. Clearly, free ligand is common to both of these binding equilibria and hence cannot be a contributor to the *change* in binding volume observed.

These changes in binding volume are in fact equal to the ΔV associated with the disproportionation equilibria associated with the coupling parameters describing the antagonism between Fru-6-P and PEP and the facilitating interaction between Fru-6-P and MgADP, summarized in Table 16.2. Further insight into the meaning and possible origin of these volume changes is revealed if one considers that ΔV_{ax} is equal to the difference in *partial* volume change (δV) introduced in the enzyme by the binding of both ligands together and the sum of each ligand individually as shown by the following

$$\Delta V_{ax} = (V_e + V_{xea}) - (V_{xe} + V_{ea}) \tag{6}$$

$$\Delta V_{ax} = (V_{xea} - V_e) - ((V_{xe} - V_e) + (V_{ea} - V_e))$$

$$\Delta V_{ax} = \delta V_{xea} - (\delta V_{xe} + \delta V_{ea})$$

where δV_{xea}, δV_{xe}, and δV_{ea} equal the *partial* volume change associated with the corresponding enzyme form as defined above. Since both $\Delta V_{F6P\text{-}PEP}$ and $\Delta V_{F6P\text{-}MgADP}$ are negative, the following inequality must therefore hold for both couplings

$$\delta V_{xea} < \delta V_{xe} + \delta V_{ea} \tag{7}$$

Consequently, the partial volume change associated with the binding of both ligands simultaneously, *ignoring any change in volume of the free ligand*, is less than the sum of the partial volume change of each ligand binding individually. This would seem to suggest that both ligands fit on the enzyme better together than they do individually. This idea is expressed in schematic fashion in Figure 16.6. It is particularly noteworthy that this would appear to be the case for the activating interaction between Fru-6-P and MgADP, as well as the inhibiting interaction between Fru-6-P and PEP. In fact, as indicated in Table 16.2, ΔV for the inhibitory interaction is actually more negative than is ΔV for the ligands that enhance one another's binding.

It is common to envision the actions of K-type allosteric inhibitors and

Figure 16.6. Schematic summarizing the fact that the partial colume of the ternary complex is smaller than the sum of the partial volumes of each ligand individually for both the inhibitor PEP and the activator MgADP of *E. coli* PFK.

activators in terms of their ability to distort or pre-form, respectively, a binding site that is complementary to the structure of the substrate and, thereby, to either antagonize or facilitate the subsequent binding of the substrate to the enzyme. Such a view would not seem to be easily consistent with the data just described. One might argue that the volume changes associated with the couplings arise from the effects of ligands on the hydration state of the enzyme at either the unfilled ligand binding sites or at other portions of the surface of the protein that are differentially affected by the different ligands when bound individually. While one cannot rule out this possibility, we would expect such effects to be modest relative to the absolute value of the ΔV values we observed.

Finally, these observations seem to speak once more to the similarity that inhibitors and activators display when their actions are examined carefully. Previously we have observed (Johnson & Reinhart, 1994a; Reinhart et al., 1989) that coupling free energies invariably comprise opposing $T\Delta S$ and ΔH terms and that the net activation or inhibition actually observed usually corresponds to a fairly small difference between these two largely compensating quantities; the nature of the allosteric effect being dependent solely on which term has the largest absolute value. In two cases we have observed the nature of an allosteric effect switch from inhibition to activation by simply increasing temperature, and hence reversing the imbalance between the ΔH and $T\Delta S$ terms, thereby implying that the actions of the ligands on a thermodynamic level can be quite similar (Braxton et al., 1994). Similarity between the actions of inhibitors and activators is once again evident in the response of the regulatory properties of PFK from *E. coli* to changes in pressure. Both substrate and allosteric ligand seem to fit better when bound simultaneously, regardless of whether the allosteric ligand is an inhibitor or activator. It is becoming increasingly apparent that the nature of an allosteric effect depends more on subtle perturbations of a host of largely compensating interactions than on a gross structural change that introduces substantially different properties, the latter idea being a common outgrowth of the prevalent use of a two-state model to explain the actions of allosteric ligands.

ACKNOWLEDGMENTS: This research was supported by grant no. GM 33216 from the National Institutes of Health (to G.D.R.).

REFERENCES

Blangy, D., & Buc, H. (1967). Allosteric transitions in *Escherichia coli* phosphofructokinase and rabbit muscle phosphorylase b. *Bull. Soc. Chim. Biol.* **49**, 1473–1478.

Blangy, D., Buc, H., & Monod, J. (1968). Kinetics of the allosteric interactions of phosphofructokinase from *Escherichia coli*, *J. Mol. Biol.* **31**, 13–35.

Braxton, B. L., Tlapak-Simmons, V. L., & Reinhart, G. D. (1994). Temperature-induced inversion of allosteric phenomena. *J. Biol. Chem.* **269**, 47–50.

Deville-Bonne, D., & Else, A. J. (1991). Reversible high hydrostatic pressure inactivation of phosphofructokinase from *Escherichia coli*, *Eur. J. Biochem.* **200**, 747–750.

Deville-Bonne, D., Bourgain, F., & Garel, J. R. (1991). pH dependence of the kinetic properties of allosteric phosphofructokinase from *Escherichia coli*. *Biochemistry* **30**, 5750–5754.

Evans, P. R., & Hudson, J. P. (1979). Structure and control of phosphofructokinase from *Bacillus stearothermolphilus*. *Nature* (London) **279**, 500–504.

Evans, P. R., Farrants, G. W., & Hudson, P. J. (1981). Phosphofructokinase: structure and control. *Phil. Trans. R. Soc. London, Ser. B* **293**, 53–62.

Evans, P. R., Farrants, G. W., & Lawrence, M. C. (1986). Crystallographic structure of allosterically inhibited phosphofructokinase at 7 Å resolution. *J. Mol. Biol.* **191**, 713–720.

Heremans, K. (1993). The behavior of proteins under pressure. *NATO ASI Ser., Ser. C* **401**, 443–469.

Johnson, J. L., & Reinhart, G. D. (1992). Magnesium-ATP and fructose 6-phosphate interactions with phosphofructokinase from *Escherichia coli*. *Biochemistry* **31**, 11510–11518.

Johnson, J. L., & Reinhart, G. D. (1994a). Influence of MgADP on phosphofructokinase from *Escherichia coli* elucidation of coupling interactions with both substrates. *Biochemistry* **33**, 2635–2643.

Johnson, J. L., & Reinhart, G. D. (1994b). Influence of substrates and MgADP on the time-resolved intrinsic fluorescence of phosphofructokinase from *Escherichia coli*: correlation of trypotophan dynamics to coupling entropy. *Biochemistry* **33**, 2644–2650.

Kotlarz, D., & Buc, H. (1977). Two *Escherichia coli* fructose-6-phosphate kinases. Preparative purification, oligomeric structure and immunological studies. *Biochim. Biophys. Acta* **484**, 35–48.

Kotlarz, D., & Buc, H. (1982). Phosphofructokinases from *Escherichia coli*. *Methods Enzymol.* **90**, 60–70.

Kundrot, C. E., & Evans, P. R. (1991). Designing an allosterically locked phosphofructokinase. *Biochemistry* **30**, 1478–1484.

Lau, F. T. K., & Fersht, A. R. (1987). Conversion of allosteric inhibition to activation in phosphofructokinase by protein engineering. *Nature* (London) **326**, 811–812.

Monod, J., Wyman, J., & Changeux, J. P. (1965). On the nature of allosteric transitions: a plausible model. *J. Mol. Biol.* **12**, 88–118.

Paladini, A. A., & Weber, G. (1981). Absolute measurements of fluorescence polarization at high pressures. *Rev. Sci. Instrum.* **52**, 419–427.

Reinhart, G. D. (1983). The determination of thermodynamic allosteric parameters of an enzyme undergoing steady-state turnover. *Arch. Biochem. Biophys.* **224**, 389–401.

Reinhart, G. D. (1985). Influence of pH on the regulatory kinetics of rat liver phosphofructokinase: a thermodynamic linked-function analysis. *Biochemistry* **24**, 7166–7172.

Reinhart, G. D. (1988). Linked-function origins of cooperativity in a symmetrical dimer. *Biophys. Chem.* **30**, 159–172.

Reinhart, G. D., Hartleip, S. B., & Symcox, M. M. (1989). Role of coupling entropy in establishing the nature and magnitude of allosteric response. *Proc. Natl. Acad. Sci. USA* **86**, 4032–4036.

Rypniewski, W. R., & Evans, P. R. (1989). Crystal structure of unliganded phospho-fructokinase from *Escherichia coli*. *J. Mol. Biol.* **207**, 805–821.

Schirmer, T., & Evans, P. R. (1990). Structural basis of the allosteric behaviour of phosphofructokinase. *Nature* **343**, 140–145.

Shirakihara, Y., & Evans, P. R. (1988). Crystal structure of the complex of phospho-fructokinase from *Escherichia coli* with its reaction products. *J. Mol. Biol.* **204**, 973–994.

Torgerson, P. M., Drickamer, H. G., & Weber, G. (1979). Inclusion complexes of poly-β-cyclodextrin: a model for pressure effects upon ligand-protein complexes. *Biochemistry* **18**, 3079–3083.

Weber, G. (1972). Ligand binding and internal equilibriums in proteins. *Biochemistry* **11**, 864–878.

Weber, G. (1975). Energetics of ligand binding to proteins. *Adv. Protein Chem.* **29**, 1–83.

Zheng, R. L., & Kemp, R. G. (1992). The mechanism of ATP inhibition of wild type and mutant phosphofructo-1-kinase from *Escherichia coli*. *J. Biol. Chem.* **267**, 23640–23645.

17

Correlation Field Splitting of Chain Vibrations: Structure and Dynamics in Lipid Bilayers and Biomembranes

PATRICK T. T. WONG

Pressure-tuning vibrational spectroscopy was first introduced to the study of structural and dynamic properties in biological systems from our laboratory about one decade ago. One of our efforts has been the search for spectral features and their pressure dependencies related to the structural and dynamic properties in biological systems. Pressure-induced correlation field splitting of the vibrational modes of methylene chains is one of the parameters that has been applied to monitor various structural and dynamic properties of a wide range of aqueous lipid bilayers and biomembranes in our laboratory. Correlation field splitting of the vibrational modes of the methylene chains in lipid bilayers is the result of vibrational coupling interactions among the ordered methylene chains with different site symmetry in the two-dimensional matrix. However, the basic theory and the characteristics of these interchain interactions in lipid bilayers still needed to be established. It was unknown whether the interchain interactions that result in the correlation field splitting take place within each lipid molecule or between neighboring molecules in the lamellar bilayers. The relative contributions of intramolecular and intermolecular interchain interactions to the correlation field splitting, and the effects of the long-range interchain interactions and interdigitation on the correlation field splitting, were also unknown. These problems have been resolved recently and are addressed in this chapter.

Our laboratory has pioneered the study of structural and dynamic properties of biological systems by means of pressure-tuning vibrational spectroscopy (Wong et al., 1982). It is now well recognized that this spectroscopic technique is one of the most powerful physical methods for the study of biological and biomedical phenomena at the molecular level with enhanced resolution (Wong, 1984, 1987a, 1987b, 1987c, 1993). The biological systems we have studied by this method include not only various aqueous biomolecular assemblies but also whole cells and intact biological tissues (Rigas et al., 1990; Wong, 1984, 1987b, 1987c, 1993; Wong et al., 1991a, 1991b, 1993). We have found that the pressure-induced changes in many spectral features and parameters in both FTIR and Raman spectra of biological systems result from modifications in structure and dynamics at the molecular level. At present, only the pressure-induced correlation field splitting of methylene chain vibrations and their relationship with the structural and dynamic properties in biomembranes and lipid bilayers are described. Correlation field splitting of the vibrational modes of methylene chains in lipid bilayers is the result of the vibrational coupling interactions among the ordered methylene chains with different site symmetry in the bilayer matrix.

Although correlation field splitting of vibrational modes has been widely used to monitor the structural and dynamic properties in biomembranes and lipid bilayers (Wong, 1984, 1987b, 1987c, 1993), there are still some unresolved problems. In common membrane lipids, there are two hydrocarbon chains in each lipid molecule. The vibrational modes of these two intramolecular chains would certainly interact with each other. It is unknown whether the interchain interactions that result in the correlation field splitting take place within each lipid molecule (intramolecular interchain interactions) or between neighboring molecules in the lamellar bilayers (intermolecular interchain interactions). The relative contributions between the intramolecular and the intermolecular interchain interactions to the correction field splitting, and the effects of the long-range interchain interactions and interdigitation on the correlation field splitting, are also unknown. To address these problems, we have studied the pressure-induced correlation field splitting in the FTIR spectra of the following lipid bilayer systems: (1) aqueous bilayer dispersion of dimyristoylphosphatidylcholine (DMPC), (2) isolated DMPC molecules in the bilayer matrix of perdeuterated DMPC-d_{54}, (3) crystalline DMPC, and (4) interdigitated phase of DMPC.

The solid DMPC/DMPC-d_{54} mixtures (1:1 and 1:2 molar ratio) were prepared by codissolving the solid components in chloroform and drying the solutions with nitrogen gas. The solid mixture samples were then dispersed into water and lyophilized for 48 hours. The dispersing and lyophilizing procedure was repeated twice. Fully hydrated DMPC bilayers and DMPC/DMPC-d_{54} mixed bilayers were prepared by dispersing about 50 mg of solid lipid in a Tris buffer made with D_2O 24 hours before analysis. The hydrated DMPC was then concentrated by centrifugation to approximately (w/w) 40%. For interdigitated DMPC dispersion, the lipids were hydrated with a buffer made with D_2O and containing 5 mg of tetracaine. The lipid dispersions were subjected to at least five freeze-thaw cycles. Anhydrous crystalline DMPC samples were prepared by loading the optical cell with crystalline DMPC, inserting the optical cell into the sample compartment of the infrared spectrophotometer, and purging with dry nitrogen gas for about 72 hours. The optical cell was then sealed by closing the space between the diamond anvils and the gasket. The relative intensity of the H_2O stretching band to the lipid CH_2 symmetric band was used to monitor the water content of the samples.

Small amounts (typically 0.1 mg) of the lipid samples were placed at room temperature, together with powdered α-quartz, into a 0.45-mm diameter hole in a 0.23-mm thick stainless steel gasket mounted on a diamond anvil optical cell, as described previously (Wong et al., 1985). Infrared spectra were collected on a Digilab FTS-40A spectrophotomer with a mercury cadmium telluride detector. The infrared beam was condensed by a sodium chloride lens system onto the diamond anvil cell. For each spectrum, 512 scans were coadded, at a spectral resolution of 4 cm^{-1}. Pressures on the sample were determined from the 695 cm^{-1} phonon band of α-quartz (Auger et al., 1988; Wong et al., 1985).

METHODOLOGY

In lipid bilayers, the vibrational interchain interactions between the opposing bilayer leaflets are extremely weak. Therefore, only the correlation field interchain interactions among the methylene chains within each bilayer leaflet are important. In an ordered zigzag methylene chain, the translational repeat unit along the chain is C_2H_4, and each repeat unit contains two CH_2 chemical groups. Most of the observed bands in the infrared and Raman spectra of methylene chains arise from the in-phase normal modes of the C_2H_4 groups along each chain.

The potential function of each internal vibration of C_2H_4 groups in an isolated methylene chain j is

$$V_j = f_j q_j^2 \tag{1}$$

where q_j is the C_2H_4 normal coordinate of the jth methylene chain and f_j is the force constant of the normal mode in an isolated chain. In terms of the normal coordinate q_j, the potential energy of an internal vibration of all the methylene chains in each lipid bilayer leaflet is given by

$$U = \sum_{j,j'} F_{jj'} q_j q_{j'} = \sum_j F_j q_j^2 + \sum_{j,j' \neq j} F_{jj'} q_j q_{j'} \tag{2}$$

The F terms are the force constants of each methylene normal mode for the entire bilayer leaflet, and thus $F_j \neq f_j$. Combining Eqs. 1 and 2, one has

$$U = \sum_j V_j + \sum_j (F_j - f_j) q_j^2 + \sum_{j,j' \neq j} F_{jj'} q_j q_{j'} \tag{3}$$

The second and third terms in this equation represent the correlation field perturbation on the intrachain vibration. The second term in Eq. 3 expresses a frequency shift of $1/2\pi (Fj - fj)^{1/2}$ resulted from the lowering of the molecular symmetry of the chains to the sites symmetry. The third term in Eq. 3 represents the coupling of the normal mode among different chains.

According to Eq. 3, the internal potential energy of the total bilayer leaflet is that of all the isolated chains plus a perturbation $\bar{U}$. Namely:

$$U = \sum_j V_j + \bar{U} \tag{4}$$

$$\bar{U} = \sum_j (F_j - f_j) q_j^2 + \sum_{j,j' \neq j} F_{jj'} q_j q_{j'} \tag{5}$$

When the perturbation potential $\bar{U}$ is expressed as a sum over pairs of chains

$$\bar{U} = 1/2 \sum_{jm} U_{jm} \tag{6}$$

it can be shown by the first-order perturbation theory (Davydov, 1971; Hexter, 1960) that the energy levels of the excited states split into n branches, where n is the number of nonequivalent chains in the repeat unit along the bilayer leaflet. The energy difference between the spectrally excited state (E) and the ground state $(E°)$ is given by

$$E - E° = \varepsilon - \varepsilon° + \frac{h}{8\pi^2 c v_o} \sum_{j}^{nN} \left(\frac{\partial^2 U_{jm}}{\partial q_m^2}\right)_o + \frac{nh}{8\pi^2 c v_o} \sum_{xjx}^{nN} B_{\alpha a}^* B_{\alpha x} \left(\frac{\partial^2 U_{jxm}}{\partial q_{jx} \partial q_m}\right)_o \tag{7}$$

where v_o is the harmonic frequency of a C_2H_4 normal mode, $\varepsilon - \varepsilon° = hv_o$, and N is the total number of repeat units in each bilayer leaflet. Chain m is on site a, and chain j_x is on site x; x runs through $a, b \ldots$ to n number of nonequivalent sites in a repeat unit. B is the transformation matrix between the symmetric coordinates of the chain repeat unit and the C_2H_4 normal coordinates of the isolated chains at the nonequivalent sites of the repeat unit along the bilayer leaflet. For a lipid bilayer leaflet with two nonequivalent chains per repeat unit $(n = 2)$, the B matrix is

$$B = \begin{bmatrix} B_{\alpha a} B_{\alpha b} \\ B_{\beta a} B_{\beta b} \end{bmatrix} = \begin{bmatrix} 1 & 1 \\ 1 & -1 \end{bmatrix} \tag{8}$$

For $n = 2$, the exciton level splits into two branches (α and β), and thus there are two repeat unit modes for each intrachain vibrational mode. By substituting Eq. 8 for Eq. 7, one may obtain

$$v_\alpha = v_o + \frac{1}{8\pi^2 c v_o} \left[\sum_{j}^{2N} \left(\frac{\partial^2 U_{jm}}{\partial q_m^2}\right)_o + \sum_{ja}^{N} \left(\frac{\partial^2 U_{jam}}{\partial q_{ja} \partial q_m}\right)_o + \sum_{jb}^{N} \left(\frac{\partial^2 U_{jbm}}{\partial q_{jb} \partial q_m}\right)_o \right] \tag{9}$$

$$v_\beta = v_o + \frac{1}{8\pi^2 c v_o} \left[\sum_{j}^{2N} \left(\frac{\partial^2 U_{jm}}{\partial q_m^2}\right)_o + \sum_{ja}^{N} \left(\frac{\partial^2 U_{jam}}{\partial q_{ja} \partial q_m}\right)_o - \sum_{jb}^{N} \left(\frac{\partial^2 U_{jbm}}{\partial q_{jb} \partial q_m}\right)_o \right] \tag{10}$$

where

$$v_{\alpha, \beta} = \frac{1}{h} (E_{\alpha, \beta} - E°)$$

Consequently, the exciton level splits into two branches (α, β). The splitting of each normal vibrational mode into α and β components in the bilayers is the so-called correlation field splitting, which is

$$v_\alpha - v_\beta = \frac{1}{4\pi^2 c v_o} \sum_{jb}^{N} \left(\frac{\partial^2 U_{jbm}}{\partial q_{jb} \partial q_m}\right)_o \tag{11}$$

The frequency shift of each normal mode of the isolated chain in the bilayers is

$$1/2(v_\alpha + v_\beta) - v_o = \frac{1}{8\pi^2 c v_o}\left[\sum_{j}^{2N}\left(\frac{\partial^2 U_{jm}}{\partial q_m^2}\right)_o + \sum_{j_a}^{N}\left(\frac{\partial^2 U_{j_a m}}{\partial q_{j_a}\partial q_m}\right)_o\right] \qquad (12)$$

U_{jm} is a function of the interchain distance and the relative orientation of the methylene chains in lipid bilayers, which are certainly varied with pressure. Therefore, both frequency shift and correlation field splitting are expected to be pressure dependent. By following their modification induced by pressure, one can monitor the structural changes in lipid bilayers. Any discontinuity of the relative orientation of the chains and of the interchain distances across the liquid crystal/gel and the gel/gel phase transitions may be detected from the discontinuity in the frequency-pressure plot. Thus, the pressure dependence of frequency will provide useful information about the existence and the nature of a pressure-induced phase transition.

If the orientations of all the methylene chains in a bilayer system are parallel to each other, the fourth term in Eqs. 9 and 10, and also the $v_\alpha - v_\beta$ value in eq. 11, will be zero. Consequently, there will be no correlation field splitting in the methylene vibrational modes. Under the following two circumstances, the correlation field splitting will be also absent in the spectra: (1) the conformation of the methylene chains is highly disordered due to the presence of a large number of gauche bonds, and thus the coupling of the vibrational modes between neighboring chains is random and weak, and (2) the methylene chains are conformationally highly ordered and fully extended, but the orientation of these fully extended chains is disordered due to reorientation fluctuations and the torsion/twisting motions of the chains, which are usually observed at low pressure or high temperature. In these cases, only broadening rather than correlation field splitting in the vibrational bands of methylene chains will be observed.

It has been demonstrated that the conformational disorder, the reorientational fluctuations, and the torsion/twisting motions of the methylene chains in lipid bilayers can be ordered and dampended by external pressure (Wong, 1984, 1987a, 1987b, 1987c, 1993). At high enough pressure, these disordered structures can be removed, and thus the correlation field splitting in the vibrational modes of the methylene chains would appear in the spectra, provided that the equilibrium orientations of neighboring chains are nonequivalent. For orientationally more disordered chains, a higher pressure is required to stop these fluctuations and motions, and thus the pressure at which the splitting appears is higher. Consequently, the order/disorder dynamics of the methylene chains in lipid bilayers can be determined by the magnitude of the correlation field splitting pressure. Moreover, the magnitude of the correlation field splitting is a measure of the degree of interchain interactions in lipid bilayers.

The types of structural and dynamic properties in aqueous lipid bilayers and biomembranes that we have studied by means of pressure-induced correlation field splitting are as follows: (1) large angle reorientational fluctuations (Auger et al., 1988; Choma & Wong, 1992; Tupper et al., 1992; Wong & Mantsch, 1988);

(2) interchain packing (Siminovitch et al., 1987a; Wong et al., 1982, 1989b; Wong & Mantsch, 1984); (3) interchain and intrachain configuration distortion (Wong & Mantsch, 1985); (4) mechanism for the formation of the lamellar subgel phase (Wong et al., 1986); (5) interdigitation (Auger et al., 1988; Siminovitch et al., 1987b, 1987c; Wong & Huang, 1989); (6) interchain interactions (Hubner et al., 1990; Siminovitch et al., 1987b; Tupper et al., 1992); (7) unsaturation-induced changes in molecular configurations in lipid bilayers (Siminovitch et al., 1987a, 1988; Wong & Mantsch, 1988); (8) the effects of anesthetics (Auger et al., 1988, 1990), toxins (Ahmed et al., 1992; Zakim & Wong, 1990), drugs (Popovic et al., 1992; Taylor et al., 1992), alkanes (Wong & Zakim, 1990), cholesterol (Wong et al., 1989a, 1989b), polypeptides (Carrier et al., 1990), proteins (Gicquand & Wong, 1994; Philp et al., 1990), fluorescent probes (Chong et al., 1989; Chong & Wong, 1993), and other exogenous molecules (unpublished work from this laboratory) on the structural and dynamic properties in the bilayer interior of various lipids; (9) the location and binding sites of these exogenous molecules in lipid bilayers (Ahmed et al., 1992; Auger et al., 1988, 1990; Carrier et al., 1990; Chong et al., 1989; Chong & Wong 1993; Gicquaud & Wong, 1994; Philp et al., 1990; Popovic et al., 1992; Taylor et al., 1992; Wong et al., 1989a,b; Wong & Zakim, 1990; Zakim & Wong, 1990); and (10) changes in the structural and dynamic properties in biomembranes of whole cells and tissues with various diseases and conditions (Rigas et al., 1990; Wong et al., 1991a,b, 1993).

RESULTS AND DISCUSSION

Intermolecular and Intramolecular Correlation

To determine the contribution of the correlation field interactions between the neighboring methylene chains within each lipid molecule to the correlation field splitting (intramolcuclear correlation), the pressure profile of the infrared spectra in the CH_2 bending region of the isolated DMPC molecules in the bilayer matrix of perdeuterated DMPC-d_{54} molecules was measured (see Figure 17.1B). The relative intensities of the shoulder bands from the choline methyl groups at 1479 and 1490 cm^{-1} in these spectra are much higher than those of the pure DMPC bilayers in Figure 17.1A because the choline methyl groups are not deuterated in the perdeuterated DMPC-d_{54} samples. The CD_2 bending mode of the methylene chains of DMPC-d_{54} is shifted to lower than 1430 cm^{-1} by the mass effect, which is outside the frequency region in Figure 17.1B. Therefore, the 1467 cm^{-1} band in Figure 17.1B is due to the δCH_2 mode of the isolated DMPC. It is clear from Figure 17.1 that the pressure behavior of the δCH_2 band of the isolated DMPC molecules (Figure 17.1B) is considerably different from that of the nonisolated DMPC molecules (Figure 17.1A). In the spectra of the isolated DMPC, the δCH_2 band exhibits as a singlet up to 35.7 kbar, and only slight broadening of the δCH_2 band is induced at high pressure.

The frequencies of the δCH_2 bands of the nonisolated DMPC and the isolated DMPC are plotted as a function of pressure in Figure 17.2. In the nonisolated DMPC bilayers, the δCH_2 band splits into two at 3.5 kbar, and the magnitude of the splitting increases gradually with increasing pressure. The frequency of the

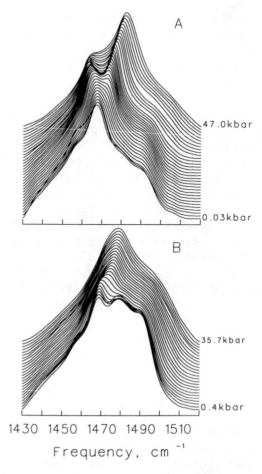

Figure 17.1. Stacked contour plots of the infrared spectra in (**A**) the δCH_2 region of the aqueous DMPC bilayers and (**B**) the isolated DMPC in perdeuterated DMPC-d$_{54}$ bilayers.

low-frequency component band decreases gradually and starts to increase slightly above 9.2 kbar with increasing pressure. The frequency of the high-frequency component band increases linearly with increasing pressure above 9.2 kbar. The frequency of the gauche δCH_2 band at 1457 cm^{-1} decrease with increasing pressure above 3.5 kbar. This band disappears at 9.2 kbar and is undetectable even in the third power derivative spectra. These results indicate that the methylene chains in DMPC bilayers become fully extended and that the gauche C–C bonds are completely removed at 9.2 kbar. Therefore, the smaller magnitude of the correlation field splitting below 9.2 kbar is due to the presence of gauche bonds in the methylene chains. As pressure increases, these disordered gauche bonds are gradually removed. Thus, the magnitude of the correlation field splitting increases gradually. Above 9.2 kbar, these gauche bonds are completely removed and the pressure-induced frequency shifts of the correlation field component bands become linear. The increase in the magnitude of the correlation field splitting above

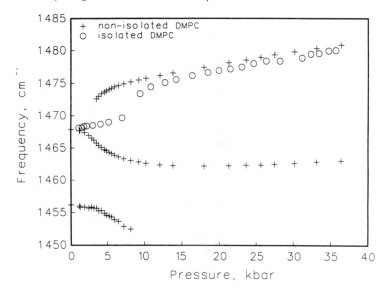

Figure 17.2. Pressure dependencies of the frequencies of the δCH_2 modes for aqueous DMPC bilayers and the isolated DMPC in perdeuterated DMPC-d_{54} bilayers.

9.2 kbar is the result of the pressure enhanced interchain interactions. As shown in Eq. 12, the interchain interactions among the orientationally equivalent chains will contribute to an increase in the mode frequency. Pressure will enhance these interchain interactions and thus increase the mode frequency. This increase in frequency with increasing pressure is observed for both the correlation field component bands in DMPC bilayers above 9.2 kbar.

The frequencies shown in Figure 17.2 were measured from the third power derivative spectra with a break point of 0.9 (Cameron et al., 1984). Even in the third power derivative spectra, the correlation field component band of the isolated DMPC is not detectable at all pressures up to 40 kbar. However, a discontinuity in the pressure range near 9.4 kbar is observed in the pressure dependence of the δCH_2 frequency of the isolated DMPC (Figure 17.2). This discontinuity pressure almost coincides with the gauche/*trans* transformation pressure (9.2 kbar) observed in the nonisolated DMPC bilayers. At this pressure, the gauche bonds are completely removed from the methylene chains, and the conformation and orientation of the methylene chains become highly ordered. Therefore, the discontinuity in the pressure dependence of the δCH_2 frequency in the isolated DMPC at 9.4 kbar is the result of disorder/order transition of the orientational and the conformational structure of the methylene chains in the isolated DMPC molecules.

It is evident from Figures 17.1 and 17.2 that the correlation field interactions are insignificant among the isolated DMPC molecules, and the correlation field splitting observed in the infrared spectra of the nonisolated DMPC bilayers is mainly contributed by the intermolecular interchain interactions among neighboring lipid molecules rather than the intramolecular interchain interactions within individual lipid molecules.

Correlation Interactions in Crystalline DMPC

X-ray single crystal studies of DMPC (Hauser et al., 1981; Pearson & Pascher, 1979) have shown that the unit cell contains four DMPC molecules arranged tail-to-tail in pairs in a bilayer configuration. Each pair consists of two crystallographically independent molecules. The lateral arrangement of the four hydrocarbon chain planes of the two DMPC molecules in each monolayer resembles the hybrid chain packing mode (Abrahamsson et al., 1978). Consequently, the orientation of the chain planes of the methylene chains within each DMPC molecule is nearly perpendicular, whereas the one between molecules is nearly parallel.

The pressure contour of the infrared spectra of the δCH_2 mode in the crystalline DMPC is shown in Figure 17.3A. The pressure profile of the correlation field splitting of the δCH_2 mode in the crystalline DMPC is comparable with that in aqueous bilayers (Figure 17.1A) except for a slight difference in the relative

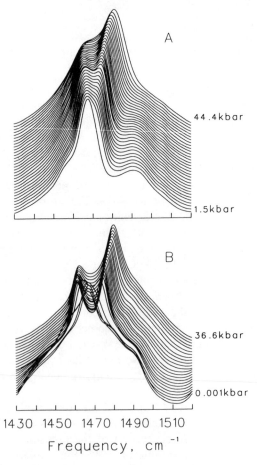

Figure 17.3. Stacked contour plots of the infrared spectra in (**A**) the δCH_2 region of the crystalline DMPC and (**B**) the interdigitated DMPC.

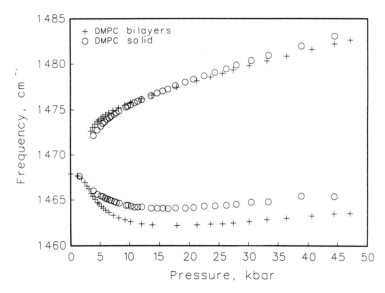

Figure 17.4. Pressure dependencies of the frequencies of the δCH_2 modes for the crystalline DMPC and aqueous DMPC bilayers.

intensities of the component bands at the corresponding pressures. The pressure dependencies of the frequencies of the δCH_2 component bands in the crystalline DMPC are compared with those in the aqueous DMPC bilayers in Figure 17.4. The correlation field splitting pressure at which the splitting of the δCH_2 band starts is about the same between these two DMPC samples. However, the magnitude of the splitting is much larger in the aqueous bilayers than in the crystal. It has been demonstrated that the molecular conformation of the glycerol backbone in DMPC crystal is essentially retained in aqueous DMPC bilayers, and the intramolecular orientation of the hydrocarbon chain planes are largely determined by the conformation of the glycerol backbone (Hauser et al., 1981). Therefore, the intramolecular orientational structure in DMPC crystal may also be retained in aqueous DMPC bilayers. In this case, the intramolecular correlation field interactions within each DMPC molecule are expected to be comparable between the crystalline state and the aqueous bilayer state. The difference in the magnitude of the correlation field splitting between these two states of DMPC shown in Figure 17.4 also indicates that this splitting is the result of the intermolecular interchain interactions among the neighboring DMPC molecules.

Figure 17.5 compares the pressure contours of the τCH_2 bands between the crystalline DMPC and the aqueous DMPC bilayers. The low-frequency correlation field band of the τCH_2 mode in the crystalline DMPC is extremely weak (see Figure 17.5A). This component band shifts from $729.3\ cm^{-1}$ at 3.9 kbar to $751.7\ cm^{-1}$ at 44.4 kbar. According to the oriented gas model (Snyder, 1961), the intensity ratio between the correlation field component bands of the τCH_2 mode is a measure of the relative orientations of the planes of the interacting methylene chains. If the correlation field splitting in DMPC is from the interactions between the methylene chains within each molecule, this intensity ratio would be close to

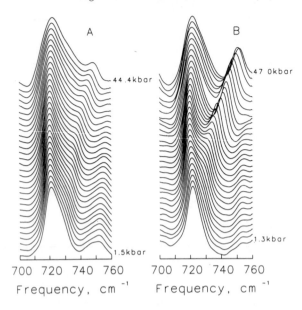

Figure 17.5. Comparison of the stacked contour plots of the infrared spectra in (**A**) the τCH_2 region between the crystalline DMPC and (**B**) the aqueous DMPC bilayers.

unity because of the nearly perpendicular orientation between the chain planes in each molecule. The peak intensity ratio of these component bands in the crystalline DMPC is extremely small. The maximum ratio, which is less than 0.2, is observed at 44.4 kbar. This result further confirms that the correlation field splitting in the spectra of the crystal is from the intermolecular rather than the intramolecular interchain interactions. The relatively low intensity ratio of the τCH_2 component bands of the aqueous DMPC bilayer shown in Figure 17.5B suggests that this correlation field splitting in aqueous DMPC bilayers is also from the intermolecular interchain interactions. The intensity ratio of the τCH_2 correlation field bands in the aqueous DMPC bilayers is 0.27 at 47 kbar, which is slightly larger than that in the crystalline DMPC. Therefore, the intermolecular arrangement in the bilayer state is not exactly the same as that in the solid state.

Short-Range and Long-Range Correlation

In Figure 17.1, the δCH_2 band of the isolated DMPC is broadened and becomes asymmetric at high pressure. This indicates that the δCH_2 band of the isolated DMPC consists of closely overlapping bands at high pressure. In the third power derivative spectra with an extremely high break point (0.98), this asymmetric δCH_2 band can be resolved into two bands at presssure above 30 kbar (see Figure 17.6). However, this splitting is much smaller than the intermolecular correlation field splitting in pure DMPC solid (Figure 17.4) and in aqueous DMPC dispersions (Figures 17.2 and 17.6). Moreover, the component band near 1470 cm^{-1} in the isolated DMPC is also observed in the third power derivative spectra of the nonisolated DMPC bilayers with a break point of 0.98 at pressure above 25 kbar

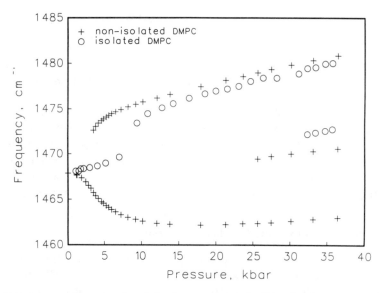

Figure 17.6. Pressure dependencies of the frequencies of the δCH_2 modes for aqueous DMPC bilayers and the isolated DMPC in perdeuterated DMPC-d$_{54}$ bilayers obtained from the third power derivative spectra with a break point of 0.98.

(Figure 17.6). Therefore, this small splitting in the δCH_2 band in the isolated DMPC is most likely the result of the short-range correlation field interactions between the two methylene chains in each DMPC molecule. The fact that the magnitude of this splitting is much smaller than that of the intermolecular correlation field interactions observed in the nonisolated DMPC bilayers suggests that the intramolecular correlation field interactions are much weaker than the intermolecular correlation field interactions. The smaller magnitude in the intramolecular correlation field interactions is consistent with the molecular structure of the DMPC molecules (Hauser et al., 1981; Pearson & Pascher, 1979). First, due to the restriction of the head group and the glycerol moiety in the DMPC molecules, the methylene groups of the two methylene chains in the fully extended DMPC molecules locate at different levels with respect to the head group. Each methylene group of the *sn-1* chain is at the position between two methylene groups of the *sn-2* chain (Hauser et al., 1981; Pearson & Pascher, 1979). Consequently, the distance between the correlated methylene groups in the two methylene chains within each DMPC molecule is large, and the correlation field interactions among them are expected to be small. Second, in the isolated DMPC, only half of the methylene groups take part in the correlation field interactions between the two methylene chains in each DMPC molecule. The methylene groups on the far end of each zigzag methylene chain in DMPC are too far away to take part in the correlation field interactions. This certainly results in a smaller magnitude of intramolecular correlation field interactions. In the presence of neighboring DMPC molecules in the nonisolated DMPC bilayers, the above restrictions are removed and all the methylene groups in the methylene chains will take place in the intermolecular correlation field interactions.

The fourth term in Eqs. 9 and 10 includes the long-range correlation field interactions with the nth neighboring molecules. In the isolated DMPC dispersions with 1:1 molar ratio between DMPC and perdeuterated DMPC-d_{54}, the first neighbors of each DMPC molecule are perdeuterated DMPC-d_{54} molecules whereas the second neighbors are nonperdeuterated DMPC molecules. If the correlation field interactions are significant among the second or higher neighbors, there would be intermolecular correlation field splitting in the spectra of the isolated DMPC similar to that observed in the nonisolated DMPC bilayers shown in Figure 17.2. The absence of this intermolecular correlation field splitting in the spectra of the isolated DMPC strongly suggest that the intermolecular correlation field interactions only take place among the nearest neighbors, and the long-range interactions with the second or higher neighboring molecules are insignificant.

Correlation Interactions in the Interdigitated Bilayers

It has been demonstrated that in the presence of charged tetracaine, DMPC molecules assemble into interdigitated bilayers (Auger et al., 1988; McIntosh et al., 1983). If the correlation field splitting is the result of the interchain interactions among neighboring DMPC molecules, then the pressure profiles of the correlation field splitting of the δCH_2 and the τCH_2 modes in the interdigitated DMPC would be dramatically different from those of the noninterdigitated DMPC bilayers. The pressure profiles of the δCH_2 and the τCH_2 bands of the interdigitated DMPC are shown in Figure 17.3B and Figure 17.7A, respectively. The pressure profiles of the corresponding δCH_2 and τCH_2 bands of the noninterdigitated DMPC bilayers are given in Figure 17.7A and Figure 17.7B, respectively. It is evident

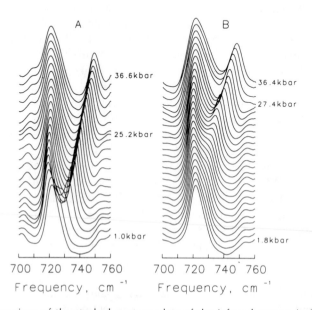

Figure 17.7. Comparison of the stacked contour plots of the infrared spectra in (**A**) the τCH_2 region between the interdigitated DMPC bilayers and (**B**) the aqueous DMPC bilayers.

from these figures that they differ significantly between the interdigitated and the noninterdigitated DMPC bilayers. The splitting of both the δCH_2 and τCH_2 bands in the spectra of the interdigitated DMPC into two well-defined bands is more abrupt and a pronounced valley between the correlation field component bands is observed. On the other hand, the correlation field component bands ($\delta' CH_2$ and $\tau' CH_2$) of the noninterdigitated DMPC bilayers appear as a broad shoulder on the high-frequenty side of the δCH_2 and τCH_2 bands, and then they steadily gain intensity with increasing pressure. Moreover, the corresponding valley between the two component bands in the noninterdigitated DMPC bilayers is comparatively shallow. The ratios between the peak height of the correlation field component band ($\tau' CH_2$ and $\delta' CH_2$) and the height of the valley in the spectra of the τCH_2 and the δCH_2 bands are compared between the interdigitated and the noninterdigitated DMPC bilayers at various pressures in Figure 17.8. The ratios of both the τCH_2 and δCH_2 modes are larger at all pressures in the interdigitated DMPC. Moreover, the $\tau' CH_2 / \tau CH_2$ and $\delta' CH_2 / \delta CH_2$ intensity ratios of the interdigitated DMPC are higher than those of the noninterdigitated DMPC at all pressures (Figure 17.9). Therefore, these ratios can be considered as parameters for determining the presence of interdigitation in lipid bilayers.

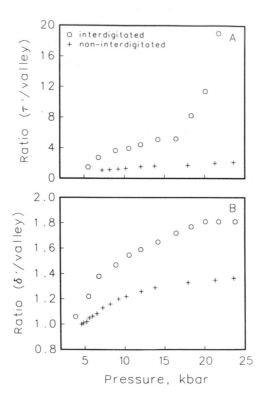

Figure 17.8. Pressure dependencies of (**A**) the ratio between the peak height of the τCH_2 band and the valley and (**B**) the ratio between the peak height of the $\delta' CH_2$ band and the valley, for the interdigitated and noninterdigitated DMPC bilayers.

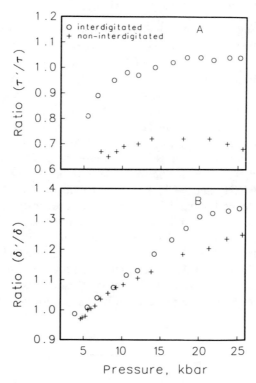

Figure 17.9. Pressure dependencies of (**A**) the peak height ratio between the $\tau'CH_2$ and the τCH_2 bands and (**B**) the peak height ratio between the $\delta'CH_2$ and the δCH_2 bands, for the interdigitated and noninterdigitated DMPC bilayers.

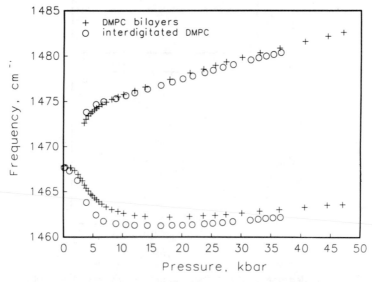

Figure 17.10. Pressure dependencies of the δCH_2 modes for the interdigitated and noninterdigitated DMPC bilayers.

Figure 17.10 shows the pressure dependencies of the frequencies of the δCH_2 correlation field component bands of the interdigitated DMPC. It is evident from this figure that the correlation field splitting is more abrupt and that the magnitude of the splitting is larger in the interdigitated DMPC at all pressures. The larger splitting indicates that the packing of the neighboring methylene chains is tighter. The methylene chain packing within each DMPC molecule is mainly governed by the conformation of the glycerol moiety of the head group (Hauser et al., 1981; Pearson & Pascher, 1979) and would not be significantly affected by interdigitation. On the other hand, the packing between neighboring molecules would certainly be strongly affected by the interdigitation. All these results in the interdigitated DMPC bilayers also provide evidence that the correlation field interactions in lipid bilayers, which give rise to the splitting of the τCH_2 and δCH_2 modes, are predominated by the vibrational interactions among the methylene chains of the neighboring molecules.

REFERENCES

Abrahamsson, S., Dahlen, B., Lofgren, H., & Pascher, I. (1978). Lateral packing of hydrocarbon chains. *Prog. Chem. Fats Other Lipids* **16**, 125–143.

Ahmed, K., Choma, C. T., & Wong, P. T. T. (1992). High pressure FTIR study of interaction of melittin with dimyristolphosphatidyl glycerol bilayers. *Chem. Phys. Lipids* **63**, 139–148.

Auger, M., Jarrell, H. C., Smith, I. C. P., Siminovitch, D. J., Mantsch, H. H., & Wong, P. T. T. (1988). Effects of the local anesthetic tetracaine on the structural and dynamic properties of lipids in model and nerve membranes: a high pressure FT-IR study, *Biochemistry* **27**, 6086–6093.

Auger, M., Smith, I. C. P., Mantsch, H. H., & Wong, P. T. T. (1990). High-pressure infrared study of phosphatidylserine bilayers and their interactions with the local anesthetic tetracaine. *Biochemistry* **29**, 2008–2015.

Cameron, D. G., & Moffatt, D. J. (1984). Deconvolution, derivation, and smoothing of spectra using Fourier transforms. *J. Test. Eval.* **12**, 78–85.

Carrier, D., Mantsch, H. H., & Wong, P. T. T. (1990). Protective effect on lipidic surfaces against pressure-induced conformational changes of poly-L-lysine. *Biochemistry* **29**, 254–258.

Choma, C. T., & Wong, P. T. T. (1992). The structure of anhydrous and hydrated dimyristoylphosphatidylglycerol: a pressure tuning infrared spectroscopic study. *Chem. Phys. Lipids* **61**, 131–137.

Chong, P. L.G., & Wong, P. T. T. (1993). Interaction of Laurdan with phosphatidylcholine liposomes: a high pressure FT-IR study. *Biochim. Biophys. Acta* **1149**, 260–266.

Chong, P. L.-G., Capes, S., & Wong, P. T. T. (1989). Effects of hydrostatic pressure on the location of PROPAN in lipid bilayers: a FT-IR study. *Biochemistry* **28**, 8358–8363.

Davydov, A. S. (1971). *Theory of Molecular Exitons*. New York, Plenum, pp. 23–111.

Gicquaud, C., & Wong, P. T. T. (1994). Mechanisms of interaction between actim and membrane lipids: a pressure tuning infrared spectroscopy study. *Biochem. J.* **303**, 769–774.

Hauser, H., Pascher, I., Pearson, R. H., & Sundell, S. (1981). Preferred conformation and molecular packing of phosphatidylethanolamine and phosphatidylcholine. *Biochim. Biophys. Acta* **650**, 21–51.

Hexter, R. M. (1960). Intermolecular coupling of vibrations in molecular crystals: a vibrational exiton approach. *J. Chem. Phys.* **33**, 1833–1841.

Hubner, W., Wong, P. T. T., & Mantsch, H. H. (1990). The effect of hydrostatic pressure on the bilayer structure of phosphatidylcholines containing ω-cyclohexyl fatty acyl chains. *Biochim. Biphys. Acta* **1027**, 229–237.

McIntosh, T. J., McDaniel, R. V., & Simon, S. A. (1983). Induction of an interdigitated gel phase in fully hydrated phosphatidylcholine containing bilayers. *Biochim. Biophys. Acta* **731**, 109–110.

Pearson, R. H., & Pascher, I. (1979). The molecular structure of lecithin dihydrate. *Nature* **281**, 499–501.

Philp, R. B., McIver, D. J., & Wong, P. T. T. (1990). Pressure distortion of an artificial membrane and the effect of ligand/protein binding. *Biochim. Biophys. Acta* **1021**, 91–95.

Popovic, P. M., Wong, P. T. T., Goel, R., Evans, W. K., Howell, S. B., Auersperg, N., & Stewart, D. J. (1992). Pressure-tuning infrared spectroscopy of cisplatin sensitive versus resistant ovarian cancer cells. *Proc. Am. Assoc. Cancer Res.* **33**, 2722.

Rigas, B., Morgello, S., Goldman, I. S., & Wong, P. T. T. (1990). Human colorectal cancers display abnormal Fourier transform infrared spectra. *Proc. Natl. Acad. Sci. USA* **87**, 8140–8144.

Siminovitch, D. J., Wong, P. T. T., & Mantsch, H. H. (1987a). Effects of *cis* and *trans* unsaturation on the structure of phospholipid bilayers: a high-pressure infrared spectroscopic study. *Biochemistry* **26**, 3277–3287.

Siminovitch, D. J., Wong, P. T. T., & Mantsch, H. H. (1987b). High-pressure infrared spectroscopy of ether- and ester-linked phosphatidylcholine aqueous dispersions. *Biophys. J.* **51**, 465–473.

Siminovitch, D. J., Wong, P. T. T., & Mantsch, H. H. (1987c). High-pressure infrared spectroscopy of lipid bilayers: new test for interdigitation. *Biochim. Biophys. Acta* **900**, 163–167.

Siminovitch, D. J., Wong, P. T. T., Berchtold, A., & Mantsch, H. H. (1988). A comparison of the effect of one and two monounsaturated acyl chains on the structure of phospholipid bilayers: a high-pressure infrared spectroscopic study. *Chem. Phys. Lipids* **46**, 79–87.

Snyder, R. G. (1961). Vibrational spectra of a crystalline n-paraffins: II. Intermolecular effects. *J. Mol. Spectrosc.* **7**, 116–144.

Taylor, K. D., Goel, R., Stewart, D. J., & Wong, P. T. T. (1992). Pressure-tuning infrared spectroscopic study of cisplatin induced changes in a phosphatidylserine model membrane. *Proc. Am. Assoc. Cancer Res.* **33**, 2677.

Tupper, S., Wong, P. T. T., & Tanphaichitr, N. (1992). Binding of Ca^{2+} to sulfogalacto-sylceramide and the sequential effects on the lipid dynamics. *Biochemistry* **31**, 11902–11907.

Wong, P. T. T. (1984). Raman spectroscopy of thermotropic and high-pressure phases of aqueous phospholipid dispersions. *Annu. Rev. Biophys. Bioeng.* **13**, 1–24.

Wong, P. T. T. (1987a). High-pressure studies of biomembranes by vibrational spectroscopy. In *High Pressure Chemistry and Biochemistry*, ed. R. van Eldik & J. Jonas. NATO series C. D. Dordrecht, Reidel, **197**, 381–400.

Wong, P. T. T. (1987b). High-pressure vibrational spectroscopy of aqueous systems: Phospholipid dispersions and proteins. In *Current Perspectives in High Pressure Biology*, ed. H. W. Jannasch, R. E. Marquis, & A. M. Zimmerman. New York, Academic Press, pp. 287–314.

Wong, P. T. T. (1987c). Vibrational spectroscopy under high pressure. In *Vibrational Spectra and Structure*, ed. J. Durig. New York, Elsevier, **16**, 357–445.

Wong, P. T. T. (1993). High pressure vibrational spectroscopic studies of aqueous biological systems: from model systems to intact tissues. In *High Pressure Chemistry, Biochemistry and Material Science*, ed. R. Winter & J. Jonas. Dordrecht, Kluwer Academic, pp. 511–543.

Wong, P. T. T., & Huang, C. (1989). Structural aspects of pressure effects on infrared spectra of mixed-chain phosphatidylcholine assemblies in D_2O. *Biochemistry* **28**, 1259–1263.

Wong, P. T. T., & Mantsch, H. H. (1984). The phase behavior of aqueous dipalmitoyl phosphatidylcholine bilayers at very high pressures. *J. Chem. Phys.* **81**, 6367–6370.

Wong, P. T. T., & Mantsch, H. H. (1985). Pressure effects on the infrared spectrum of 1,2-dipalmitoyl phosphatidylcholine bilayers in water. *J. Chem. Phys.* **83**, 3268–3274.

Wong, P. T. T., & Mantsch, H. H. (1988). Reorientational and conformational ordering processes at elevated pressure in 1,2-dioeoyl phosphatidylcholine: a Raman and infrared spectroscopic study. *Biophys. J.* **54**, 781–790.

Wong, P. T. T., & Zakim, D. (1990). Structural and dynamic properties of a two-dimensional solution of hexadecane in aqueous lipid bilayers: a pressure-tuning infrared spectroscopic study. *J. Phys. Chem.* **94**, 5052–5056.

Wong, P. T. T., Murphy, W. H., & Mantsch, H. H. (1982). Pressure effects on the Raman spectra of phospholipid membranes: pressure induced phase transitions and structural changes in 1,2-dimyristoyl 3-sn-phosphatidylcholine water dispersions. *J. Chem. Phys.* **76**, 5230–5237.

Wong, P. T. T., Moffatt, D. J., & Baudais, F. L. (1985). Crystalline quartz as an internal pressure calibrant for high-pressure infrared spectroscopy. *Appl. Spectrosc.* **39**, 733–735.

Wong, P. T. T., Mushayakarara, E., & Mantsch, H. H. (1986). Mechanism of subgel phase formation of DPPC: a high pressure Raman spectroscopic study. *Biophys. J.* **49**, 315a.

Wong, P. T. T., Capes, S., & Mantsch, H. H. (1989a). Hydrogen bonding between anhydrous cholesterol and phosphatidylcholine: an infrared spectroscopic study. *Biochim. Biophys. Acta* **980**, 37–41.

Wong, P. T. T., Chagwedera, T. E., & Mantsch, H. H. (1989b). Effect of cholesterol on structural and dynamic properties of tripalmitoyl glyceride: a high pressure infrared spectroscopic study. *Biophys. J.* **56**, 845–850.

Wong, P. T. T., Cadrin, M., & French, S. W. (1991a). Abnormal molecular vibrations in mouse liver tumor: pressure-tuning FTIR spectroscopy. *Exp. Molec. Pathology* **55**, 269–284.

Wong, P. T. T., Wong, R. K., Caputo, T. A., Godwin, T. W., & Rigas, B. (1991b). Infrared spectroscopy of exfoliated human cervical cells: evidence of extensive structural changes during carcinogenesis. *Proc. Natl. Acad. Sci. USA* **88**, 10988–10992.

Wong, P. T. T., Chagwedera, T. E., & Mantsch, H. H. (1991c). Pressure tuning of frequencies and resonance in the infrared and Raman spectra of n-pentadecane. *J. Mol. Str.* **247**, 31–46.

Wong, P. T. T., Goldstein, S. M., Thomas, R. C., Godwin, A., Pivik, C., & Rigas, B. (1993). Distinct infrared spectroscopic patterns of human basal cell carcinoma of the skin. *Cancer Res.* **53**, 762–765.

Zakim, D., & Wong, P. T. T. (1990). A high pressure infrared spectroscopic study of the solvation of bilirubin in lipid bilayer. *Biochemistry* **29**, 2003–2007.

18

High-Pressure Effects on the Structure and Phase Behavior of Model Membrane Systems

ROLAND WINTER, ANNE LANDWEHR, THOMAS BRAUNS,
JÖRG ERBES, CLAUS CZESLIK, and OLIVER REIS

Phospholipids, which provide valuable model systems for lipid membranes, display a variety of polymorphic phases, depending on their molecular structure and on environmental conditions. High hydrostatic pressure has been used as a physical parameter to study the thermodynamic properties and phase behavior of these systems. High pressure is also a characteristic feature of certain natural membrane environments. In the first part of this article, we review our recent work on the temperature- and pressure-dependent phase behavior of phospholipid systems differing in lipid conformation and headgroup structure. In the second part, we report on the determination of the (T, x, p) phase diagrams of binary phospholipid mixtures. An additional section deals with effects of incorporating ions, small amphiphilic molecules, and steroids into the bilayer on the experimental temperature- and pressure-dependent phase behavior of lipid systems. Finally, we discuss lamellar to nonlamellar thermotropic and barotropic phase transformations, which occur for a number of lipids, such as phosphatidylethanolamines, monoacylglycerides, and lipid mixtures. It has been suggested that nonlamellar lipid structures might play an important role as transient and

---

Abbreviations: MA-myristic acid; MO-monoolein; ME-monoelaidin; TTC-tetracaine; DMPC-1,2-dimyristoly-sn-glycero-3-phosphatidylcholine (di-Cl4:0); DPPC-1,2-dipalmitoly-sn-glycero-3-phosphatidylcholine (di-C16:0); DSPC-1,2-distearoyl-sn-glycero-3-phosphatidylcholine (di-C18:0); DOPC-1,2-dioleoyl-sn-glycero-3-phosphatidylcholine (di-C18:1, cis); DOPE-1,2-dioleoyl-sn-glycero-3-phosphatidylethanolamine (di-C18:1 cis); POPC-1-palmitoyl-2-oleoyl-sn-glycero-3-phosphatidylcholine (C6:0, C18:1, cis); DEPC-1,2-dielaidoyl-sn-glycero-3-phosphatidylcholine (di-C18:1, trans); DPPE-1,2-dipalmitoyl-sn-glycero-3-phosphatidylethanolamine (di-C16:0); DMPS-1,2-dimyristoyl-sn-glycero-3-phosphatidylserin (di-C14:0); egg-PE-egg-yolk phosphatidylethanolamine.

local intermediates in a number of biochemical processes. High-pressure small-angle x-ray (SAXS) and neutron (SANS) scattering, differential scanning calorimetry (DSC), high-pressure differential thermal analysis (DTA), and p, V, T measurements have been used as experimental methods for the investigation of these systems.

Lipid bilayer dispersions, in particular the phosphatidylcholines and phosphatidylethanolamines, are the workhorses for the investigation of biophysical properties of membrane lipids because they constitute the basic structural component of biological membranes. They exhibit a rich lyotropic and thermotropic phase behavior (Cevc & Marsh, 1987; Marsh, 1991; Yeagle, 1992). Most fully hydrated saturated phospholipid bilayers exhibit two principal thermotropic lamellar phase transitions, corresponding to a gel to gel ($L_{\beta'}-P_{\beta'}$) transition and a gel to liquid-crystalline ($P_{\beta'}-L_{\alpha}$) main transition at a temperature T_m. In the fluid-like L_{α} phase, the hydrocarbon chains of the lipid bilayers are conformationally disordered, whereas in the gel phases the hydrocarbon chains are more extended and relatively ordered.

In addition to these thermotropic phase transitions, several pressure-induced lamellar to lamellar phase transitions have been observed (for example, see Chong & Weber, 1983; Braganza & Worcester, 1986; Wong et al., 1988; Wong & Mantsch, 1989; Winter & Pilgrim, 1989; Winter & Thiyagarajan, 1990; Driscoll et al., 1991; Scarlata, 1991; Peng & Jonas, 1992). Besides the physicochemical interest in high-pressure phase behavior and structure of amphiphilic molecules, high pressure is also of considerable physiological interest. For example, pressure studies on lipid systems are of interest in understanding the physiology of deep sea organisms, the sensitivity of excitable cell membranes to pressure ("high-pressure nervous syndrome"), and the antagonistic effect of pressure on anesthetic action (Zimmerman, 1978; Rostain et al., 1989; Balny et al., 1992). Furthermore, studies of pressure effects on lipid systems often led to the discovery of new phases and helped to elucidate the mechanisms underlying lipid phase transformations.

Although most lipids in excess water exist in lamellar bilayer phases, certain lipids also can form nonbilayer, hexagonal (H_{II}), or cubic liquid-crystalline phases (Lindblom & Rilfors, 1989; Seddon, 1990; Tate et al., 1991; Seddon & Templer, 1993). Many naturally derived lipids, including phosphatidylethanolamines and monoacylglycerides, can form a hexagonal phase. The H_{II} phase is formed by a periodic hexagonal lattice of cylindrical pores of water of about 40 Å diameter, being surrounded by a monolayer of lipid molecules. Many of the cubic liquid-crystalline phases are now known to consist of bicontinuous regions of water and hydrocarbon, which can be described by infinite periodic minimal surfaces (IPMSs). An IPMS is an intersection-free surface that is periodic in three dimensions and has zero mean curvature everywhere.

The nonlamellar phases, which occur in a number of membrane lipids probably play an important functional role as local and transient intermediates (Mariani et al., 1988; Seddon, 1990; Tate et al., 1991). For example, the cubic phases seem to be more or less directly involved in membrane fusion and fat digestion. Evidence has also consolidated in support of the hypothesis that microorganisms control the lipid composition of their membranes to maintain them close to a composition at which nonlamellar structures would begin to appear (Cronan & Vagelos, 1975; Rilfors et al., 1978).

RESULTS AND DISCUSSION

Thermotropic and Barotropic Phase Transitions of Single Component Phospholipid Dispersions

As an example of the volumetric properties of a saturated phospholipid, Figure 18.1a shows the temperature dependence of the apparent specific lipid volume V_L^* of DMPC in water. The change of V_L^* near 14 °C corresponds to a small volume change in course of the $L_{\beta'}-P_{\beta'}$ transition. The main transition at $T_m = 23.9$ °C is accompanied by a well-pronounced 3% change in volume (see, for example, Venneman et al., 1986; Wiener et al., 1988; Raudino et al., 1980; Böttner et al., 1994), which is mainly due to changes of the chain cross-sectional area, because chain disorder increases drastically at the transition. Assuming that the volume of the headgroup is independent of temperature in the temperature range covered (Tardieu, et al., 1973), a mean volume increase of 1.5 Å^3 can be calculated for the methylene groups at the main transition. Figure 18.1b shows the pressure dependence of V_L^* at 30 °C. Increasing pressure triggers the phase transformation from the L_α into the gel phase, as can be seen from the rather abrupt decrease of the lipid volume at 270 bar. The volume change, ΔV_m at the main transition decreases slightly with increasing temperature and pressure along the main transition line, as can be seen in Figure 18.2. It appears that the compressibility of the $P_{\beta'}$ gel phase is substantially lower than that of the liquid-crystalline phase (for $T = 30$ °C: $\chi_T(P_{\beta'}) = 5 \ (\pm 2)10^{-5}$ bar^{-1} and $\chi_T(L_\alpha) = 13 \ (\pm 2)10^{-5}$ bar^{-1}). The (T, p) phase diagram for the main transition of DMPC-water is displayed in Figure 18.3. Up to pressures of about 1 kbar, a linear increase of the phase transition curve with $21 \ (\pm 1)$°C kbar^{-1} is observed, whereas for higher pressures, dT_m/dp slightly decreases. Similar deviations from linearity of $T_m(p)$ at higher pressures have also been observed for DPPC and DSPC dispersions (Ceh, 1991; Winter & Böttner, 1993).

Assuming the main transition to be of first order, the entropy and enthalpy changes at the transition can be calculated by means of the Clapeyron equation. An enthalpy change for DMPC of $\Delta H_m = 26$ kJ mol^{-1}, and an entropy change of $\Delta S_m = 86$ Jmol^{-1} K^{-1} is obtained. Recent high-pressure DTA experiments (Landwehr & Winter, 1994) on this system revealed that ΔH_m does not change significantly up to about 1.5 kbar. By applying the pressure-jump technique in combination with time-resolved x-ray diffraction, Caffrey et al. (1991) showed that the pressure-induced main transition of DMPC dispersions appears to proceed by a two-state mechanism with no nonlamellar intermediates occurring. The transition in either the pressurization or depressurization direction appears to take place in two steps. The first involves the actual phase conversion itself, which is fast, occurring on the time scale of seconds. The second involves a slow relaxation of the new phase toward its final equilibrium state.

Figure 18.4 shows an example of the high-pressure phase behavior of an unsaturated phospholipid and displays the high-pressure DTA measurements of

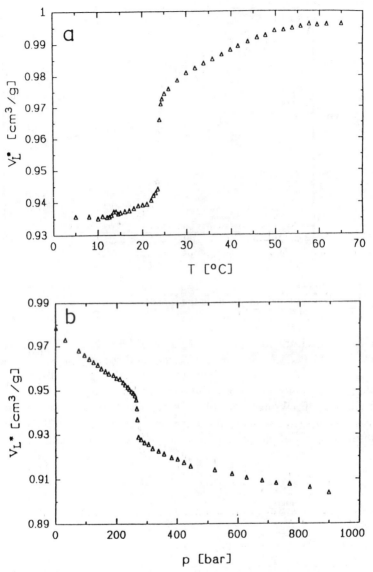

Figure 18.1. Apparent specific lipid volume of DMPC in water (**a**) as a function of temperature at ambient pressure and (**b**) as a function of pressure at $T = 30\,°C$.

DEPC-water as a function of pressure. In this lipid with two *trans* double-bonds, one can observe a drastic increase of the L_α-gel transition temperature with increasing pressure ($dT_m/dp = 19(\pm 2)°C/kbar$). Figure 18.4 points to a constant ΔH_m value over the whole pressure and temperature range covered.

In the lower pressure regime, a common value for the L_α-gel transition slope of about $22(\pm 2)°C/kbar$ has been observed for the saturated phosphatidylcholines DMPC, DPPC, and DSPC. Similar values have been found for

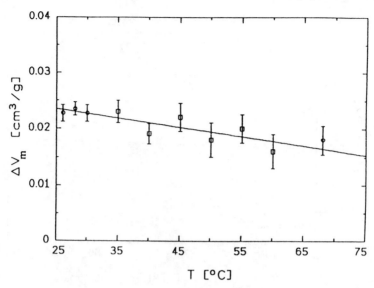

Figure 18.2. Pressure-induced volume changes at the main transition of DMPC dispersions taken at different temperatures along the L_α/gel transition curve.

the mono-*cis*-unsaturated POPC ($dT_m/dp = 21(\pm 1)$ °C/kbar), the phosphatidylserine DMPS ($dT_m/dp = 17(\pm 2)$ °C/kbar), and for the phosphatidylethanolamine DPPE ($dT_m/dp = 23(\pm 1)$ °C/kbar) (Winter & Pilgrim, 1989; Winter et al., 1989; Reis, 1994). Only the slopes of the main transition lines of *cis*-unsaturated DOPC and DOPE (Wong et al., 1988; Wong and Mantsch, 1989; Winter et al., 1990) have been found to be significantly smaller (see Figure 18.5). Obviously, the transition slopes of phospholipids are not affected appreciably by differences

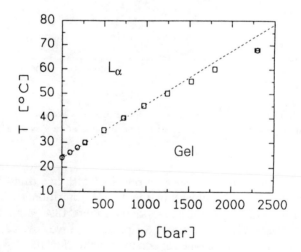

Figure 18.3. Temperature-pressure phase diagram for the gel to liquid-crystalline transition of DMPC water dispersions as determined by volumetric measurements.

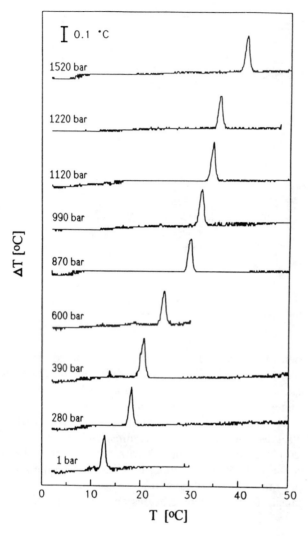

Figure 18.4. DTA traces of a DEPC water dispersion at different pressures (heating scan rate 1 °C/minute).

in the hydrocarbon chain length or the nature of the choline headgroup (ethanolamine or serine); these affect mainly the transition temperature. However, the presence of double bonds in the chain region influences the transition slope. The introduction of *cis* double bonds leads to the lowest transition temperatures and smallest transition slopes, presumably because the cis double bonds impose a kink in the linearity of the acyl chains, thus creating significant free volume in the bilayer, which reduces the ordering effect of high pressure.

Further pressure-induced gel phases have been observed in single-component phospholipid dispersions, such as an interdigitated high-pressure gel phase in DPPC and DSPC (Braganza & Worcester, 1986; Winter & Pilgrim, 1989), as

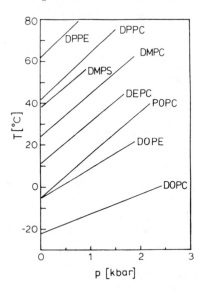

Figure 18.5. Temperature-pressure phase diagram for the liquid-crystalline to gel transition of different phospholipid bilayer systems.

identified by neutron diffraction experiments on oriented multilamellar vesicles (Braganza & Worcester, 1986). The (T, p) phase diagram for DSPC-water obtained from recent high-pressure neutron diffraction and DTA experiments is shown in Figure 18.6. Interestingly, in addition to the isothermal, pressure-induced liquid-crystalline to gel transformation, an interdigitated hydrocarbon phase with d-spacing of 50 Å is found at higher pressures, and the pressure at which this transition takes place varies nonlinearly with temperature. The partially inter-digitated gel phase probably forms under conditions of temperature and pressure at which the pretransition vanishes, and its phase line seems to merge with that of the main transition at high temperatures and pressures. No interdigitated gel

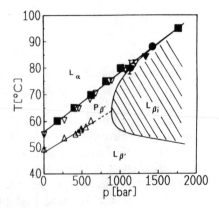

Figure 18.6. (T, p) phase diagram of the system DSPC-H$_2$O in excess water.

phase has been observed for DMPC multilayers up to 3 kbar (Winter & Pilgrim, 1989).

These studies clearly demonstrate that by regulating the lipid composition of the cell membranes through changes in lipid chain length, degree of unsaturation, and headgroup structure, biological organisms are provided with a mechanism for efficiently modulating the physical state of their membranes in response to changes in the external environment.

Effect of Additives on the High-Pressure Phospholipid Phase Behavior

Considerable experimental and theoretical attention has been devoted to the nature of pure phospholipid membrane phase transitions and how they are affected by the incorporation of other species interacting with these membranes. Such processes are of interest because they intimately reflect the molecular interactions of the membrane and may thus help in understanding membrane systems and function on a molecular level.

Salt Effect

It is well known that the effect of inorganic ions on the main transition temperature strongly depends on the nature of the ions (Caffrey & Feigenson, 1984; Hauser, 1991). Because of the formation of rather stable complexes between the divalent ion and the phosphate group, the effect is especially dramatic when Ca^{2+} ions are adsorbed on negatively charged membranes. DSC measurements on DMPC/Ca^{2+} dispersions revealed that increasing Ca^{2+} concentration leads to an increase in main transition temperature (Zorn & Nimtz, 1990; Jacobs, 1992), with little change in transition enthalpy, and to an increase of the $L_{\beta'}-P_{\beta'}$ gel-to-gel transition temperature, until both transitions merge around 50 mol% Ca^{2+} (see Figure 18.7).

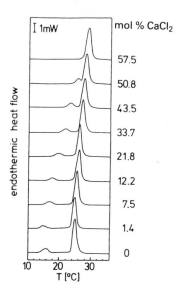

Figure 18.7. DSC traces of a DMPC water dispersion at different concentrations of $CaCl_2$.

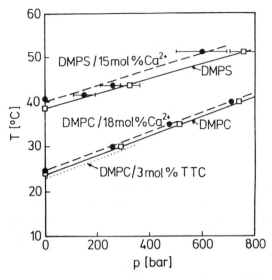

Figure 18.8. Influence of $CaCl_2$ or tetracaine on the main transition of DMPC or DMPS as a function of temperature and pressure.

Volumetric measurements on Ca^{2+}/DMPC dispersions showed that the thermodynamic properties are only slightly affected by addition of 20 mol% Ca^{2+}. Compared to pure DMPC dispersions, the main transition is shifted 30 bar toward smaller pressures at constant temperature. Otherwise, the transition slope (dT_m/dp) is parallel to that of pure DMPC up to pressures of about 1.7 kbar, and the volume change ΔV_m, at the main transition is of comparable magnitude. A similar behavior has been observed for the negatively charged lipid DMPS with addition of 15 mol% Ca^{2+}. The corresponding (T_m/p) phase diagrams for the main transition of the two lipid systems with and without Ca^{2+} are presented in Figure 18.8.

Local Anesthetics

In this section, we present data on the influence of the local anesthetic tetracaine on the thermodynamic properties and the temperature- and pressure-dependent phase behavior of the model biomembrane DMPC (Winter et al., 1991; Böttner et al., 1992). As tetracaine (TTC) can be viewed as a model system for a large group of amphiphilic molecules, these results also provide insight into the general understanding of the physicochemical action of this kind of amphiphilic molecule on membranes. From volumetric measurements on a sample containing 3 mol % TTC, it has been found that the main transition at ambient pressure shifts to a lower temperature. The expansion coefficient α_p increases drastically relative to that of the pure lipid system in the gel phase, and the incorporation of the anesthetic into the DMPC bilayer causes a decrease of ΔV_m relative to that of the pure lipid system, by about 15%. Also depicted in Figure 18.8, the addition of 3 mol% TTC shifts the pressure-induced liquid-crystalline to gel-phase transition

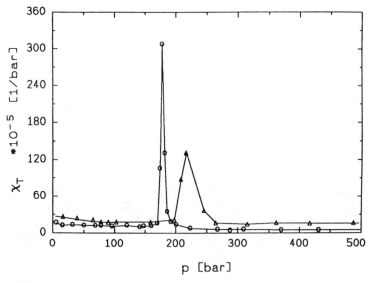

Figure 18.9. Isothermal compressibility χ_T of a pure DMPC (O) and 3 mol% TTC/DMPC ($\triangle$) dispersion at, for example, $T = 28\,°C$ and pH = 9.

toward somewhat higher pressures. Figure 18.9 shows the isothermal compressibilities of pure and 3 mol% TTC containing DMPC at $T = 28\,°C$. The addition of 3 mol% TTC yielded larger values for the compressibilities of both lipid phases, but there is no apparent difference in the coefficients of compressibility of the gel and liquid-crystalline phases. While χ_T is reduced drastically at the main transition point, it is enhanced in the neighborhood of the transition. The relative root mean square fluctuations of the lipid volume $\delta V_{L,\,rel} = (\langle \delta V_L^* \rangle^2)^{1/2}/V_L^*$ in the liquid-crystalline phase of pure DMPC are about 7%; they increase to about 30% at the main transition and decrease to 4% in the gel phase. These relative volume fluctuations of the lipids in membranes are large in comparison to volume fluctuations of other biochemical systems such as proteins (Gekko & Hasegawa, 1986). This fact may be relevant to understanding the dynamics, structure, and, therefore, the function of membrane-bound proteins. The addition of 3 mol% of the local anesthetic leads to an increase of the relative lipid volume fluctuations up to about 9% in both lipid phases. In a computer simulation study (Mouritsen, 1991), it could be shown that foreign molecules, such as anesthetics, in membranes might strongly couple to the thermal density and concentration fluctuations of the lipid system near its gel to liquid-crystalline phase transition, thus leading to a strong enhancement of $\delta V_{L,\,rel}$ in the neighborhood of T_m, which has indeed been observed experimentally. These findings might be of further biochemical relevance, since, in lipid bilayer membranes, strong density or concentration fluctuations are related to the transmembrane permeability of ions and small molecules. Although the biochemical action of local anesthetics is still controversial as to whether or not the action is lipid mediated, it is clear that they do strongly perturb the lipid bilayer system and change its phase behavior and thermomechanical properties.

Cholesterol Effect

The introduction of cholesterol into lipid membranes can have a drastic effect on the phase behavior of membranes [see, for example, Yeagle, 1992]. DSC studies on the system POPC/cholesterol-water revealed that an increase in the cholesterol concentration leads to a drastic reduction of ΔH_m until the main transition vanishes at cholesterol contents higher than about 30 mol%. Volumetric studies showed that the addition of 5 mol% cholesterol shifts the main transition of POPC toward higher pressures by 150 bar. The volume change at the main transition stays approximately constant up to 5 mol% cholesterol, but decreases at higher cholesterol concentrations. Concurrent with the decrease in size of ΔV_m, the width of the transition then substantially increases (Brauns & Winter, unpublished). Figure 18.10 displays the lamellar d-spacing of pure and 28 mol% cholesterol containing POPC at 10 °C as a function of pressure (T_m(POPC) = -5 °C at 1 bar). The d-spacing in the L_α state is increased drastically by the addition of cholesterol, whereas at higher pressures its value is only slightly smaller than the corresponding value in the gel state of pure POPC. These observations indicate that—in agreement with fluorescence depolarization studies—cholesterol has the effect of increasing the conformational order in the fluid-like state, as has been observed in saturated phospholipid systems; this, in turn, should increase the bilayer thickness. At pressures corresponding to the gel state of pure POPC, the addition of cholesterol has only a small effect on the lipid chain order parameter. This clearly demonstrates the drastic effect cholesterol can impose on the structure of lipid membranes and on their temperature- and pressure-dependent phase behavior.

Binary Phospholipid Mixtures

Although studies of the phase behavior of single-component liposomes are valuable, they are not very realistic models for biological membranes, which contain different types of phospholipids. As a first step toward understanding these complex systems, studies of binary phospholipid mixtures can be carried out.

Here, we present data on the temperature- and pressure-dependent phase diagram of the lipid mixtures DMPC/DPPC and DMPC/DSPC in excess water. The phase diagrams have been constructed by specifying the onset and completion temperatures of the phase transitions by an inspection of the shapes of DTA traces. The onset and completion temperatures were corrected for the contributions to the total transition widths, which stem from the finite widths of the transition curves of the two pure lipids. As an example, Figure 18.11 shows DTA traces of a DMPC/DPPC dispersion at $x_{DMPC} = 0.1$ as a function of pressure. The results clearly show the pressure-dependent shift of the phase coexistence region to higher temperatures. The (T, x) phase diagram of the DMPC/DPPC mixture at ambient pressure is shown in Figure 18.12. A similar phase diagram has been derived for this system with data from other experimental techniques (Mabrey & Sturtevant, 1976; Lee, 1977; Schmidt & Knoll, 1985). Figure 18.12 also shows the theoretical curve corresponding to an isomorphous system with ideal mixing behavior in the fluid and gel phase. Although the overall features of the phase diagram are

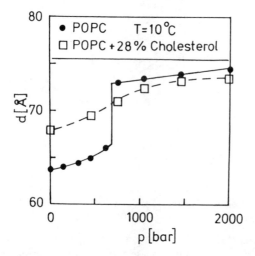

Figure 18.10. Lamellar repeat distance d of a POPC and POPC/28 mol% cholesterol dispersion at $T = 10\,°C$ as a function of pressure.

reproduced by the theoretical curve, clear evidence for deviations from ideal mixing are observed. The agreement can be improved by assuming nonideal mixing in both phases. The nonideality can be accounted for in a very simple way by using regular solution theory, where the enthalpy of mixing is assumed to have the simple form $\Delta H_{mix} = \rho_0 x_A x_B$. The quantity ρ_0 is a measure of the strength of interaction between lipid A and lipid B with mole fractions x, and can be expressed simply in terms of lipid-lipid interaction energies U_{ij}: $\rho_0 = Z(U_{AB} - 0.5(U_{AA} + U_{BB}))$, where Z is the coordination number of nearest neighbors. If the repulsive interaction between unlike lipids outweighs the interaction between like lipids, there is a tendency toward phase separation. The theoretical curve for DMPC/DPPC with an interaction parameter ρ_0 of 2 kJ mol^{-1} in the fluid and 4 kJ mol^{-1}

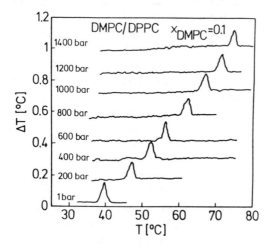

Figure 18.11. DTA traces of a DMPC/DPPC mixture at $x_{DMPC} = 0.1$ as a function of pressure.

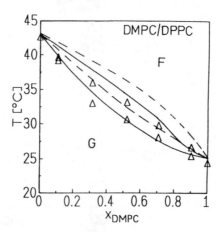

Figure 18.12. (T, x) phase diagram of DMPC/DPPC dispersions (△-experimental DTA data; dashed curves-ideal mixing phase diagram; solid curves-transition curves calculated on the basis of regular solution theory).

in the gel phase is added in Figure 18.12. Clearly, an improvement is observed. For the DMPC/DSPC dispersion, the deviations from ideal mixing are considerably greater. Obviously, increasing disparity in chain length of the two lipids leads to increasing nonideality. In a statistical-mechanical calculation of lipid phase separation in DMPC/DSPC dispersions by von Dreele (1978), it has been found that deviations from ideal mixing in this system are repulsive in nature, tending toward phase separation, and are considerably greater in the gel phase, as expected from the higher packing density in this phase.

Recently, an extended Monte Carlo computer simulation study applying a ten-state Pinks model has been used to describe the phase diagram of two-component phospholipid bilayer membranes of saturated phospholipids (Jørgensen et al., 1993). Figure 18.13 shows the experimental results for DMPC/DSPC dispersions in comparison to the theoretical calculated phase diagram. In DMPC/DSPC-water, the gel-fluid coexistence region has become very broad. Quite reasonable aggreement is found between experiment and theoretical calculation. For both mixtures, DMPC/DPPC and DMPC/DSPC, strong deviations from nonideal mixing behavior are found in the calculations. The simulations showed that in the fluid thermodynamic one-phase region of the phospholipids, these phases are clearly not uniform or random in structure. Instead, distinct local lateral structures consisting of correlated lipids are observed. These local structures become more pronounced as the degree of nonideal mixing increases. They also depend on the distance from the phase boundaries and are most pronounced near equimolar concentrations. Figures 18.14 and 18.15 exhibit the (T, x) phase diagram for mixtures of DMPC/DPPC and DMPC/DSPC, respectively, up to pressures of 1500 bar as determined by high-pressure DTA. No change in shape of the fluid/gel coexistence regions is observed for either mixture in the whole temperature and pressure range covered. Thus, increasing pressure does not lead to significant changes in the nonideality parameter or even to changes in miscibility in the fluid or gel state of the lipid mixtures.

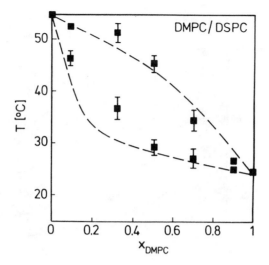

Figure 18.13. Comparison of the (T, x) phase diagram of DMPC/DSPC-water with computer simulation data (Jørgensen et al., 1993).

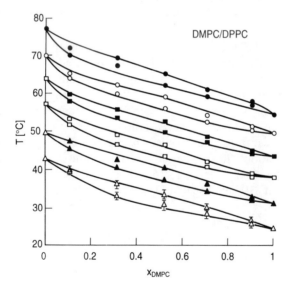

Figure 18.14. (T, x, p) phase diagram of DMPC/DPPC dispersions as obtained from high-pressure DTA experiments. △ -1 bar; ▲ -300 bar; □ -600 bar, ■ -900 bar; ○ -1200 bar; ● - 1500 bar.

Lamellar to Nonlamellar Lipid Phase Transitions

This final section discusses the temperature- and pressure-dependent phase behavior of three examples of lipid systems that also exhibit stable or metastable nonlamellar structures: a phosphatidylethanolamine (DOPE), monoacylglycerides (MO and ME), and a binary mixture (DMPC/MA).

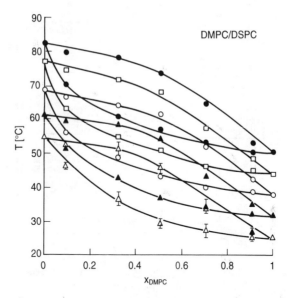

Figure 18.15. (T, x, p) phase diagram of DMPC/DSPC dispersions as obtained from high-pressure DTA experiments. △-1 bar; ▲-300 bar; ○-600 bar; □-900 bar, ●-1200 bar.

In contrast to DOPC, the corresponding *cis*-unsaturated phospholipid with ethanolamine as its headgroup (DOPE) exhibits, in addition to the lamellar gel to liquid-crystalline main transition, a bilayer to nonbilayer phase transformation at higher temperatures. From DSC studies, it is known that DOPE water dispersions undergo a lamellar L_β to lamellar L_α transition at $T_m \approx -6\,°C$ and a lamellar L_α to inverted hexagonal (H_{II}) thermotropic transition at $T_h \approx 10$–$13\,°C$. Whereas the main transition at ambient pressure is accompanied by an enthalpy change ΔH_m of about 20–30 kJ mol^{-1}, ΔH_h of the L_α–H_{II} transition is only 2 kJ mol^{-1} (Hahn, 1993). Figure 18.16 displays the DTA trace taken of a DOPE water dispersion at 810 bar. The first DTA peak at 2 °C corresponds to the lamellar

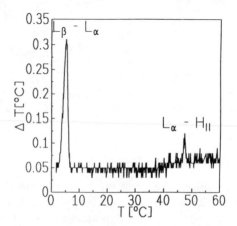

Figure 18.16. DTA trace of a DOPE water dispersion at 810 bar (heating scan rate 1 °C/minute).

L_β–L_α transition, whereas the smaller endothermic peak at 47 °C is due to the L_α–H_{II} transition. The enthalpy change connected with the hexagonal transition is only about 14% of that for the main transition. The transition enthalpy is significantly lower than that of the gel to liquid-crystalline transition, because the transition takes place wholly in a fluid-like state. Obviously, increasing pressure not only shifts the lamellar to lamellar transition, but also the lamellar to hexagonal transition to higher temperatures.

From high-pressure SANS experiments on DOPE dispersions we found that the lattice constant of the H_{II} phase varies drastically with temperature and pressure (for example, $a(50 \, °C) = 68 \, Å$, $(da/dp)_{T=50 \, °C} = 5.3 \, Å/kbar$, and $(da/dT)_{p=1 \, bar} = -0.2 \, Å/°C$). Higher pressures force a closer packing of the chains, which presumably results in a reduction of the number of gauche bonds within the chains so that T_m and T_h increase with increasing pressure. The application of hydrostatic pressure in the presence of excess water leads to an increase in the lattice constant of the hexagonal phase, probably due to an increase in the diameter of the water cores and, hence, in the hydration level associated with the lipid headgroups.

Figure 18.17 displays the corresponding (T, p) phase diagram for the two endothermic transitions of the DOPE-water system (Landwehr et al., 1993). As can be clearly seen, the transition slope of the lamellar chain-melting transition $(dT_m/dp = 14 \, °C/kbar)$ is significantly lower than that of saturated phospholipids. For the smaller pressures, the slope of the L_α–H_{II} transition $(dT_h/dp = 40 \, °C/kbar)$ is almost three times steeper than the slope of the lamellar chain-melting transition. At higher pressures, dT_h/dp decreases slightly. The phase diagram is in qualitative agreement with the one obtained recently by x-ray and volumetric measurements (So et al., 1993). Using high-pressure FTIR in combination with the diamond anvil technique, Wong and Mantsch found an indication of an additional lamellar to lamellar transition at 9 kbar (Wong & Mantsch, 1989). A similarly steep slope for the L_α–H_{II} transition has also been observed for egg-PE by turbidity and (p, V, T) measurements (Yager & Chang, 1983; Hahn, 1993). This L_α–H_{II} transition is the most pressure-sensitive lipid phase transition found to date.

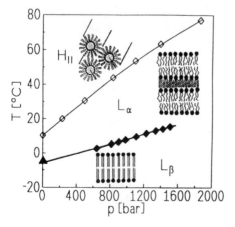

Figure 18.17. (T, p) phase diagram for the L_β–L_α and L_α–H_{II} phase transition of DOPE-water.

The data here clearly indicate that the bilayer form is stabilized at high pressure. Therefore, biochemical processes involving the H_{II} phase as an intermediate state should be slowed down by the application of hydrostatic pressure (Chang & Yager, 1983).

Recently, it has been shown that a cubic primitive structure can be induced in DOPE by subjecting the sample to an extensive temperature cycling process between $-5\,°C$ to $15\,°C$, that is, at temperatures close to the transition region of the L_{α} and H_{II} phase (Shyamsunder et al., 1988). By changing the thermal history of the sample, even two cubic phases may be induced, possibly of space group Pn3m and Im3m (Erbes et al., 1994). They exhibit limited temperature stability, however. The cubic phase Im3m disappears at about $45\,°C$, whereas the cubic primitive phase is stable up to about $75\,°C$. Siegel (1986) has suggested that formation of these cubic structures may arise from topological defects (inverted micellar intermediates and interlamellar attachments) of the membrane surface generated by cycling through the $L_{\alpha}-H_{II}$ transition. One might speculate that the number of these created interfacial defects increases as one repeatedly crosses the $L_{\alpha}-H_{II}$ transition, until these defects finally coalesce and form a new phase that corresponds to a bicontinuous cubic structure. Andersson et al. (1988) showed that in certain situations, the topology of bicontinuous cubic phases can result in less frustration and hence a lower free energy than either the lamellar or H_{II} phase. The occurrence of these phases may be obscured by large kinetic barriers, however.

Recently, it has also been shown (So et al., 1993) that at higher pressures the cubic phase probably forms within a few hours when the temperature and pressure are kept in the region of the $H_{II}-L_{\alpha}$ transition. In similar experiments on DOPE dispersions using high-pressure SANS, we observed that the change from the hexagonal to the cubic lattice at $83.5\,°C$ and 2145 bar occurs with a characteristic time constant of about 10%/hour.

As a second example, two monoacylglycerides, monoolein (MO) and monoelaidin (ME), were chosen for investigation of their high-pressure phase behavior because they exhibit mesomorphic phases with one-, two-, and three-dimensional periodicity in a rather easily accessible temperature and pressure range. Furthermore, their physical and chemical properties have received considerable interest because of their importance as intermediates in lipid digestion and metabolism; hence their applications in food industry. Figure 18.18 displays the SAXS traces of a MO dispersion in the temperature range from 72 to $96\,°C$. In the lower temperature region, the Bragg reflections of a cubic structure are seen and they can tentatively be indexed on the space group Pn3m with lattice spacing 86 Å at $45\,°C$. At about $87\,°C$, the cubic phase transforms into a H_{II} phase with lattice constant 55 Å. The phase transition enthalpy change from the cubic to the H_{II} phase in MO is about 0.3 kJ mol^{-1}, which is a typical value for phase transition enthalpies between phases with fluid-like character. Recently, it has been shown by time-resolved x-ray diffraction that all of the MO transitions examined appear to be two state (Caffrey, 1987). The thermal lattice expansion coefficients in both phases are negative: -0.36 Å/$°C$ for the cubic and -0.14 Å/$°C$ for the H_{II} phase. From high-pressure SANS experiments, the lattice parameter, and therefore the volume per unit cell, is found to be strongly pressure dependent. For the cubic structure, a value of $(da/dp)_{T=50\,°C} = 8$ Å/kbar has been found. Somewhat higher

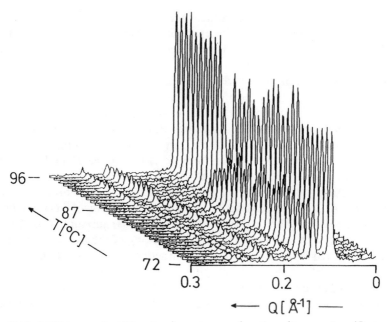

Figure 18.18. SAXS traces of a MO water dispersion as a function of temperature (Q momentum transfer)

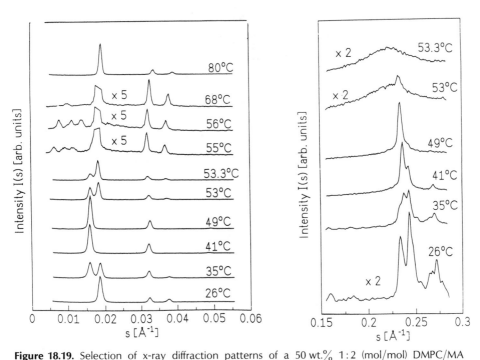

Figure 18.19. Selection of x-ray diffraction patterns of a 50 wt.% 1:2 (mol/mol) DMPC/MA dispersion at different temperatures (scan rate 0.5 °C/minute).

have been measured for the pressure dependence of the lattice parameter of the corresponding cubic phase of ME (for example, $(da/dp)_{T=58\,°C} = 12$ Å/kbar). The (T, p) phase diagrams of MO and ME in excess water are described in Czeslik et al. (1995).

Mixtures of diacylphosphatidylcholines with fatty acids of the same chain-length tend to form 1:2 (mol/mol) stoichiometric compounds. For the palmitic chain system, it has been shown that the low-temperature phase is a lamellar phase, but the high-temperature fluid phase is an inverted hexagonal one. For the 1:2 (mol/mol) DMPC/MA system, however, an additional isotropic phase of cubic symmetry is observed at higher temperatures (see, for example, Hyde & Andersson, 1984; Heimburg et al., 1990; Seddon et al. 1990). Figure 18.19 exhibits a selection of x-ray diffraction patterns of a 50 wt.% DMPC/MA dispersion (scan rate ca. 0.5 °C/minute). At 26 °C, the diffraction pattern characteristic of a crystalline lipid structure (Cryst.)—probably a phase-separated lamellar mixture of DMPC/MA and MA—occurs, with at least partially rigid packing of the hydrocarbon chains, as might be inferred from the wide-angle diffraction region. At about 32 °C, the first two orders of the lamellar L_β gel phase appear with repeat spacing of 63 Å. Above 49 °C, a hexagonal phase H_{II} with fluid-like chains develops with lattice spacing 63 Å. Around 55 °C, additional Bragg reflections, which index on a cubic

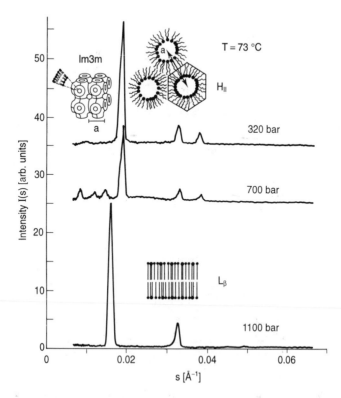

Figure 18.20. Scattering intensity $I(s)$ of a 50 wt.% 1:2 (mol/mol) DMPC/MA dispersion at $T = 73$ °C at selected pressures ($s = (2/\lambda)\sin\theta$; λ wavelength of radiation; 2θ scattering angle).

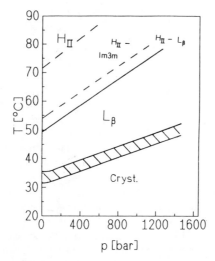

Figure 18.21. (T, p) phase diagram of the system $1:2$ (mol/mol) DMPC/MA-50 wt.% water as obtained from DTA, x-ray, and neutron diffraction experiments in the heating direction (additional metastable phases, such as a further cubic phase, are encountered upon cooling the sample, in particular at elevated pressures).

lattice, appear at low momentum transfers, possibly with space group Im3m and with lattice constant 216 Å. This cubic phase seems to coexist with the hexagonal phase up to about 70 °C. It thus appears that the behavior of this shorter chainlength phosphatidylcholine/fatty acid mixture is rather complex, with co-existing H_{II} and cubic phases occurring over a limited temperature range. The induction of inverse bicontinuous cubic phases at shorter chainlengths may be a general feature of those lipid systems that have a tendency to form the inverse hexagonal H_{II} phase (Seddon et al., 1990). As an example of a pressure-dependent study, Figure 18.20 shows the scattering intensity I(s) of DMPC/MA at $T = 73$ °C for three selected pressures. The reflections at $p = 1100$ bar correspond to a lattice constant of 62 Å of the L_β phase. At $p = 700$ bar, coexistence between the H_{II} ($a = 61$ Å) and cubic phase ($a = 164$ Å) is observed. The cubic phase finally disappears at still lower pressures. From the combined results of high-pressure DTA and x-ray and neutron diffraction experiments, a tentative (T, p) phase diagram has been constructed for this system (see Figure 18.21). As can be clearly seen, not only the Cryst. to L_β transition shifts to higher temperatures with increasing pressure ($dT/dp = 12$ °C/kbar), but so does the L_β–H_{II}/cubic transition, which, however, has the much steeper slope of 22 °C/kbar.

A theoretical unterstanding of the form of the (T, p) phase diagram of the various lipid systems requires a detailed understanding of all the complex interactions involved, such as interfacial, hydration, and van der Waals forces, steric repulsion, and hydrogen bonding. At present, considerable effort is devoted to model these contributions.

ACKNOWLEDGMENTS: We are grateful to the DFG and the Fonds der Chemischen Industrie for financial support. We thank Dr. G. Rapp for his help with the synchrotron-X-ray diffraction experiments at the EMBL outstation at DESY in Hamburg.

REFERENCES

Andersson, S., Hyde, S. T., Larsson, K., & Lidin, S. (1988). Minimal surfaces and structures: from inorganic and metal crystals to cell membranes and biopolymers. *Chem. Rev.* **88**, 221–242.

Balny, C., Hayashi, R., Heremans, K., & Masson, P., eds. (1992). *High Pressure and Biotechnology.* vol. 224. Colloque INSERM, **224**, John Libbey Eurotext.

Böttner, M., Ceh, D., Jacobs, U., & Winter, R. (1994). High pressure volumetric measurements on phospholipid bilayers. *J. Phys. Chem.* **184**, 205–218.

Böttner, M., Christmann, M.-H., & Winter, R. (1992). The influence of local anaesthetics on the pressure dependent phase behaviour of model membranes. In *The Structure and Conformation of Amphiphilic Membranes*, ed. R. Lipowski, D. Richter, & K. Kremer. Berlin, Springer Proceedings in Physics, **66**, 65–69.

Braganza, L. F., & Worcester, D. L. (1986). Hydrostatic pressure induces hydrocarbon chain interdigitation in single-component phospholipid bilayers. *Biochemistry* **25**, 2591–2596.

Caffrey, M. (1985). Kinetics and mechanisms of the lamellar gel/lamellar liquid-crystal and lamellar/inverted hexagonal phase transition in phosphatidylethanolamine: a real-time x-ray diffraction study using synchrontron radiation. *Biochemistry* **24**, 4826–4844.

Caffrey, M. (1987). Kinetics and mechanism of transitions involving the lamellar, cubic, inverted hexagonal, and fluid isotropic phases of hydrated monoacylglycerides monitored by time-resolved x-ray diffraction. *Biochemistry* **26**, 6349–6363.

Caffrey, M., & Feigenson, G. W. (1984). Influence of metal ions on the phase properties of phosphatidic acid in combination with natural and synthetic phosphatidylcholines: an x-ray diffraction study using synchrotron radiation. *Biochemistry* **23**, 323–331.

Caffrey, M., Hogan, J., & Mencke, A. (1991). Kinetics of the barotropic ripple ($P_{\beta'}$) lamellar liuid crystal (L_α) phase transition in fully hydrated dimyristoylphosphatidylcholine (DMPC) monitored by time-resolved x-ray diffraction. *Biophys. J.* **60**, 456–466.

Ceh, D. (1991). PVT-Messungen an Modell-Biomembranen. Diploma thesis, University of Marburg.

Cevc, G., & Marsh, D. (1987). *Phospholipid Bilayers.* New York, Wiley.

Chang, E. L., & Yager, P. (1983). Effect of high pressure on a lipid non-bilayer phase. *Mol. Cryst. Liq. Cryst.* **98**, 125–129.

Chong, P.-L. G., & Weber, G. (1983). Pressure dependence of 1,6-diphenyl-1,3,5-hexatriene fluorescence in single-component phosphatidylcholine liposomes. *Biochemistry* **22**, 5544–5550.

Cronan, J. E., & Vagelos, P. R. (1975). Metabolism and function of the membrane phospholipids of *Escherichia coli. Biochim. Biophys. Acta* **265**, 25.

Czeslik, C., Winter, R., Rapp, G., & Bartels, K. (1995). Temperature- and pressure-dependent phase behavior of monoacylglycerides monoolein and monoelaidin. *Biophys. J.* **68**, 1423–1429.

Driscoll, D. A., Jonas, J., & Jonas, A. (1991). High pressure ^{2}H nuclear magnetic resonance study of the gel phases of dipalmitoylphosphatidylcholine. *Chem. Phys. Lipids* **58**, 97–104.

Erbes, J., Czeslik, C., Hahn, W., Rapp, G., & Winter, R. (1994). On the existence of bicontinuous cubic phases in dioleoylphosphatidylethanolamine. *Ber. Bunsenges. Phys. Chem.* **98**, 1287–1293.

Gekko, K., & Hasegawa, Y. (1986). Compressibility-structure relationship of globular proteins. *Biochemistry* **25**, 6563–6571.

Hahn, W. (1993). Biophysikalische Untersuchungen an invertiert kubischen und invertiert hexagonalen Phasen der Phospholipide DOPE und Egg-PE. Diploma thesis, University of Marburg.

Hauser, H. (1991). Effect of inorganic cations on phase transitions. *Chem. Phys. Lipids* **75**, 309–325.

Heimburg, T., Ryba, N. J. P., Würz, U., & Marsh, D. (1990). Phase transition from a gel to a fluid phase of cubic symmetry in dimyristoylphosphatidylcholine/myristic acid (1:2, mol/mol) bilayers. *Biochim. Biophys. Acta* **1025**, 77–81.

Hyde, S. T., Andersson, S., Ericsson, B., & Larsson, K. (1984). A cubic structure consisting of a lipid bilayer forming an infinite periodic minimum surface of the gyroid type in the glycerolmonooleat-water system. *Z. Kristallogr.* **168**, 213–219.

Jacobs, U. (1992). Biophysikalische Untersuchungen zum Einfluß von $CaCl_2$ auf die Modellbiomembran DMPC. Diploma thesis, University of Marburg.

Jørgensen, K., Sperotto, M. M., Mouritsen, O. G., Ipsen, J. H., & Zuckermann, M. J. (1993). Phase equilibria and local structure in binary lipid bilayers. *Biochim. Biophys. Acta* **1152**, 135–145.

Landwehr, A., & Winter, R. (1994). High-pressure differential thermal analysis of lamellar to lamellar and lamellar to non-lamellar lipid phase transitions. *Ber. Bunsenges. Phys. Chem.* **98**, 214–218.

Landwehr, A., Böttner, M., Hahn, W., & Winter, R. (1994). The effect of high pressure on model biomembranes. In *High-Pressure Science and Technology–1993*, ed. S. C. Schmidt, J. W. Shaner, G. A. Samara, and M. Ross. New York, AIP Press, pp. 1435–1438.

Lee, A. G. (1977). Lipid phase transitions and phase diagrams: II. Mixtures involving lipids. *Biochim. Biophys. Acta* **472**, 285–344.

Lindblom, G., & Rilfors, L. (1989). Cubic phases and isotropic structures formed by membrane lipids—possible biological relevance. *Biochim. Biophys. Acta* **988**, 221–256.

Mabrey, S., & Sturtevant, J. M. (1976). Investigation of phase transitions of lipids and lipid mixtures by high sensitivity differential scanning calorimetry. *Proc. Natl. Acad. Sci. USA* **73**, 3862–3866.

Mariani, P., Luzzati, V., & Delacroix, H. (1988). Cubic phases of lipid-containing systems. Structure analysis and biological implications. *J. Mol. Biol.* **204**, 165–189.

Marsh, D. (1991). General features of phospholipid phase transitions. *Chem. Phys. Lipids* **57**, 109–120.

Mouritsen, O. G. (1991). Theoretical models of phospholipid phase transitions. *Chem. Phys. Lipids* **57**, 179–194.

Peng, X., & Jonas, J. (1992). High-pressure ^{31}P NMR study of dipalmitoylphosphatidylcholine bilayers. *Biochemistry* **31**, 6383–6390.

Raudino, A., Zuccarello, F., La Rosa, C., & Buemi, G. (1990). Thermal expansion and compressibility coefficients of phospholipid vesicles. Experimental determination and theoretical modeling. *J. Phys. Chem.* **94**, 4217–4223.

Reis, O. (1994). Die Struktur und thermodynamische Eigenschaften binärer Phospholipidmischungen. Diploma thesis, University of Bochum.

Rilfors, L., Wieslander, A., & Stahl, S. (1978). Lipid and protein composition of membranes of Bacillus metaterium variants in the temperature range 5 to 70 °C. *J. Bacteriol.* **135**, 1043.

Rostain, J. C., Martinez, E., & Lemaire, C., Eds. (1989). *High Pressure Nervous Syndrome— 20 Years Later*. Marseille, ARAS-SNHP Publications.

Scarlata, S. F. (1991). Compression of lipid membranes as observed at varying membrane positions. *Biophys. J.* **60**, 334–340.

Schmidt, G., & Knoll, W. (1985). Densitometric characterization of aqueous lipid dispersions. *Ber. Bunsenges. Phys. Chem.* **89**, 36–43.

Seddon, J. M. (1990). Structure of the inverted hexagonal (H_{II}) phase, and non-lamellar phase transitions of lipids. *Biochim. Biophys. Acta* **1031**, 1–69.

Seddon, J. M., & Templer, R. H. (1993). Cubic phases of self-assembled amphiphilic aggregates. *Phil. Trans. R. Soc. Lond. A* **344**, 377–401.

Seddon, J. M., Hogan, J. L., Warrender, N. A., & Pebay-Peyroula, E. (1990). Structural studies of phospholipid cubic phases. *Progr. Colloid Polym. Sci.* **81**, 189–197.

Shyamsunder, E., Gruner, S. M., Tate, M. W., Turner, D. C., So, P. T. C., & Tilcock, C. P. S. (1988). Observation of inverted cubic phases in hydrated dioleoylphosphatidylethanolamine membranes. *Biochemistry* **27**, 2332–2336.

Siegel, D. (1986). Inverted micellar intermediates and the transitions between lamellar, cubic, and inverted hexagonal lipid phases: I. Mechanism of the L_α–H_{II} phase transitions. *Biophys. J.* **49**, 1155–1170.

So, P. T. C., Gruner, S. M., & Erramilli, S. (1993). Pressure-induced topological phase transitions in membranes. *Phys. Rev. Lett.* **70**, 3455–3458.

Tardieu, A., Luzzati, V., & Reman, F. C. (1973). Structure and polymorphism of the hydrocarbon chains of lipids: a study of lecithin-water phases. *J. Mol. Biol.* **75**, 711–733.

Tate, M. W., Eikenberry, E. F., Turner, D. C., Shyamsunder, E., & Gruner, S. M. (1991). Nonbilayer phases of membrane lipids. *Chem. Phys. Lipids* **57**, 147–164.

Vennemann, N., Lechner, M. D., Henkel, T., & Knoll, W. (1986). Densitometric characterization of the main phase transition of dimyristoyl-phosphatidylcholine between 0.1 and 40 MPa. *Ber. Bunsenges. Phys. Chem.* **90**, 888–891.

von Dreele, P. H. (1978). Estimation of lateral species separation from phase transitions in nonideal two-dimensional lipid mixtures. *Biochemistry* **17**, 3939–3943.

Wiener, M. C., Tristram-Nagle, S., Wilkinson, D. A., Campbell, L. E., & Nagle, J. F. (1988). Specific volumes of lipids in fully hydrated bilayer dispersions. *Biochim. Biophys. Acta* **938**, 135–142.

Winter, R., & Böttner, M. (1993). Volumetric properties of model biomembranes. In *High Pressure Chemistry, Biochemistry and Materials Science*, ed. R. Winter & J. Jonas. Dordrecht, Kluwer Academic, pp. 545–560.

Winter, R., & Pilgrim., W.-C. (1989). A SANS study of high pressure phase transitions in model biomembranes. *Ber. Bunsenges. Phys. Chem.* **93**, 708–717.

Winter, R., & Thiyagarajan, P. (1990). High-pressure phase transitions in model biomembranes. *Progr. Colloid. Polym. Sci.* **81**, 216–221.

Winter, R., Christmann, M.-H., Böttner, M., Thiyagarajan, P., & Heenan, R. K. (1991). The influence of the local anaesthetic tetracaine on the temperature and pressure dependent phases behavior of model biomembranes. *Ber. Bunsenges. Phys. Chem.* **95**, 811–820.

Winter, R., Xie, C.-L., Jonas, J., Thiyagarajan, P., & Wong, P. T. T. (1989). High-pressure small-angle neutron scattering (SANS) study of 1, 2-dielaidoly-*sn*-glycero-3-phosphocholine bilayers. *Biochim. Biophys. Acta* **982**, 85–88.

Wong, P. T. T., & Mantsch, H. H. (1989). X-ray diffraction study of the effects of pressure on bilayer to nonbilayer lipid membrane phase transitions. *J. Chem. Phys.* **90**, 1295–1296.

Wong, P. T. T., Siminovitch, D. J., & Mantsch, H. H. (1988). Structure and properties of model membranes: new knowledge from high-pressure vibrational spectroscopy. *Biochim. Biophys. Acta* **47**, 139–171.

Yager, P., & Chang, E. L. (1983). Destabilization of a lipid non-bilayer phase by high pressure. *Biochim. Biophys. Acta* **731**, 491–494.

Yeagle, P. (1992). *The Structure of Biological Membranes.* Boca Raton, CRC Press.
Zimmerman, A. M. (1978). *High Pressure Effects on Cellular Processes.* New York, Academic Press.
Zorn, R., & Nimtz, G. (1990). Influence of Ca^{2+} and Mg^{2+} ions on model membranes. *Ber. Bunsenges. Phys. Chem.* **94**, 573–578.

19

Membrane-Free Volume Variation with Bulky Lipid Concentration by Regular Distribution: A Functionally Important Membrane Property Explored by Pressure Studies of Phosphatidylcholine Bilayers

PARKSON LEE-GAU CHONG

We have used the dips in the ratio of excimer to monomer fluorescence intensity (E/M) as the index of lipid regular distribution to examine the effect of pressure on lipid lateral organization in the liquid-crystalline state of dimyristoylphosphatidylcholine (DMPC)/1-palmitoyl-2-(10-pyrenyl)decanoyl-sn-glycerol-3-phosphatidylcholine (Pyr-PC) bilayers. In the pressure range of 0.001–0.7 kbar at 30 °C, E/M dips remain discernible, suggesting that lipid regular distribution appears favorably in the liquid-crystalline state. In the same pressure range, E/M decreases steadily with increasing pressure at noncritical mole fractions; in contrast, E/M changes little with pressure at critical mole fractions. This result reveals an important physical principle underlying lipid regular distribution—that is, that membrane free volume reaches a local minimum at critical mole fractions of bulky lipids (for example, cholesterol and Pyr-PC). Using the activity of phospholipase A2 and the membrane fluidity inferred from diphenylhexatriene fluorescence polarization, we demonstrate that regular distribution of bulky lipids has a

---

Abbreviations: DMPC = L-α-dimyristoylphosphatidylcholine; DPH = diphenylhexatriene; DPPC = L-α-dipalmitoylphosphatidylcholine; MLV = multilamellar vesicles; LUV = large unilamellar vesicles; Pyr-PC = 1-palmitoyl-2-(10-pyrenyl)decanoyl-sn-glycerol-3-phosphatidylcholine; T_m = main phase transition temperature; X_{PyrPC} = mole fraction of Pyr-PC in DMPC; Y'_{PyrPC} = critical mole fraction of Pyr-PC at which the acyl chains of DMPC form regularly distributed hexagonal superlattices; Y_{PyrPC} = critical mole fraction of Pyr-PC at which the pyrene-containing acyl chains form regularly distributed hexagonal superlattices.

regulatory role in membrane properties or functions. Our data show that the activity of snake venom phospholipase A2 can be significantly modulated by minute changes (for example, 0.5 mol%) in the concentration of bulky lipids on either side of a critical mole fraction as a result of membrane free volume variation. The approach employed in this study illustrates the usefulness of high-pressure fluorescence methodology for obtaining new information about membranes at ambient conditions.

METHODOLOGY

High-pressure fluorescence spectroscopy has emerged as a powerful tool in biophysical research during the past 15 years (Paladini & Weber, 1981; Weber, 1992). Previous fluorescence studies of membranes at high pressure were focused on dynamic aspects, particularly on membrane fluidity. These dynamic studies have included probe rotations in lipid membranes (Chong & Weber, 1983; Chong et al., 1983; Mateo et al., 1993 and references cited therein), probe lateral diffusion (Flamm et al., 1982; Muller & Galla, 1983, 1987; Turley & Offen, 1985, 1986; Eisinger & Scarlata, 1987; Macdonald et al., 1988; Kao et al., 1992), spontaneous intermembrane transfer (Mantulin et al., 1984), and the relationship of membrane dynamics to the activity of membrane-bound proteins (Chong et al., 1985; Verjovski-Almeida et al., 1986; Jona & Martonosi, 1991).

While membrane dynamics at high pressure has been extensively studied by fluorescence probe techniques, as well as other physical techniques (for example, NMR, light scattering), little has been done with regard to high-pressure fluorescence studies of lipid lateral organization. This situation has arisen mainly because of the lack of a clear understanding of lipid lateral distribution at the molecular level. Although sometimes a macroscopic lipid organization can be visualized by microscopy and scattering techniques, the molecular details of lipid lateral distribution are usually not known. To conduct molecular analysis of lipid lateral organization, studies on simple binary mixtures are needed. In theory, lipids in two-component membranes can be laterally organized in three ways: domain separation, random distribution, and regular distribution (Von Dreele, 1978). A regular distribution is a lateral organization in which the guest molecules are maximally separated in the lipid matrix.

Pyrene fluorescence has been useful in studying lipid lateral distribution. For example, the abrupt change in the ratio of the excimer to monomer fluorescence intensity (E/M) of 1-palmitoyl-2-(10-pyrenyl)decanoyl-sn-glycerol-3-phosphatidyl-choline (Pyr-PC) in lipid bilayers was interpreted in terms of temperature-induced changes in lipid lateral distribution (Galla & Hartmann, 1980; Jones & Lentz, 1986). From using oxygen quenching to determine the excimer formation constant, Chong and Thompson (1985) concluded that binary mixtures composed of Pyr-PC and 1-palmitoyl-2-oleoyl-L-α-phosphatidylcholine (POPC) form a randomly distributed liquid-crystalline system in the temperature range of 15–55 °C, whereas binary mixtures of pyrene-labeled sphingomyelin/POPC form domains. On the basis of E/M and fluorescence lifetime phase/modulation data, Hresko et al. (1986) suggested that Pyr-PC is randomly distributed in both L-α-dimyristoylphos-phatidylcholine (DMPC) and L-α-dipalmitoylphosphatidylcholine (DPPC) at temperatures outside the phase transition regions. On the basis of lipid transfer data, Roseman and Thompson (1980) also suggested that Pyr-PC is randomly distributed in the DMPC matrix at temperatures greater than the main phase transition temperature of DMPC.

In addition to random and domain distributions, regular distributions in lipid membranes of Pyr-PC has also been suggested. Somerharju et al. (1985) showed that the plot of E/M versus the mole fraction of Pyr-PC in egg yolk phosphatidyl-choline and in DPPC exhibits several linear regions separated by kinks. The appearance of E/M kinks was interpreted in terms of the regular distribution of Pyr-PC into hexagonal superlattices. The hexagonal superlattice model (Somerharju et al., 1985; Mustonen et al., 1987; Virtanen et al., 1988) proposed that (1) the acyl chains of the phospholipid molecules form a hexagonal host lattice, (2) the pyrene-containing acyl chains are guest elements, which are bulky and cause steric perturbation in the host lattice, and (3) the guest elements maximally separate to minimize the total energy. According to Ruocco and Shipley (1982), the acyl chains of phospholipids can be arranged into a quasi two-dimensional hexagonal lattice. For a given acyl chain, its position can be described by two coordinates (n_a, n_b) once the origin and the principal axes, a and b, have been defined. In the regular distribution, Pyr-PC molecules may establish a hexagonal superlattice within the host lattice. The critical Pyr-PC mole fraction, $Y_{PyrPC}(n_a, n_b)$, at which the pyrene-containing acyl chains form regularly distributed hexagonal superlattices can be calculated by the equation (Kinnunen et al., 1987; Virtanen et al., 1988)

$$Y_{PyrPC}(n_a, n_b) = 2/(n_a^2 + n_a n_b + n_b^2) \tag{1}$$

Although the E/M kinks observed by Somerharju et al. (1985) seem to agree with Eq. 1, only a few E/M kinks were observed.

By using a three-state photophysical model and a global analysis of the E/M and phase/modulation lifetime data, we show that the lateral distribution parameter w deviates negatively from ideal mixing, which suggests that Pyr-PC molecules form regular distributions in the Pyr-PC/DMPC binary mixtures (Sugar et al., 1991). The three-state photophysical model overcomes the inadequacy of the two-state model constructed by Birks et al. (1963); thus the new conclusion is significant. However, a negative w value does not indicate whether the regular distribution is in the form of a hexagonal superlattice. Nevertheless, the approach of the three-state model and the global analysis provides additional evidence, independent of the E/M kinks (Somerharju et al., 1985), that lipids in Pyr-PC/DMPC mixtures are regularly distributed.

The most compelling evidence for lipid regular distribution comes from our recent observations of E/M dips in Pyr-PC/DMPC binary mixtures. A series of dips, in addition to kinks, were observed in the plot of E/M versus the mole fraction of DMPC, X_{PyrPC}, at 30 °C (Tang & Chong, 1992). The E/M dips were interpreted in terms of the extended hexagonal superlattice model (Tang & Chong, 1992). According to this model, the E/M dips/kinks below 66.7 mol% Pyr-PC (excluding the dip/kink at 33.3 mol%) were formed as a result of the pyrene-containing acyl chains being regularly distributed in the DMPC lipid matrix. The dip positions in this concentration region agree with Eq. 1. On the other hand, the dips above

66.7 mol% plus the dip/kink at 33.3 mol% are described by Eq. 2 (Tang & Chong, 1992):

$$Y'_{PyrPC}(m_a, m_b) = 1 - (2/(m_a^2 + m_a m_b + m_b^2)) \tag{2}$$

where Y'_{PyrPC} is the critical mole fraction of Pyr-PC at which the acyl chains of DMPC form regularly distributed hexagonal superlattices, and m_a and m_b are the projections along axes a and b, respectively, for an acyl chain of DMPC in the superlattice. The good agreement between the observed dips and the Y or Y' values predicted from Eqs. 1 and 2 provides compelling evidence that lipids in the Pyr-PC/DMPC binary mixtures are regularly distributed.

Imperfect superlattices due to thermal fluctuations, impurities, and/or variations in membrane curvature are expected to occur under ambient conditions. The coexistence of regular regions with irregular regions explains why E/M does not go to zero at critical Pyr-PC mole fractions (Tang & Chong, 1992). This idea has been corroborated by computer simulations (Sugar et al., 1994), which further indicate that the ratio of regular areas (where regular patterns can be recognized from the snapshot of simulated lateral distributions) to irregular areas (where regular patterns cannot be recognized) reaches a local maximum at critical mole fractions and a minimum between two neighboring critical mole fractions.

It is believed that the deformation in the hexagonal lattice caused by the bulky pyrene ring generates a long-range repulsive interaction (Sugar et al., 1994; Chong et al., 1994) between pyrene-containing acyl chains. This long-range repulsion is the physical origin for the maximal separation of Pyr-PC molecules in the membrane, which leads to the drop of the E/M value at critical mole fractions.

In the present study, we used the E/M dips as the index of lipid regular distribution to examine the effect of pressure on lipid lateral organization in Pyr-PC/DMPC multilamellar vesicles. Fluorescence measurements under pressure were made isothermally on an SLM high-pressure optical cell mounted in an SLM DMX-1000 fluorometer (SLM Instruments, Urbana, IL). The excitation wavelength was 342 nm. E/M was determined using the intensities at 478 nm and 378 nm, respectively. The lipid concentration used for fluorescence measurements was $\sim 2 \times 10^{-6}$ M. The concentration of Pyr-PC was determined from an excitation coefficient at 342 nm equal to 42,000 M^{-1} cm^{-1} (in methanol) (Somerharju et al., 1985). The phospholipid concentration was determined by the method Bartlett (1959). Vesicles were prepared according to the procedures described in Chong et al. (1994).

Our results suggest that regular distribution persists in the liquid-crystalline state of Pyr-PC/DMPC mixtures and that the membrane free volume varies with lipid fractional composition, with a local minimum at the critical mole fraction and a maximum between two adjacent critical mole fractions. This periodic variation in free volume has been shown to modulate membrane fluidity and the activity of snake venom phospholipase A2.

RESULTS AND DISCUSSION

Figure 19.1 shows the effects of pressure on E/M dips in Pyr-PC/DMPC multilamellar vesicles at 30 °C (Chong et al., 1994). E/M dips are clearly discernible

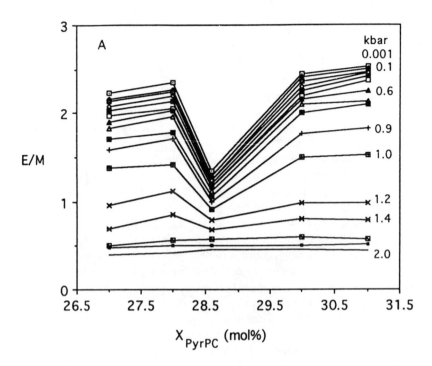

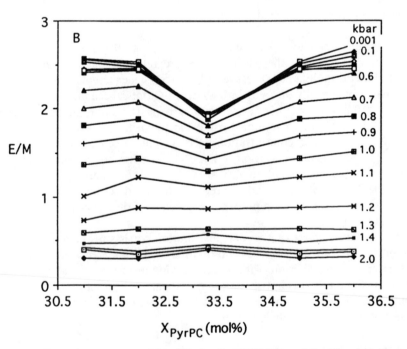

Figure 19.1. Effect of pressure on E/M dips for Pyr-PC/DMPC multilamellar vesicles in the neighborhood of a critical Pyr-PC mole fraction: (**A**) 28.6 mol%, (**B**) 33.3 mol%, (**C**) 50 mol%, and (**D**) 66.7 mol%. Temperature = 30 °C (Chong et al., 1994).

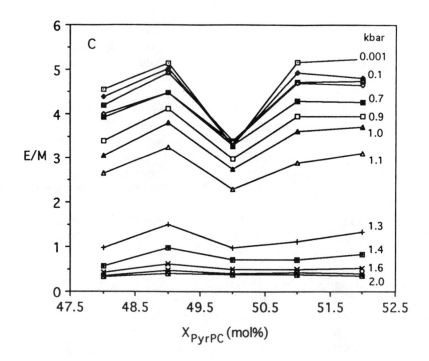

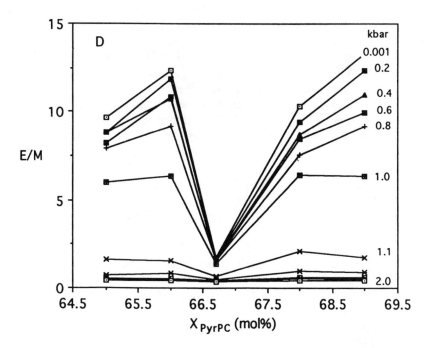

Figure 19.1 Continued

at 28.6, 33.3, 50, and 66.7 mol% Pyr-PC in the pressure range 0.001–0.7 kbar. At much higher pressures, E/M dips are not detected. The dip positions agree with the critical mole fractions Y or Y' predicted from Eqs. 1 and 2. In the pressure range 0.001–0.7 kbar, the depth of the dip decreases only slightly with increasing pressure. These results suggest that lipid regular distribution persists in the liquid-crystalline state.

Figure 19.2A shows the effects of pressure on the E/M values of Pyr-PC in DMPC multilamellar vesicles at 30 °C (Chong et al., 1994). E/M undergoes an abrupt decrease at elevated pressures with a 50% change in E/M at about 1.1 kbar. This abrupt decrease is believed to be due to the pressure-induced phase transition from the liquid-crystalline state to the gel state. At pressures below 0.7 kbar, the slope of E/M versus pressure at noncritical mole fractions (for example, 48, 49, 52, 53 mol%) is significantly greater than that at the critical mole fraction (that is, 50 mol%).

A similar trend is seen in Pyr-PC/DMPC mixtures in the neighborhood of 33.3 mol% (Figure 19.2B) (Chong et al., 1994). An abrupt decrease in E/M occurs between 0.6 and 1.2 kbar. Below 0.5 kbar, E/M decreases steadily with increasing pressure at noncritical mole fractions (for example, 31, 32, 35, 36 mol%); within the same pressure range, E/M at the critical mole fraction increases only slightly with pressure. The curves in Figure 19.2 exhibit a sharp transition at around 1.0 kbar, with a break point at low and high pressures.

Using the low-pressure and high-pressure break points obtained from various Pyr-PC mole fractions, a pressure-composition phase diagram-like curve is constructed for Pyr-PC in DMPC at 30 °C (Figure 19.3).

Normally, the onset and completion temperatures determined from a differential scanning calorimetric curve represent the beginning and completion of the trans-gauche isomerization in the lipid acyl chains, whereas the breakpoints in the plot of E/M versus temperature (Chong et al., 1994) reflect the lateral reorganization of the lipids. So the curve shown in Figure 19.3 can be considered only as the pressure-composition "phase diagram" with regard to lipid lateral organization. Note that this diagram may differ from the pressure-composition phase diagrams determined by other techniques. The temperature-pressure-composition phase diagram of Pyr-PC/DMPC is not available in the literature. However, at 30 °C and at pressures below 0.7 kbar, pure DMPC is in the liquid-crystalline state according to the phase diagram constructed by Winter and Pilgrim (1989). Pyr-PC is likely to lower the T_m of the Pyr-PC/DMPC mixture to below that of DMPC (24 °C) since its T_m is lower (14.5 °C) (Somerharju et al., 1985). This suggests that at 30 °C and at pressures below 0.7 kbar, Pyr-PC/DMPC is most likely to be in the liquid-crystalline state. The results in Figure 19.1 show that, in the pressure range 0.001–0.7 kbar, the E/M dips remain discernible; the depth of the dip decreases only slightly with increasing pressure, and the dip position remains virtually unchanged. These pressure results, combined with temperature results (Chong et al., 1994), suggest that lipid regular distribution appears favorably in the liquid-crystalline state.

The pressure data (Figures 19.2A & 19.2B) in the liquid-crystalline state (< 0.7 kbar at 30 °C) show a sharp contrast between membranes at critical mole fractions (33.3 mol% and 50 mol%) and membranes at noncritical mole fractions.

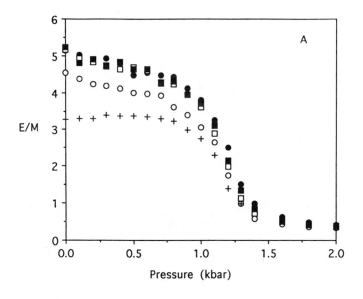

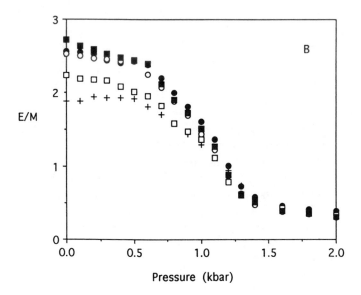

Figure 19.2. Effect of pressure on E/M in Pyr-PC/DMPC multilamellar vesicles at 30 °C: (**A**) mol%
Pyr-PC: 48 (○), 49 (●), 50 (+), 52 (□), and 53 (■); (**B**) mol% Pyr-PC: 31 (○), 32
(●), 33.3 (+), 35 (□), 36 (■) (Chong et al., 1994).

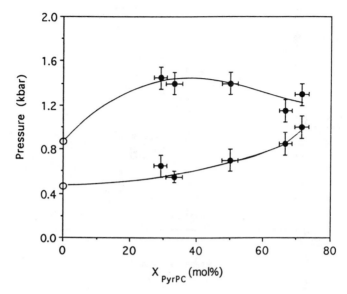

Figure 19.3. The pressure-composition phase diagram with regard to lipid lateral distribution for Pyr-PC/DMPC multilamellar vesicles at 30 °C. The upper and lower curves represent the low-pressure and high-pressure break points, respectively, in the plot of E/M versus pressure. The data (○) for pure DMPC are derived from the phase transition of pure DMPC determined by infrared spectroscopy (Chong et al., 1989).

At noncritical Pyr-PC mole fractions, E/M initially decreases steadily with increasing pressure, then decreases abruptly. In contrast, at critical mole fractions, E/M changes little with pressure before the abrupt decrease in E/M occurs. The steady decrease in E/M with pressure observed at noncritical mole fractions below 0.7 kbar is similar to the results that have been previously reported in various other membrane systems (Flamm et al., 1982; Muller & Galla, 1983, 1987; Turley & Offen, 1985, 1986; Eisinger & Scarlata, 1987; Macdonald et al., 1988; Kao et al., 1992). In those studies, the decrease in E/M with pressure was interpreted in terms of a pressure-induced decrease in membrane fluidity/membrane free volume which in turn gives rise to a decreased lateral diffusion. Note that the lateral diffusion of lipids in membranes has been shown (Galla et al., 1979; Vaz & Hallman, 1983; Peters & Beck, 1983; Vaz et al., 1985; King & Marsh, 1986) to follow the free volume model (Cohen & Turnbull, 1959; Galla et al., 1979; Muller & Galla, 1983; Vaz & Hallmann, 1983) and that membrane free volume $V_f = V\alpha(T - T_m)$, where V is the average volume of the lipid above the phase transition temperature T_m, and α is the thermal expansion coefficient (Galla et al., 1979).

The striking difference in the pressure dependence of E/M between membranes at critical and noncritical mole fractions can be understood in terms of changes in lipid lateral organization from a noncritical to a critical mole fraction. Previous studies have established that increased pressure reduces membrane volume in the liquid-crystalline state of lipid bilayers (Liu & Kay, 1977; Tosh & Collings, 1986). This pressure-induced decrease in volume should have two effects on Pyr-PC/DMPC systems: (1) pressure reduces the lateral diffusion of lipids via the reduction in membrane free volume, thus decreasing the E/M value, and (2) pressure shortens the intermolecular distance, thus increasing the E/M value. At critical mole fractions, membranes possess less free volume as a result of the higher ratio of regular area to irregular area. In this case, a decrease in E/M due to factor (1) may be offset by an increase in E/M due to factor (2) such that, overall, the E/M changes very little with pressure (50 mol% Pyr-PC in Figure 19.2A and 33.3 mol% Pyr-PC in Figure 19.2B at pressures below 0.7 kbar). At noncritical mole fractions, membranes possess more free volume because of the lower ratio of regular to irregular area. In this case, factor (1) should dominate and, consequently, E/M decreases steadily with increasing pressure.

Our pressure data suggest that membrane free volume is less at critical Pyr-PC mole fractions than at noncritical mole fractions. This indicates that membrane free volume varies with Pyr-PC mole fraction in a periodic manner, since several critical mole fractions occur over a wide range of Pyr-PC concentrations (Tang & Chong, 1992). This result parallels those from computer simulations which show that the ratio of regular region to irregular region reaches a local maximum at critical Pyr-PC mole fractions and a local minimum between two neighboring critical mole fractions (Sugar et al., 1994). These results are completely consistent with the concept of lipid regular distribution because the regular region requires a more stringent lateral arrangement to maintain the hexagonal superlattice, whereas the irregular region allows membrane defects or void space due to imperfect interfacial contact between regular regions.

Pyr-PC-like molecules (for example, cholesterol and dehydroergosterol), which are bulkier than bilayer phospholipids, do exist in biological membranes. In fact, we have recently shown that the lipids in dehydroergosterol/DMPC mixtures are also regularly distributed in a way similar to those in Pyr-PC/DMPC mixtures (Chong, 1994). Although we have not carried out pressure studies on sterol/DMPC mixtures, the physical principles of lipid regular distribution derived from Pyr-PC/DMPC systems can be extended to sterol/PC mixtures. That is to say that membrane free volume in sterol/PC mixtures should also reach a local minimum at critical mole fractions and a maximum between two consecutive critical sterol mole fractions.

The periodic variation of membrane free volume with the mole fraction of bulky lipids is of great biological significance because many membrane properties or functions such as membrane fluidity, lipid lateral diffusion (Galla et al., 1979; Vaz & Hallman, 1983; Peters & Beck, 1983; Vaz et al., 1985; King & Marsh, 1986), membrane fusion (Hui et al., 1981), the activity of phospholipase A2 (Upreti & Jain, 1980), and spontaneous transfer of lipids between membranes (Wimley & Thompson, 1991) have been previously suggested to be related to membrane free volume or membrane defects.

The biophysical significance of membrane free volume variation with bulky lipid concentration has been tested by using the commonly used membrane probe, diphenylhexatriene (DPH), in cholesterol/DMPC multilamellar vesicles. It is well established that the steady-state polarization of DPH fluorescence reflects mainly the molecular order of lipid acyl chains. If membrane free volume reaches a local minimum at a critical sterol mole fraction, then DPH polarization should be higher at critical cholesterol mole fractions than at their neighboring noncritical mole

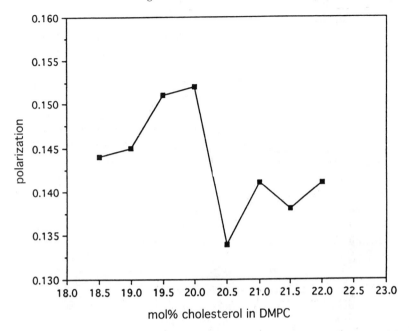

Figure 19.4. The steady-state fluorescence polarization of DPH in cholesterol/DPMC multilamellar vesicles at 33 °C as a function of the cholesterol mole fraction. Excitation wavelength = 355 nm. Emission was collected through a KV417 cutoff filter. Vesicles were prepared as previously described (Chong et al., 1994).

fractions. As shown in Figure 19.4, the steady-state DPH polarization does indeed vary with cholesterol mole fraction and exhibits a local maximum at 20 mol% cholesterol. This concentration is the critical mole fraction predicted for regularly distributed cholesterol in lipid membranes (Chong, 1994).

The results reported here suggest that there is a need to reevaluate membrane properties in the liquid-crystalline state of binary mixtures containing bulky groups such as cholesterol. Physical properties such as compressibility and membrane fluidity need to be examined at small concentration intervals, with special attention focused near critical mole fractions. A general trend from low sterol mole fraction to high sterol mole fraction represents a *global* change in the physical property of interest. Thus, the increased DPH polarization that accompanies an increase in cholesterol content from 0 to 50 mol%, as previously shown by many investigators, is a global effect of lipid acyl chain ordering by cholesterol. However, many critical mole fractions occur between 0 and 50 mol% (Chong, 1994). At a critical mole fraction, a change in physical property is expected to occur (for example, the DPH polarization at around 20 mol% cholesterol in Figure 19.4); this phenomenon is a *local* concentration effect. At physiological conditions, cholesterol content in the membrane may not undergo a huge global change; instead, small changes in cholesterol concentration near the critical mole fraction may play a far more significant role in modulating membrane functions.

Whether membrane free volume variation with bulky lipid mole fraction has a biological significance has been tested by using Pyr-PC/DMPC large unilamellar vesicles (average diameter ~400 nm) as the substrate for snake venom phospho-lipase A2. Phospholipase A2 (EC 3.1.1.4) is a water-soluble enzyme that catalyzes the hydrolysis of the *sn*-2 fatty acid ester of phosphatidylcholine to lysophospha-tidylcholine and fatty acid. It has been proposed that phospholipase A2 binding is enhanced by membrane defects (Upreti & Jain, 1980). According to our pressure data, membrane free volume/membrane defect reaches a local minimum at a critical Pyr-PC mole fraction. Thus, it is expected that the activity of phospholipase A2 is lowest at critical mole fractions. Figure 19.5 shows the activity of *Crotalus*

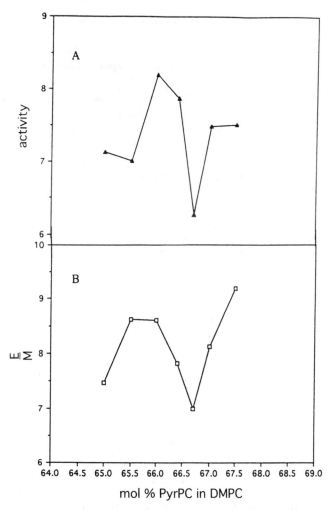

Figure 19.5. (**A**) Activity of phospholipase A2 as a function of Pyr-PC mole fraction in DMPC large unilamellar vesicles (LUV, diameter = ~400 nm) at 37 °C. The activity is expressed in terms of the percentage change in monomer fluorescence per second. The sample volume in the cuvette was 1.4 ml at a total lipid concentration of 1.9–2.1 µM. The amount of phospholipase A2 added was 94.5 ng in 5 ml buffer (0.1 M Tris, 1 mM CaCl$_2$, pH = 7.4 (at 37 °C). (**B**) Plot of E/M versus Pyr-PC mole fraction in DMPC(LUV) at 37 °C.

durissus terrificus venom phospholipase A2 (obtained from Sigma) as a function of Pyr-PC mole fraction at 37 °C. The activity is expressed in terms of the percentage change in monomer fluorescence per second according to the method of Radvanyi et al. (1989). In brief, once the hydrolysis is initiated by adding an aliquot of enzyme in 0.1 M Tris, 1 mM CaCl$_2$, pH 7.4 (at 37 °C) into the substrate, the percentage increase in monomer fluorescence of Pyr-PC is monitored at 378 nm as a function of time. The slope of the intensity change in the first minute represents the initial hydrolysis rate of the substrate lipids. It is observed that this hydrolysis activity reaches a local minimum at the critical mole fraction 66.7 mol% (Figure 19.5, ▲) where the E/M versus mole fraction profile (Figure 19.5, □) also shows a minimum at this critical mole fraction. This correlation suggests that the activity of phospholipase A2 is modulated by bulky lipid fractional concentration via the free volume changes due to the extent of the regularly distributed area.

As demonstrated in Figure 19.5, a critical mole fraction acts like a switch which increases significantly the activity of phospholipase A2 by minute changes (for example, 1 mol%) of Pyr-PC content on either side of the critical mole fraction. Since many critical concentrations occur over a wide range of Pyr-PC mole fraction (Tang & Chong, 1992), the activity of phospholipase A2 should, in principle, be modulated up and down periodically by Pyr-PC content according to the relative size of the regular area.

In conclusion, our pressure data reveal an important physical principle underlying lipid regular distribution—that is, that membrane free volume reaches a local minimum at critical mole fractions of bulky lipids. On the basis of the activity of phospholipase A2 and the membrane fluidity inferred from DPH fluorescence polarization, we demonstrate that lipid regular distribution has a regulatory role in membrane properties or functions. The approach employed in this study illustrates the usefulness of high-pressure fluorescence methodology in obtaining information on membranes at ambient conditions.

ACKNOWLEDGMENTS: We thank Ruth Rusch, Khanh Truong, and Daxin Tang for technical assistance. This research was supported in part by the Navy Research Office and in part by the American Heart Association.

REFERENCES

Bartlett, G. R. (1959). Phosphorus assay in column chromatography. *J. Biol. Chem.* **234**, 466–468.

Birks, J. B., Dyson, D. J., & Munro, I. H. (1963). Excimer fluorescence II: lifetime studies of pyrene solutions. *Proc. R. Soc. London. Ser. A.* **275**, 575–588.

Chong, P. L.-G. (1994). Evidence for regular distribution of sterols in liquid crystalline phosphatidylcholine bilayers. *Proc. Natl. Acad. Sci. USA.* **91**, 10069–10073.

Chong, P. L.-G., & Thompson, T. E. (1985). Oxygen quenching of pyrene-lipid fluorescence in phosphatidylcholine vesicles: a probe for membrane organization. *Biophys. J.* **47**, 613–621.

Chong, P. L.-G., & Weber, G. (1983). Pressure dependence of 1,6-diphenyl-1,3,5-hexatrene fluorescence in single-component phosphatidylcholine liposomes. *Biochemistry* **22**, 5544–5550.

Chong, P. L.-G., Capes, S., & Wong, P. T. T. (1989). Effects of hydrostatic pressure on the location of PRODAN in lipid bilayers: a FTIR study. *Biochemistry* **28**, 8358–8363.

Chong, P. L.-G., Cossins, A. R., & Weber, G. (1983). Differential polarized phase fluorometric study of the effects of high hydrostatic pressure upon the fluidity of cellular membranes. *Biochemistry* **22**, 409–415.

Chong, P. L.-G., Fortes, P. A. G., & Jameson, D. M. (1985). Mechanisms of inhibition of (Na, K)-ATPase by hydrostatic pressure studied with fluorescent probes. *J. Biol. Chem.* **260**, 14484–14490.

Chong, P. L.-G., Tang, D., & Sugar, I. P. (1994). Exploration of physical principles underlying lipid regular distribution: effects of pressure, temperature and radius of curvature on E/M dips in pyrene-labeled PC/DMPC binary mixtures. *Biophys. J.* **66**, 2029–2038.

Cohen, M. H., & Turnbull, D. (1959). Molecular transport in liquid and glasses. *J. Chem. Phys.* **31**, 1164–1169.

Eisinger, J., & Scarlata, S. F. (1987). The lateral fluidity of erythrocyte membranes: temperature and pressure dependence. *Biophys. Chem.* **28**, 273–281.

Flamm, M., Okubo, T., Turro, N. J., & Schachter, D. (1982). Pressure dependence of pyrene excimer fluorescence in human erythrocyte membranes. *Biochim. Biophys. Acta* **687**, 101–104.

Galla, H.-J., & Hartmann, E. (1980). Excimer-forming lipids in membrane research. *Chem. Phys. Lipids* **27**, 199–219.

Galla, H.-J., Hartmann, W., Theilen, U., & Sackmann, E. (1979). On two-dimensional passive random walk in lipid bilayers and fluid pathways in biomembranes. *J. Membr. Biol.* **48**, 215–236.

Hresko, R. C., Sugar, I. P., Barenholz, Y., & Thompson, T. E. (1986). Lateral distribution of a pyrene-labeled phosphatidylcholine in phosphatidylcholine bilayers: fluorescence phase and modulation study. *Biochemistry* **25**, 3813–3823.

Hui, S. W., Stewart, T. P., & Boni, L. T. (1981). Membrane fusion through point defects in bilayers. *Science* **212**, 921–923.

Jona, I., & Martonosi, A. (1991). The effect of high pressure on the conformation, interactions and activity of the Ca^{2+}-ATPase of sarcoplasmic reticulum. *Biochim. Biophys. Acta* **1070**, 355–373.

Jones, M. E., & Lentz, B. R. (1986). Phospholipid lateral organization in synthetic membranes as monitored by pyrene-labeled phospholipids: effects of temperature and prothrombin fragment 1 binding. *Biochemistry* **25**, 567–574.

Kao, Y. L., Chang, E. L., & Chong, P. L.-G. (1992). Unusual pressure dependence of the lateral motion of pyrene-labeled phosphatidylcholine in bipolar lipid vesicles. *Biochem. Biophys. Res. Comm.* **188**, 1241–1246.

King, M. D., & Marsh, D. (1986). Free volume model for lipid lateral diffusion coefficients. Assessment of the temperature dependence in phosphatidylcholine and phosphatidylethanolamine bilayers. *Biochim. Biophys. Acta.* **862**, 231–234.

Kinnunen, P. K. J., Tulkki, A., Lemmetyinen, H., Paakkola, J., & Virtanen, A. (1987). Characters of excimer formation in Langmuir-Blodgett assemblies of 1-palmitoyl-2-pyrenedecanoylphosphatidylcholine and dipalmitoylphosphatidylcholine. *Chem. Phys. Lett.* **136**, 539–545.

Liu, N.-I., & Kay, R. L. (1977). Redetermination of the pressure dependence of the lipid bilayer phase transition. *Biochemistry* **16**, 3484–3486.

Macdonald, A. G., Wahle, K. W. J., Cossins, A. R., & Behan, M. K. (1988). Temperature, pressure and cholesterol effects on bilayer fluidity; a comparison of pyrene excimer/monomer ratios with the steady-state fluorescence polarization of diphenylhexatriene in liposomes and microsomes. *Biochim. Biophys. Acta* **938**, 231–242.

Mantulin, W. W., Gotto, A. M., & Pownall, H. J. (1984). Effect of hydrostatic pressure on the transfer of a fluorescent phosphatidylcholine between apolipoprotein-phospholipid recombinants. *J. Am. Chem. Soc.* **106**, 3317–3319.

Mateo, C. R., Tauc, P., & Brochon, J.-C. (1993). Pressure effects on the physical properties of lipid bilayers detected by trans-parinaric acid fluorescence decay. *Biophys. J.* **65**, 2248–2260.

Muller, H.-J., & Galla, H.-J. (1983). Pressure variation of the lateral diffusion in lipid bilayer membranes. *Biochim. Biophys. Acta* **733**, 291–294.

Muller, H.-J., & Galla, H.-J. (1987). Chain length and pressure dependence of lipid translational diffusion. *Eur. Biophys. J.* **14**, 485–491.

Mustonen, P., Virtanen, J. A., Somerharju, P. J., & Kinnunen, P. K. J. (1987). Binding of cytochrome *c* to liposomes as revealed by the quenching of fluorescence from pyrene-labeled phospholipids. *Biochemistry* **26**, 2991–2997.

Paladini, A. A., & Weber, G. (1981). Absolute measurements of fluorescence polarization at high pressures. *Rev. Sci. Instrum.* **52**, 419–427.

Peters, R., & Beck, K. (1983). Translational diffusion in phospholipid monolayers measured by fluorescence microphotolysis. *Proc. Natl. Acad. Sci. USA* **80**, 7183–7187.

Radvanyi, F., Jordan, L., Russo-Marie, F., & Bon, C. (1989). A sensitive and continuous fluorometric assay for phospholipase A2 using pyrene-labeled phospholipids in the presence of serum albumin. *Anal. Biochem.* **177**, 103–109.

Roseman, M. A., & Thompson, T. E. (1980). Mechanism of the spontaneous transfer of phospholipids between bilayers. *Biochemistry* **19**, 439–444.

Ruocco, M. J., & Shipley, G. G. (1982). Characterization of the subtransition of hydrated dipalmitoylphosphocholine bilayers: kinetics, hydration and structural studies. *Biochim. Biophys. Acta* **691**, 309–320.

Somerharju, P. J., Virtanen, J. A., Eklund, K. K., Vainio, P., & Kinnunen, P. K. J. (1985). 1-Palmitoyl-2-pyrenedecanoyl glycerophospholipids as membrane probes: evidence for regular distribution in liquid-crystalline phosphatidylcholine bilayers. *Biochemistry* **24**, 2773–2781.

Sugar, I. P., Zeng, J., & Chong, P. L.-G. (1991). Use of Fourier transforms in the analysis of fluorescence data: 3. Fluorescence of pyrene-labeled phosphatidylcholine in lipid bilayer membrane: a three-state model. *J. Phys. Chem.* **95**, 7524–7534.

Sugar, I. P., Tang, D., & Chong, P. L.-G. (1994) Monte Carlo simulation of lateral distribution of molecules in a two-component lipid membrane: effect of long-range repulsive interactions. *J. Phys. Chem.* **98**, 7201–7210.

Tang, D., & Chong, P. L.-G. (1992). E/M dips: evidence for lipids regularly distributed into hexagonal super-lattices in pyrene-PC/DMPC binary mixtures at specific concentrations. *Biophys. J.* **63**, 903–910.

Tosh, R. E., & Collings, P. J. (1986). High pressure volumetric measurements in dipalmitoyl-phosphatidylcholine bilayers. *Biochim. Biophys. Acta* **859**, 10–14.

Turley, W. D., & Offen, H. W. (1985). Fluorescence detection of gel-gel phase transitions in DMPC vesicles at high pressures. *J. Phys. Chem.* **89**, 3962–3964.

Turley, W. D., & Offen, H. W. (1986). Lipid microviscosity of DMPC vesicles at high pressures: dipyrenylpropane excimer fluorescence. *J. Phys. Chem.* **90**, 1967–1970.

Upreti, G. C., & Jain, M. K. (1980). Action of phospholipase A2 on unmodified phosphatidylcholine bilayers: organizational defects are preferred sites of action. *J. Membrane Biol.* **55**, 113–123.

Vaz, W. L. C., & Hallmann, D. (1983). Experimental evidence against the applicability of the Saffman-Delbruck model to the translational diffusion of lipids in phosphastidyl-choline bilayer membranes. *FEBS Lett.* **152**, 287–290.

Vaz, W. L. C., Clegg, R. M., & Hallmann, D. (1985). Translational diffusion of lipids in liquid crystalline phase phosphatidylcholine multilayers: a comparison of experiment with theory. *Biochemistry* **24**, 781–786.

Verjovski-Almeida, S., Kurtenbach, E., Amorim, A. F., & Weber, G. (1986). Pressure-induced dissociation of solubilized sarcoplasmic reticulum ATPase. *J. Biol. Chem.* **261**, 9872–9878.

Virtanen, J. A., Somerharju, P., & Kinnunen, P. K. J. (1988). Prediction of patterns for the regular distribution of soluted guest molecules in liquid crystalline phospholipid membranes. *J. Mol. Electr.* **4**, 233–236.

Von Dreele, P. H. (1978). Estimation of lateral species separation from phase transitions in nonideal two-dimensional lipid mixtures. *Biochemistry* **17**, 3939–3943.

Weber, G. (1992). Effects of temperature and pressure on molecular associations and on single peptide chain proteins. In *Protein Interactions*. London, Chapman and Hall, pp. 199–270.

Wimley, W. C., & Thompson, T. E. (1991). Transbilayer and interbilayer phospholipid exchange in dimyristoylphosphatidylcholine/dimyristoylphosphatidylethanomine large unilamellar vesicles. *Biochemistry* **30**, 1702–1709.

Winter, R., & Pilgrim, W.-C. (1989). A SANS study of high pressure phase transitions in model biomembranes. *Ber. Bunsenges. Phys. Chem.* **93**, 708–717.

20

NMR Studies of the Order
and Dynamics of
Dipalmitoylphosphatidylcholine
Bilayers as a Function of Pressure

ANA JONAS, XIANGDONG PENG,
BAO-SHIANG LEE, STEPHANIE SCHWER, and JIRI JONAS

We have used ^{2}H NMR methods to examine the order and dynamics of dipalmitoylphosphatidylcholine (DPPC) in multilamellar and small unilamellar vesicles in water as a function of pressure. Multipulse ^{2}H NMR techniques were used with selectively deuterated DPPC on both chains at positions C-2, C-9, or C-13, to obtain lineshapes, spin-lattice relaxation times (T_1), and spin-spin relaxation times (T_2) at 50 °C from 1 bar to 5.2 kbar pressure. This pressure range allowed us to explore the phase behavior of DPPC from the liquid crystalline (LC) phase through various gel phases (GI, GII, GIII, GX), including the interdigited Gi phase. Pressure has an ordering effect on all chain segments in all the phases. In the LC phase, the order parameter (S_{CD}) decreases from C-2 > C-9 > C-13, while in the gel phases S_{CD} decreases from C-9 > C-13 > C-2, indicating that in the gel phases the middle segments of the chains are more restricted in their motions than the ends. In the LC phase, T_1 and T_2 values for all segments decrease with pressure and have an order from C-13 > C-9 > C-2. These results suggest that similar conformational motions and molecular rotational motions occur in the LC state in all segments, but have increased amplitudes and frequencies toward the methyl ends. At the phase transitions, discontinuities and abrupt reversal of the slopes for the T_1 or T_2 dependences on pressure indicate major changes in motional modes and rates for DPPC molecules in the different structures. In the second part of this study, we have measured the lateral diffusion of DPPC in sonicated vesicles in D_2O as a function of pressure. The spin-lattice relaxation rate in the rotating frame $T_{1\rho}^{-1}$ was plotted as a function of the square root of the spin-locking field angular frequency $(\omega_1)^{1/2}$, and the lateral diffusion coefficient (D) was calculated from the slope. Pressure effects are observed on lateral diffusion in the LC phase ($D = 5.4 - 2 \times 10^{-9}$ cm^2 seconds, from 1 to 300 bar) but are negligible in the Gl phase ($D \approx 1.0 \times 10^{-9}$ cm^2 seconds, from 400 to 800 bar).

Phospholipid bilayers and monolayers have important roles in nature as components of cell membranes and lipoproteins. In cell membranes, phospholipid bilayers constitute the permeability barriers between aqueous compartments. In lipoproteins, phospholipid monlayers solubilize lipids for transport in blood. Aside from their structural interfacial roles, aggregated phospholipids modulate the functions of associated proteins and enzymes by direct binding and by physical effects. Therefore, synthetic phospholipid bilayers, in multilamellar or small bilayer vesicle forms, have become models for the study of the structural and dynamic propoerties of natural phospholipid aggregates.

Since the 1970s, NMR methods, in particular ^{2}H NMR, have been applied very effectively to the investigation of the biophysical properties of phospholipid bilayers. Seelig and Seelig (1974) and Seelig (1977) first described the determination of order parameters in different segments of the acyl chains of dipalmitoylphosphatidylcholine (DPPC) by ^{2}H NMR. They demonstrated clearly that the acyl chains of DPPC become progressively more disordered from the glycerol backbone toward the methyl ends. In addition, they confirmed that both chains have nonequivalent conformations in the liquid crystalline (LC) bilayers near the glycerol backbone. Subsequent studies exploited the dramatic line-shape changes of ^{2}H NMR spectra of perdeuterated DPPC acyl chains to follow the thermotropic phase transitions of this phospholipid (Davis, 1979), and measured the spin-lattice (T_1) (Brown et al., 1979) and spin-spin relaxation times (T_2) (Bloom & Sternin, 1987; Mayer et al., 1988) of selectively deuterated DPPC to assess the structural dynamics of bilayers. Although attempts have been made in these studies to develop theoretical interpretations of the relaxation times in terms of the correlation times (τ_c) of the involved motions, a general theoretical model is not available yet because of the great complexity of the molecular motions in bilayers. Current NMR pulse sequences allow one to probe motions with τ_c from 10^{-11} to 10^{-4} seconds. For phospholipid bilayers, this time range encompasses *trans-gauche* isomerizations and segmental chain motions, rotation and wobbling of the entire molecule around the bilayer normal axis, and lateral diffusion (Bloom et al., 1991a,b).

High-pressure studies of phospholipid bilayers are relatively recent. Wong and coworkers (Wong & Mantsch, 1985; Wong, 1987) used IR and Raman spectroscopy to observe various gel phases (GII to GV) induced by high pressure. Their studies revealed several new gel phases that are not readily accessible at decreased temperatures. Neutron diffraction studies at high pressures revealed the existence of an interdigitated gel phase (Gi) of DPPC above 40 °C (Winter & Pilgrim, 1989; Prasad et al., 1987).

Our laboratory has combined NMR methods with high-pressure and variable-temperature measurements to access diverse gel phases in phospholipid bilayers and to study the order and dynamics of phospholipids in bilayers as a function of volume changes. Eventually this will allow the testing of theoretical models for

Table 20.1. NMR studies on pressure effects on model membranes

System	Experiment	Results	Reference
DPPC	Natural abundance ^{13}C T_1, T_2	Phase transitions	Jonas et al. (1988)
DMPC, POPC	^1H 2D-NOESY	NOE build-up curves	Jonas et al. (1990)
DPPC-d_{62} ($\pm$TTC)	^2H lineshapes	Phase diagram: order parameters: pressure reversal of the effects of TTC	Driscoll et al. (1991a, 1991b)
DPPC ($\pm$TTC)	^{31}P lineshapes, T_1	Structure and dynamics of the head group; phase diagram	Peng and Jonas (1992)
DPPC-d_2 (2, 2); (9, 9); (13, 13)	^2H lineshapes, T_1, T_2	Order parameters; chain motions	This study, Peng et al. (1995a)
DPPC ($\pm$TTC)	^1H T_1, 2D-NOESY	Dynamics; location of TTC; spin-diffusion	Peng et al. (1995b)
DPPC-d_{62}-cholesterol	2H lineshapes	Phase diagram	Samarasinghe and Jonas (unpublished)
DPPC	^1H $T_{1\rho}$	Lateral diffusion	This study, Lee et al. (1995)

Pressure range from 0.1 MPa (1 bar) to 500 MPa (5 kbar); temperature range from 7 °C to 75 °C; high resolution ^1H NMR at 300 MHz.

NMR relaxation times in the absence of confounding temperature effects. Table 20.1 lists the studies that have been performed in our laboratory on phospholipid bilayers using high-pressure NMR methods. Especially relevant to the work presented in this article is the study by Driscoll et al. (1991) of perdeuterated DPPC (DPPC-d_{62}) multilamellar vesicles. Deuterium NMR lineshapes were the basis for the construction of an extensive temperature-pressure phase diagram for this system, which confirmed the existence of GII, GIII, GIV, and Gi gel phases reported by others and revealed a new high-pressure, low-temperature phase (GX). Since the complex ^2H NMR spectra of DPPC-d_{62} did not allow accurate relaxation measurements, in the present study we expanded that work to selectively deuterated DPPC. This time our objective was to examine the order (S_{CD}) and dynamics (T_1 and T_2) of specific segments (C-2, C-9, and C-13) of the DPPC acyl chains under increasing pressures in the liquid crystalline state and several gel-phases including the Gi phase.

In addition to the ^2H NMR work, we report here the measurement of lateral diffusion coefficients of DPPC molecules in small unilamellar vesicles by using the ^1H NMR rotating frame spin-lattice relaxation method (Burnett & Harmon, 1972; Fisher & James, 1978). This approach permits the relatively slow lateral diffusion rate of DPPC to be assessed in the liquid-crystalline phase and in two gel phases under increasing pressures.

METHODOLOGY

^{2}H NMR Studies of Order and Relaxation

The objectives were to assess the order and motions of specific segments of DPPC molecules in multilamellar vesicles as a function of pressure. The deuterium-labeled DPPC samples were synthesized by Avanti Polar Lipids (Birmingham, Ala.) with two ^{2}H labels on either the 2, 9, or 13 carbons of both acyl chains. The DPPC samples (35% by weight) were dispersed in ^{2}H-depleted water above 41 °C. Deuterium NMR experiments were performed on a home-built NMR spectrometer with a wide-bore (130 mm) Oxford 4.2 T superconducting magnet. The spectrometer is interfaced to a GE 293D pulse programmer and a GE/Nicolet 1280 computer system. The contribution by Peng et al. (Chapter 8 in this volume) describes the high-pressure NMR probe and sample cells used in our laboratory and their performance characteristics. During the experiments, temperature was controlled at 50 ± 0.2 °C, and pressure was varied between 1 and 5 kbar (± 20 bar). Deuterium NMR spectra were obtained at a 27.6 MHz spectrometer frequency. The typical multiphase sequences were (1) the quadrupole echo sequence, $(\pi/2)_x - t_1 - (\pi/2)_y - t_2$–echo (Davis, 1979) and (2) the inversion recovery sequence, $(\pi)_{\text{comp.}} - \tau - (\pi/2)_x - t_1 - (\pi/2)_y - t_2$–echo. The t_1 and t_2 values were adjusted to 30–50 μs, and the π and $\pi/2$ pulses were replaced by composite pulses to compensate for B_1 field inhomogeneities and pulse length inaccuracy. The quadrupole splitting values (Δv_Q) were obtained with the quadrupole echo sequence. The spin-lattice relaxation times (T_1) were measured from the inversion recovery sequence, and the delay time (τ) was varied depending on the T_1 values of the samples. The spin-spin relaxation times (T_2) were determined with the quadrupole echo sequence; the times t_1 and t_2 were varied according to the T_2 values of different samples and conditions.

^{1}H NMR Study of Lateral Diffusion

The NMR rotating-frame spin-lattice relaxation method (Burnett & Harmon, 1972; Fisher & James, 1978), which has been used successfully in our laboratory in studies of pressure effects on the diffusion in highly viscous liquids (Walker et al., 1988), can be used in measuring lateral diffusion of the phospholipid molecules in DPPC vesicles. The benefit of this method is that the diffusion coefficient is found directly from measured quantities without estimating parameters or perturbing the bilayer with probes. If intermolecular dipolar interactions modulated by the translational motion contribute significantly to the proton relaxation, the rotating-frame spin-lattice relaxation rate ($1/T_{1\rho}$) is a function of the square root of the spin-locking field angular frequency ($\omega_1^{1/2}$), according to

$$\frac{1}{T_{1\rho}} = C\omega_1^{1/2} + \frac{1}{T_2} \tag{1}$$

where C is a constant that contains the lateral diffusion coefficient and $1/T_2$ is the spin-spin relaxation rate. By taking the derivative of the above equation with respect to $\omega_1^{1/2}$ and applying the resulting equation to motion in only two dimensions, an equation is obtained by which the lateral diffusion coefficient can be determined directly (Fisher & James, 1978)

$$\frac{d(1/T_{1\rho})}{d(\omega_1^{1/2})} = -3\sqrt{3}\gamma^4\hbar^2 n/40D^{3/2} \tag{2}$$

where γ is the gyromagnetic ratio, h is Planck's constant divided by 2π, n is the spin density (spins/ml), and D is the lateral diffusion coefficient. However, two experimental requirements must be met before this equation will hold (Burnett & Harmon, 1972). The first is that $\omega_1 = \gamma H_1 > \gamma H_{loc}$, where ω_1 is the angular frequency of the spin-locking field, H_1 is the strength of the applied spin-locking rf field, and H_{loc} is the strength of the local dipolar field. The second requirement is that $(\omega_1\sigma^2/D)^{1/2} \leq 2$, where σ is the molecular diameter.

Lateral diffusion coefficients of the phospholipid molecules in sonicated DPPC vesicles were measured by using the proton NMR rotating frame spin-lattice relaxation method at temperatures ranging from 50 to 70 °C and pressures ranging from 1 to 5 kbar. The DPPC was obtained from Avanti Polar Lipids (Birmingham, Ala.) and was dispersed in D_2O at a concentration of 15% by weight. The DPPC dispersion was purged with N_2 at 55 °C and sonicated in a bath sonicator until clear.

The ^1H NMR experiments were carried out with the same equipment described for the ^2H NMR experiments, but with a spectrometer frequency of 180 MHz, 4K data points in the FID, a spectral width of 800 Hz, and a variable acquisition delay on a time scale of seconds. After every change of pressure and temperature, samples were allowed to equilibrate for several hours before spectra were collected. The choline methyl proton resonance was chosen for the relaxation measurements because it is well resolved from the other proton resonances and can be observed even above the main LC to GI phase transition up to 1.5 kbar pressures.

Rotating frame spin-lattice relaxation times ($T_{1\rho}$) were measured by applying a $\pi/2$ pulse, followed immediately by a spin-locking pulse of variable length τ, which was phase shifted by 90° from the first pulse. For each $T_{1\rho}$ measurement, 10 different spin-locking pulse lengths were used. The peak intensities from the resulting 10 spectra were entered into the data-reduction routine of the software. The intensities were plotted versus the pulse lengths, and the value of $T_{1\rho}$ was obtained by a least squares fit to the exponentially decaying magnitude by $M(t) = M_0 \exp\{-\tau/T_{1\rho}\}$, where M_0 is the magnetization immediately following the initial $\pi/2$ pulse.

To calculate the lateral diffusion coefficient D, the angular frequency of the spin-locking field (ω_1) had to be known. It was determined from the expression $\omega_1 = \gamma H_1 = \pi/t$, where H_1 is the strength of the spin-locking pulse, and t is the time of application of H_1 needed to rotate the magnetization vector through 180°. An attenuator connected to the pulse amplifier was used to produce six to nine different 180° pulse lengths in the range of 16 μs to 80 μs for each measurement of the lateral self-diffusion coefficient.

RESULTS

Lineshape Studies

Figure 20.1 shows ^{2}H NMR spectra of DPPC deuterated on the C-2 positions of the acyl chains. The lineshapes in the liquid crystalline (LC) phase (1 bar to 500 bar) are characteristic of axially symmetric powder patterns. These spectra have sharp-edged peaks and shoulders at about twice the frequency difference between the edge and the center of the spectrum. The DPPC labeled on C-2 has spectra in the LC phase with three quadrupole splittings. The largest quadrupole splitting is observed for deuterons on the sn-1 chain, and the smaller two belong to the nonequivalent deuterons on the sn-2 chain as described by Seelig and Seelig (1974). These three quadrupole splittings persist throughout the LC phase under increasing pressures indicating that the bent conformation of the sn-chain is retained. DPPC labeled in the C-9 segments exhibits one quadrupole splitting for 1 bar through 250 bar but two splittings from 375 bar to 500 bar, whereas DPPC labeled in the C-13 segments exhibits two splittings throughout the LC phase (not shown). As the pressure is increased to 750 bar, the bilayer enters into the GI phase, and the sharp edges for all three samples disappear as observed in the

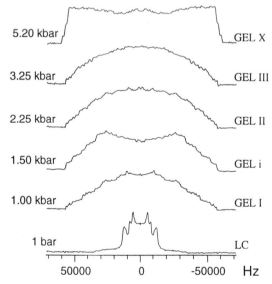

Figure 20.1. ^{2}H NMR spectra of DPPC molecules labeled on the C-2 segments of the acyl chains. A multilamellar dispersion of 35% by weight of DPPC in water was measured at pressures from 1 bar to 5.2 kbar. Spectra were recorded with a spectral width of 150 kHz and data size of 2K points. The π/2 pulse length was 6 μs and line broadening was 50 Hz for the LC phase and 300 Hz for the gel phases. A total of 8,000–40,000 scans was accumulated depending on the sensitivity of the samples; the recycle delay was chosen to be greater than 5 T_1 for a full recovery of the magnetization. All FIDs were left-shifted to the echo maximum, and the data were transformed starting right on the top of the echo.

rippled thermotropic phase (Ruocco et al., 1985). With further increases in pressure up to the interdigitated Gi phase, the intensities of the parallel edges of the spectra increase and then decrease again in the chain-tilted GII phase. The intensities of the parallel edges of the spectra increase progressively with increasing pressure. In the closely packed GX phase, prominent intensities at the outer edges were observed. The splittings of the parallel edges in the various gel phases change very little, but the Δv_Q values increase dramatically at the main phase transition.

Order Parameter Studies

Since the residual quadrupole splittings can be determined unambiguously in the LC phase, the C-D bond-order parameter S_{CD} can be calculated according to Seelig (1977) from

$$\Delta v_Q = \frac{3}{4}\left(\frac{e^2qQ}{h}\right)S_{CD} \tag{3}$$

$$S_{CD} = \frac{1}{2}\langle 3\cos^2\theta - 1\rangle \tag{4}$$

where (e^2qQ/h) is the static quadrupole splitting constant, which has been determined to be 168 kHz for paraffinic CD bonds, and θ is the angle between the CD bond vector and the bilayer normal. However, in the gel phase the sharp powder pattern disappears; therefore, to estimate the average order parameters we measured the first moment (M1) of the spectra. The first moment is related to the average order parameter S_{CD} by the relationship given by Davis (1979)

$$M1 = \frac{\pi}{\sqrt{3}}\left(\frac{e^2qQ}{h}\right)S_{CD} \tag{5}$$

Figure 20.2 shows the calculated S_{CD} values as a function of pressure for the three DPPC samples.

Inspection of the plots indicates that the values of S_{CD} change abruptly at the different phase transitions. As expected, the order parameter decreases in the LC state of the DPPC samples as follows: C-2 > C-9 > C-13, reflecting the much larger range of motions near the methyl end of the molecule. Strikingly, in the gel phases the relationship between the order paramaters in the different segments changes to C-9 > C-13 > C-2. This indicates that the middle segments of the acyl chains are more ordered than the extremes, as illustrated in Figure 20.3. Furthermore, the order is higher in the Gi phase in all segments of the DPPC acyl chains relative to the adjacent bilayer phases. In GIII and GX, the order parameters cannot be measured accurately, but they approach 0.5, the maximal value corresponding to extended, all-*trans* conformations of the chains.

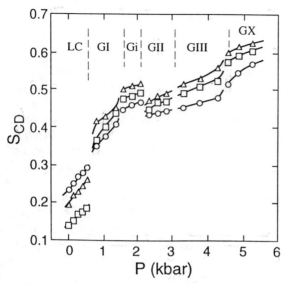

Figure 20.2. Pressure dependence of the deuterium order parameter, S_{CD}, at 50 °C, for the DPPC samples labeled in the C-2 segments (○), C-9 segments (△), and C-13 segments (□).

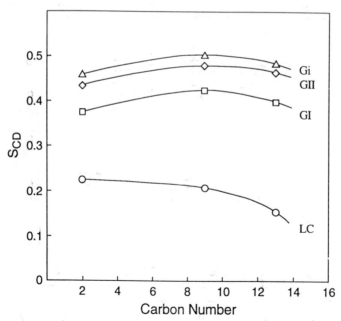

Figure 20.3. Order parameters, S_{CD}, as a function of the deuterium label positions along the acyl chain (sn-1). The S_{CD} values at 50 °C, for the LC (○), GI (□), Gi (△), and GII (◇) phases were taken from the middle pressure for each phase shown in Figure 20.2.

Relaxation Studies

The T_1 values in the LC phase were measured from the decay of the peak intensity. No distinct orientation dependence of T_1 values across the echo spectra was observed, and this is in agreement with the results of Brown and Davis (1981) and Perly et al. (1985). In the LC phase, two T_1 values were obtained for the DPPC samples labeled on C-2 at each pressure. A larger value was found for the sn-1 chain deuterons, while a smaller value was observed for the two deuterons of the sn-2 chain. For the other two samples, only average T_1 values were determined due to limited resolution. As indicated in Figure 20.4, each sample shows three discontinuities in T_1 values, occurring at phase transitions LC/GI, GI/Gi, and Gi/GII, respectively. In addition, one local maximum occurs at the phase transition GII/GIII, and another local minimum occurs near the phase transition GIII/GX. As a function of pressure, all three samples have almost the same pressure dependence with the order in magnitude of T_1 values being C-13 > C-9 > C-2 at each pressure in each phase. The T_1 values decrease with increasing pressure in each phase, except in the Gel II phase where they increase with pressure.

In the spin-spin relaxation time measurements, we found that T_2 is greater for the peaks than for the shoulders of the spectra. This kind of orientation dependence of T_2 across the echo spectra was also found in other lipid systems (Perly et al., 1985; Bloom & Sternin, 1987; Watnick et al., 1987). Thus, the spin-spin relaxation times were measured from the decay of peak intensity in the LC phase. As shown in Figure 20.5, the T_2 values for the three samples in the LC and GI

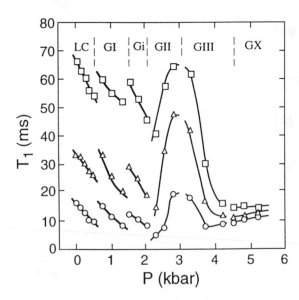

Figure 20.4. The spin-lattice relaxation time, T_1, as a function of pressure, at 50 °C, for the DPPC samples labeled in the C-2 segments (○), C-9 segments (△), and C-13 segments (□).

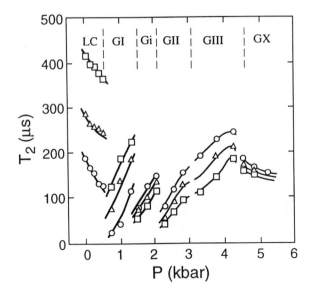

Figure 20.5. The spin-spin relaxation time, T_2, as a function of pressure, at 50 °C, for the DPPC samples labeled in the C-2 segments (○), C-9 segments (△), and C-13 segments (□).

phases vary in magnitude from C-13 > C-9 > C-2 at each pressure. However, in the other high-pressure gel phases, from Gi to GX, the order of T_2 values is inverted: C-2 > C-9 > C-13. T_2 values decrease as pressure is increased in the LC and GX phases, whereas they increase with increasing pressure in the other gel phases. In addition, a local maximum in T_1 is observed near the phase transition between the GII and GX phases.

Lateral Diffusion Studies

Reciprocal values of rotating spin-lattice times $(T_{1\rho}^{-1})$ were plotted versus $\omega_1^{1/2}$ (Figure 20.6). The slope of the least-squares-fitted line was used to calculate the lateral diffusion coefficient (Figure 20.7) as a function of pressure at two temperatures, 50 and 60 °C.

The diffusion coefficient decreases continuously with increasing pressure from 1 bar (5.4 × 10⁻⁹ cm² seconds) to 300 bar (3.6 × 10⁻⁹ cm² seconds) at 50 °C in the LC phase. A sharp decrease in the value of the lateral diffusion coefficient occurs at the pressure of the LC to GI phase transition. From 500 bar to 800 bar in the GI phase, the values of the lateral diffusion coefficient (~1 × 10⁻⁹ cm² seconds) appear constant. Another abrupt decrease in the value of the lateral diffusion coefficient occurs at the pressure of the GI-Gi phase transition. Above 1 kbar in the Gi phase, the values of the lateral diffusion coefficient (~1 × 10⁻¹⁰ cm² seconds) are relatively constant. Except for the difference in the phase transition pressures, identical behavior was observed at both 60 °C (shown in Figure 20.7) and 70 °C (data not shown). From the temperature dependence of

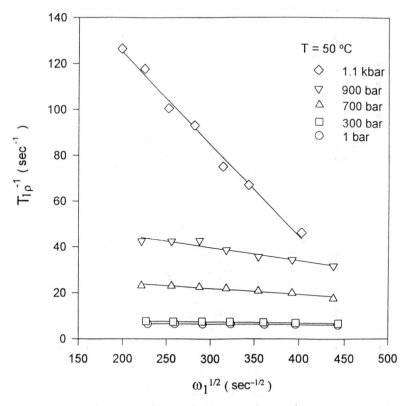

Figure 20.6. Proton NMR rotating frame spin-lattice relaxation rate $(T_{1\rho})^{-1}$ as a function of the square root of the spin-locking field angular frequency $(\omega_1)^{1/2}$ for the choline methyl groups of sonicated DPPC vesicles at 50 °C and different pressures. Symbols represent the experimental results and the lines the least-squares fitted data. From 1 bar to 300 bar the vesicles are in the LC phase, from 500 bar to 800 bar in the GI phase, and above 1 kbar in the GI phase.

the lateral diffusion coefficients in the LC state (1 bar) and GI phase (plateau of D values), the activation energies for lateral diffusion were calculated to be 3.5 kcal mol^{-1} and 6.0 kcal mol^{-1}, respectively.

DISCUSSION

The ^{2}H NMR lineshapes shown in Figure 20.1 clearly reveal the LC to GI phase transition. Transitions between the gel phases are less evident from lineshapes alone, but all other data, including first moments, S_{CD} values and T_1 and T_2 data confirm the expected gel phases (GI, Gi, GII, GIII, GX) that can be accessed by pressure at 50 °C (Driscoll et al., 1991a,b). The transition pressures observed in that study (Driscoll et al., 1991a,b) are lower than those reported here due to the isotope effects of the perdeuterated chains of DPPC-d$_{62}$.

The ^{2}H NMR spectra in the LC phase have the characteristic axially symmetric powder pattern indicative of significant intra- and intermolecular

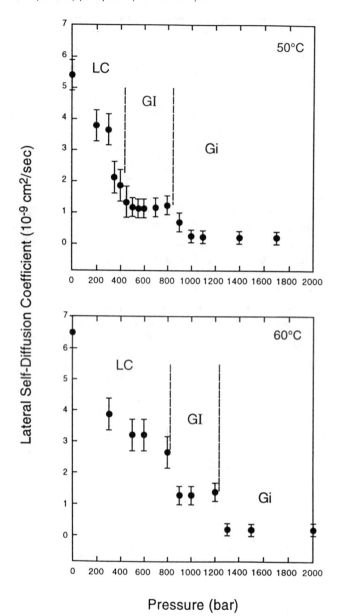

Figure 20.7. Lateral self-diffusion coefficients of DPPC in sonicated vesicles as a function of pressure at 50 °C and 60 °C. Data were calculated from the slopes of the lines using the equation for D given in the text (Eq. 2).

motions. The three quadrupole splittings for the DPPC labeled on C-2 result from the distinct conformations of the *sn*-1 *and sn*-2 chains near the glycerol backbone and the nonequivalent state of the two deuterons on the bent *sn*-2 chain (Seelig & Seelig, 1974). The same conformational differences persist with increasing pressures in the LC state, but cannot be resolved in the gel phases. Near the main LC to

GI transition, the other two labeled segments C-9 and C-13, also have distinct quadrupole splittings for the two chains, while at 1 bar the C-9 segments appear equivalent . The increased resolution for the C-9 quadrupole splittings in the LC phase with pressure reflects the increased order and decreased mobility in the central chain region which magnify the conformational and dynamic differences between the two acyl chains.

Above the LC to GI phase transition, the spectra broaden dramatically and acquire considerable spectral intensity at the shoulders, near ± 60 kHz. Since Davis and Jeffrey (1977) have shown that the quadrupole splittings for all-*trans* conformation of ethylene segments in a hydrocarbon chain give spectral peaks at ± 63 kHz, the shoulders near ± 60 kHz observed in our spectra indicate the presence of increasing populations of *trans* conformers in the gel phases.

The order parameters (Figures 20.2 and 20.3) of all three segments have the expected relative values in the LC state: C-2 > C-9 > C-13. This indicates increasing motional amplitudes from the backbone toward the methyl end of acyl chains. With pressures from 1 bar to 500 bar, the order parameters in the LC state essentially double. Beyond the LC to GI phase transition, the order parameters increase dramatically and change in relative order to C-9 > C-13 > C-2. Evidently, in all the gel phases the middle C-9 segments are most restricted in their motions, while the extremes of the chains have larger motional amplitudes. This is not surprising for the methyl ends, which are free, but is unexpected for the C-2 segments near the glycerol backbone. However, it is possible that water, which remains fluid at the highest pressures attained in this study, can have a major effect on the order and dynamics near the glycerol backbone (Cevc & Marsh, 1987).

The order parameters in the Gi phase are higher than those in the two adjacent gel phases. This indicates that the motions are most restricted for all chain segments in this state. The chains are rigid and have the highest content of *trans*-conformers. The S_{CD} values in excess of 0.5 seen in the GIII and GX phases are theoretically not meaningful; they probably arise from the observed distortion of the lineshapes due to the orientation dependence across the powder spectra of these phases. Similar distortions were seen in thermotropic gel phases (Perly et al., 1985; Bloom & Sternin, 1987; Watnick et al., 1987).

The range of molecular motions that can be detected by ^{2}H NMR relaxation studies have correlation times from about 10^{-11} to 10^{-4} seconds. This includes three major types of motions: (1) transitions between different conformations, (2) reorientations about the direction normal to the bilayer plane, and (3) diffusion in the plane of the bilayer (Bloom et al. (1991). Spin-lattice relaxation time, T_1, measurements can access internal conformational motions and rotational motions of individual molecules; spin-spin relaxation times, T_2, reveal, in addition, slower motions. Typically, for phospholipid bilayers $T_2 \approx 10^{-2} \; T_1$ (Bloom et al., 1991).

There are few reports of theoretical analyses of T_1 and T_2 ^{2}H NMR values for phospholipid bilayers in small vesicle and multilamellar vesicle forms. Brown et al. (1979) derived expressions for $1/T_1$, as a function of τ_C, the correlation time for *trans-gauche* isomerization of the chain segments and the corresponding order parameters. They concluded that in the LC state order contributes relatively little to T_1. The T_1 values mostly reflect fast segmental motions with $\tau_C \approx 10^{-10}$ seconds near the glycerol backbone, and $\tau_C \approx 10^{-11}$ seconds near the methyl group. Brown et al. (1979) observed similar motions in small vesicles and multilamellar vesicles, and noted that the motions for segments C-3 to C-9 are correlated. Our results in the LC state at 1 bar agree well with those of Brown et al. (1979), and show progressive parallel decreases in T_1 with pressure for all segments, reflecting the decreasing rates of *trans-gauche* isomerizations. In our results, transitions between the LC and GI and between the various gel phases are characterized by discontinuities and changes in the slope of T_1 versus pressure. Interpretation of these results in terms of contributing motions and correlation times would require comprehensive computer simulation and modeling beyond the scope of this study. However, Mayer et al. (1988) have analyzed ^{2}H NMR relaxation data for dimyristoylphosphatidylcholine in the LC and two thermotropic phases by using a dynamic NMR model based on the stochastic Liousville equation. They observed discontinuities and slope changes for T_1 and T_2 values at the phase transitions and within the $L_{\beta'}$ phase. Their analysis suggested that the molecular dynamics in this system are characterized by superimposed inter- and intramolecular motions, including overall reorientation of phospholipid molecules and *trans-gauche* isomerization of chain segments. They also concluded that the *sn-2* chains undergo "two-site rotational jumps ... in the plane of the membrane." Similar motions can be expected to exist in the DPPC bilayers subjected to pressure. According to Bloom and Sternin (1987), T_2 may also reflect slower lateral diffusion of molecules along curved surfaces. The change from a negative slope to a positive slope in the pressure dependence of T_1 and T_2 indicates the transition from a fast to a slow correlation time regime (Mayer et al., 1988). This and the discontinuities in T_1 and T_2 at the various phase transitions indicate complex changes in the rates and modes of the motions as the structure of the bilayer changes.

Finally, we used a different pulse sequence to obtain proton spin-lattice relaxation times in the rotating frame ($T_{1\rho}$) for DPPC molecules in sonicated small unilamellar vesicles. This relaxation time can be used to calculate lateral diffusion coefficients (Burnett & Harmon, 1972; Fisher & James, 1978). The calculated diffusion coefficients, 5.4×10^{-9} cm^2 seconds (50 °C) to 7.3×10^{-9} cm^2 seconds (70 °C) agree well with the previously reported values at 1 bar measured by the same technique (Fisher & James, 1978), but are one or two orders of magnitude lower than values obtained by EPR (Devaux and McConnell, 1972) or fluorescence recovery after photobleaching (Vaz et al., 1985) techniques. Although it is possible that the values differ because of the introduction of perturbing probes in the latter methods, it is also known that the values of diffusion coefficients depend on the extent of hydration of the bilayer, on the vesicle type, and in some cases on the sample preparation (Fahey & Webb, 1978; Kuo & Wade, 1979; McCowan et al., 1981).

Our values for the diffusion coefficient ($\sim 10^{-9}$ cm^2 seconds) in the GI phase are essentially independent of pressure and about one order of magnitude higher than the value measured by fluorescence recovery after photobleaching in the thermotropic $P_{\beta'}$ phase (Fahey & Webb, 1978). These differences probably arise from differences in the methods and conditions employed, but could also reflect inherent differences in the diffusion coefficients in the gel phases induced by pressure or temperature. The lateral self-diffusion coefficient decreases with increasing pressure in the LC phase because a pressure increase in the LC bilayer decreases the fluidity of the bilayer by decreasing the volume, which increases molecular interactions and order in the bilayer (Chong & Weber, 1983). Another way to explain this pressure dependence is by means of the free volume theory (Vaz et al., 1985): translational diffusion of a lipid molecule in the bilayer occurs only when a free volume larger than a certain critical size appears in the vicinity of the lipid molecule. The smaller the overall volume, the lower the probability for a molecule to move into a free volume of critical size and the slower the diffusion. Increasing the pressure decreases the volume, and the lateral self-diffusion is reduced. Since with increasing pressure, as revealed by the ^{2}H NMR measurements described here, the order of the DPPC chains continues to increase and the motional states continue to decrease in the GI phase while the diffusion coefficient remains almost constant, our results are not consistent with the hypothesis that the lateral diffusion is related to the fluidity of the bilayer in the gel phase.

ACKNOWLEDGMENTS: This work was supported by NIH Grant GM-42452.

REFERENCES

Bloom, M., & Sternin, E. (1987). Transverse nuclear spin relaxation in phospholipid bilayer membranes. *Biochemistry* **26**, 12101–2105.

Bloom, M., Morrison, C., Sternin, E., & Thewalt, J. L. (1991). Physical properties of the fluid lipid-bilayer component of cell membranes—a perspective. *Q. Rev. Biophys.* **24**, 293–397.

Brown, M. F., & Davis, J. H. (1981). Orientation and frequency dependence of the deuterium spin-lattice relaxation in multilamellar phospholipid dispersions: implications for dynamic models of membrane structures. *Chem. Phys. Lett.* **79**, 1431–435.

Brown, M. F., Seelig, J., & Häberlen, U. J. (1979). Structural dynamics in phospholipid bilayers from deuterium spin-lattice relaxation time measurements. *J. Chem. Phys.* **70**, 15045–5053.

Burnett, L. J., & Harmon, J. F. (1972). Self-diffusion in viscous liquids: pulse NMR measurements. *J. Chem. Phys.* **57**, 11293–1297.

Cevc, G., & Marsh, D. (1987). *Phospholipid Bilayers.* New York, Wiley, pp. 57–97.

Chong, P. L.-G., & Weber, G. (1983). Pressure dependence of 1, 6-diphenyl-1, 3, 5-hexatriene fluorescence in single-component phosphatidylcholine liposomes. *Biochemistry* **27**, 399–404.

Davis, J. (1979). Deuterium magnetic resonance study of the gel and liquid crystalline phases of dipalmitoylphosphatidylcholine. *Biophys. J.* **27**, 1339–358.

Davis, J. H., & Jeffrey, K. R. (1977). The temperature dependence of chain disorder in potassium palmitate-water: a deuterium NMR study. *Chem. Phys. Lipids* **20**, 187–104.

Devaux, P. F., & McConnell, H. M. (1972). Lateral diffusion in spin-labeled phosphatidylcholine multilayers. *J. Am. Chem. Soc.* **94**, 14475–4481.

Driscoll, D. A., Jonas, J., & Jonas, A. (1991a). High pressure ^{2}H nuclear magnetic resonance study of the gel phases of dipalmitoylphosphatidylcholine. *Chem. Phys. Lipids* **58**, 97–104.

Driscoll, D. A., Samarasinghe, S., Adamy, S., Jonas, J., & Jonas, A. (1991b). Pressure effects on dipalmitoylphosphatidylcholine bilayers measured by ^{2}H nuclear magnetic resonance. *Biochemistry* **30**, 13322–3327.

Fahey, P. F., & Webb, W. W. (1978). Lateral diffusion in phospholipid bilayer membranes and multilamellar liquid crystals. *Biochemistry* **17**, 13046.

Fisher, R. W., & James, T. L. (1978). Lateral diffusion of the phospholipid molecule in dipalmitoylphosphatidylcholine bilayers: an investigation using nuclear spin-lattice relaxation in the rotating frame. *Biochemistry* **17**, 1177–1183.

Jonas, J., Xie, C.-L., Jonas, A., Grandinetti, P. J., Campbell, D., & Driscoll, D. (1988). High-resolution ^{13}C NMR study of pressure effects on the main phase transition in 1-α-dipalmitoyl phosphatidylcholine vesicles. *Proc. Natl. Acad. Sci. USA* **85**, 4115–4117.

Jonas, J., Winter, R., Grandinetti, P. J., & Driscoll, D. (1990). High-pressure 2D NOESY experiments on phospholipid vesicles. *J. Magn. Reson,* **87**, 1536–547.

Kuo, A.-L., & Wade, C. G. (1979). Lipid lateral diffusion by pulsed nuclear magnetic resonance. *Biochemistry* **18**, 12300–2308.

Lee, B.-S., Schwer, S., Jonas, A., & Jonas, J. (1995). High pressure proton NMR study of lateral self-diffusion of phosphatidylcholines in sonicated unilamellar vesicles. *Chem. Phys. Lipids* **78**, 103–117.

Mayer, C., Muller, K., Weisz, K., & Kothe, G. (1988). Deuteron NMR relaxation studies of phospholipid membranes. *Liq. Crystals* **3**, 1797–806.

McCowan, J. T., Evans, E., Diehl, S., & Wiles, H. C. (1981). Degree of hydration and lateral diffusion in phospholipid multilayers. *Biochemistry* **20**, 13134–3138.

Peng, X., & Jonas, J. (1992). High pressure ^{31}P NMR study of dipalmitoylphosphatidyl-choline bilayers. *Biochemistry* **31**, 16383–6390.

Peng, X., Jonas, A., & Jonas, J. (1995a). High pressure ^{2}H-NMR study of the order and dynamics of selectively deuterated dipalmitoyl phosphatidylcholine in multilamellar aqueous dispersions. *Biophys. J.* **68**, 11137–1144.

Peng, X., Jonas, A., & Jonas, J. (1995b). One and two dimensional ^{1}H NMR studies of pressure and tetracaine effects on sonicated phospholipid vesicles. *Chem. Phys. Lipids* **75**, 159–69.

Perly, B., Smith, I. C. P., & Jarrel, H. C. (1985). Acyl chain dynamics of phosphatidyle-thanolamines containing oleic acid and dihydrostearolic acid: ^{2}H NMR relaxation studies. *Biochemistry* **24**, 4659–4665.

Prasad, S. K., Shashidhar, R., Gaber, B. P., & Chandrasekhar, S. C. (1987). Pressure studies on two hydrated phospholipids—1,2-dimyristoyl-phosphatidylcholine and 1,2-dipalmitoylphosphatidylcholine. *Chem. Phys. Lipids* **43**, 227–235.

Ruocco, M. J., Makriyannis, A., Siminovitch, D. J., & Griffin, R. G. (1985). Deuterium NMR investigation of ether- and ester-linked phosphatidylcholine bilayers. *Biochemistry* **24**, 4844–4851.

Seelig, A., & Seelig, J. (1974). The dynamic structure of fatty acyl chains in a phospholipid bilayer measured by deuterium magnetic resonance. *Biochemistry* **13**, 4839–4845.

Seelig, J. (1977). Deuterium magnetic resonance; theory and application to lipid membranes. *Q. Rev. Biophys.* **10**, 353–418.

Vaz, W. L. C., Clegg, R. M., & Hallmann, D. (1985). Translational diffusion of lipids in liquid crystalline phase phosphatidylcholine multibilayers: a comparison of experiment with theory. *Biochemistry* **24**, 781–786.

Walker, N. A., Lamb, D. M., Adami, S. T., Jonas, J., & Dare-Edwards, M. P. (1988). Self-diffusion in compressed highly viscous liquid z-ethylhexyl-benzoate. *J. Phys. Chem.* **92**, 3675–3679.

Watnick, O., Dea, P., Nayeem, A., & Chan, S. I. (1987). Cooperative lengths and elastic constants in lipid bilayers: the chlorophyll a/dimyristollccithin system. *J. Chem. Phys.* **86,** 5789–5800.

Winter, R., & Pilgrim, W.-C. (1989). A SANS study of high pressure transitions in model biomembranes. *Ber. Bunsenges, Phys. Chem.* **93,** 708–717.

Wong, P. T. T. (1987). High pressure studies of biomembranes by vibrational spectroscopy. In *High Pressure Chemistry and Biochemistry,* ed. R. van Eldik & J. Jonas, Dordrecht, D. Reidel, pp. 381–400.

Wong, P. T. T., & Mantsch, H. H. (1985). Effects of hydrostatic pressure on the molecular structure and endothermic phase transitions of phosphatidylcholine bilayers: a Raman scattering study. *Biochemistry* **24,** 4091–4096.

21

The Effects of Increased Viscosity on the Function of Integral Membrane Proteins

SUZANNE F. SCARLATA

For many years the idea that the activity of integral membrane proteins is regulated by the fluidity of the lipid matrix was popular and appeared to be quite rational. However, as information about the effect of viscosity on the function of different membrane proteins became available, the correlation between the two became increasingly unclear. The purpose of this article is to readdress this issue in light of our recent pressure and temperature studies. This chapter is divided into seven parts: (1) the effect of viscosity on enzyme activity; (2) the effect of viscosity on the local motions of proteins; (3) characterization of membrane viscosity; (4) demonstration of changes in protein-lipid contacts brought about by changes in viscosity; (5) an example of a protein in which the viscosity appears to stabilize a particular conformational state; (6) relations between membrane viscosity and protein function; and (7) conclusions.

BACKGROUND OF VISCOSITY EFFECTS ON ENZYME ACTIVITY

The effect of viscosity (η) on the rate (k) of a chemical reaction was first given by Kramers (1940):

$$k = A/\eta e^{-Ea/RT} \qquad (1)$$

In this expression, viscosity will affect the rate of a reaction by limiting the rate of diffusion of reactants. Viscosity will thus modify the frequency factor (A) and should not affect the activation energy. This expression has been applied to aqueous soluble enzymes (for example, Gavish, 1979; Gavish & Werber, 1979; Somogyi et al., 1984), and it appears that, in general, enzymes obey Kramers's relation, although in some cases the exponent of η is less than one.

331

Viscosity can affect enzymatic rates not only by limiting the diffusion of substrates but also by damping internal motions of the protein chains. It seems reasonable that a high enough viscosities, the protein would be damped sufficiently so that large activation energies will be required for the backbone motions that allow substrates and products to diffuse into and out of the active site. This viscosity-induced increase in activation energy was shown by studies of the reassociation of carbon monoxide and dioxygen to the heme site of myoglobin after flash photodissociation (Austin et al., 1975; Beece et al., 1980). These workers found four major activation barriers to oxygen diffusion in myoglobin and that the flux across these could be damped successively by increasing the viscosity. Their studies clearly demonstrated that at high viscosities the energy pathway of enzymatic reactions will be altered.

EFFECT OF VISCOSITY ON THE LOCAL MOTIONS OF PROTEINS

Weber and coworkers (Weber et al., 1984; Scarlata et al., 1984; Rholam et al., 1984) then investigated the extent to which the external viscosity could damp movements in the interior of proteins. Specifically, they determined whether the external viscosity could damp the small, local motions inside proteins that couple together and are ultimately responsible for conformational changes. Their approach was to monitor the rotational motions of tryptophan and tyrosine side chains of proteins under viscous conditions. The behavior of the rotational motion of these residues can be determined by the response of the fluorescence anisotropy (A) and lifetime (τ) through the Perrin equation

$$(A_0/A) - 1 = RT\tau/\eta V \tag{2}$$

where A_0 is the anisotropy in the absence of rotational motion, R is the gas constant, T is the absolute temperature, V is the rotational volume (assuming a sphere), and η is the viscosity. They then expanded the viscosity, empirically, about an arbitrary temperature (T_0)

$$\eta = \eta_0 e^{-b(T_0 - T)} \tag{3}$$

where η_0 is the viscosity at T_0, and b is defined as the thermal coefficient of the viscosity. Inserting this expression into the Perrin equation, while rearranging and taking the natural logarithm, we obtain

$$\ln[A_0/A - 1] - \ln[RT\tau/V] = \ln \eta_0 - \mathbf{b}(T_0 - T) \tag{4}$$

If we let $T_0 = 273$ K, then $(T_0 - T)$ will just refer to the temperature in degrees Celsius. The left side of this equation could be grouped together into a parameter Y. Thus, if Y is plotted against T, then a straight line should be obtained with a slope b.

Initially, a series of isolated fluorophores were monitored in different solvents (Weber et al., 1984). It was found that the thermal coefficient of the viscosity, **b**, was solely dependent on the particular solvent and did not depend on the probe used in its measurement. For example, in 80% glycerol-buffer, the naphthalene-based probe, PRODAN (6-propionyl (dimethylamino) naphthalene) gave a value identical to that of indole-based (tryptophan) and phenol-based (tryosine) probes. Also, PRODAN, which can form two strong hydrogen bonds and undergoes isotropic rotations in polar solvents such as butanol, gives the same **b** value as perylene, which cannot hydrogen bond and undergoes anisotropic rotational motion. Moreover, these fluorescence-determined, microscopic values of **b** match those determined macroscopically by flow viscometry. Thus, the magnitude of **b** depends solely on the thermal viscosity behavior of the solvent.

This strict solvent dependence of **b** was then used to gain insight into the environment inside proteins (Scarlata et al., 1984; Rholam et al., 1984). These studies were done by immersing a series of peptides and proteins in a viscous solvent (80% glycerol-buffer) to eliminate slower rotations of the proteins and by monitoring the rotational motions of the tyrosines and tryptophan residues as a function of temperature. Interestingly, all peptides and proteins tested displayed two values of **b**: one at low temperatures and high viscosities that is equal to the solvent, and a second, reduced value, at high temperatures and lower viscosities. The transition between these two viscosity regimes occurs very abruptly, over a 1–2° range of temperature. Examples of a typical Y versus T plot are shown in Figure 21.1 in which the upper set of data are those collected for the free amino acid tyrosine and the lower set of data are those for neurophysin. Neurophysin has a single tyrosine residue that is internally located, but similar behavior is observed for other proteins containing multiple fluorophores.

The interpretation of this behavior at high viscosities is as follows: the motions of the Tyr and Trp residues are very damped. Decreasing the external viscosity allows an expansion of the amplitude of protein motions, which is then transmitted to the fluorophore. This expansion allows further depolarization of the probe due to an increase in the volume element in which the probe can rotate. The increase continues until the rotational amplitude of the probe becomes limited by its neighboring residues. At this point, any increase in rotational motion is determined by the surrounding protein matrix. Thus, at low temperature the rotations of the probe molecules are effectively noninteracting with or uncoupled from the neighboring residues, and the viscosity must decrease for them to become coupled. This interpretation is supported by computer simulations that can mimic the plots purely by changing the rotational space allowed to the fluorophore and which show that the rate of rotation must be unrealistically slow to perturb this behavior.

Most of the proteins and peptides studied undergo the transition between coupled and uncoupled motions at viscosities close to 1 poise (see Scarlata et al., 1984; Rholam et al., 1984), and many physical techniques suggest that the viscosity of membranes are in this range (Radda & Smith, 1970; Cone, 1972; Rudy & Gitler, 1972; Shinitzky & Barenholz, 1978). This raises the question of whether the function of integral membrane proteins may be regulated by the coupling and uncoupling of small scale motions through changes in the viscosity and lipid packing.

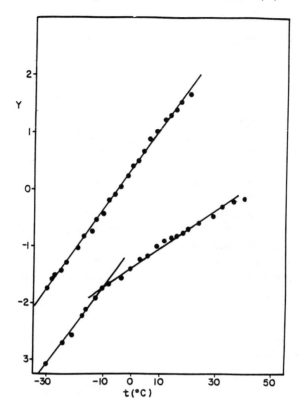

Figure 21.1. The Y versus temperature plot of free tyrosine in 80% glycerol-buffer (upper data) and for neurophysin under the same conditions. (Reprinted with permission from M. Rholam, S. Scarlata, and G. Weber, 1984. *Biochemistry* **23,** 6794. Copyright 1984 American Chemical Society.)

CHARACTERIZATION OF MEMBRANE VISCOSITY

One of the first problems encountered in dealing with lipid membranes is that their anisotropic nature precludes their characterization in bulk terms and that their viscosity can only be described microscopically. Lipid membranes have large viscosity gradients in the planes parallel and perpendicular to the hydrocarbon chains and also as a function of membrane depth (Thulborn & Sawyer, 1978; Tilley et al., 1979).

To determine the dependence of the thermal coefficient of the viscosity (**b**) on membrane position, we measured the change in fluorescence anisotropy and lifetime of a series of anthroyloxy fatty acid probes which can be located at progressively deeper positions in dioeoylphosphatidylcholine (DOPC) bilayers (Thulborn & Sawyer, 1978). We found that, although the local viscosity (as measured by fluorescence anisotropy) decreases with increasing depth, the thermal coefficient of the viscosity is invariant (Scarlata, 1989). Also, the anthroyloxy probes can be excited into distinct bands in which the resulting depolarization is either from half in-plane and half out-of-plane motions (381 nm) or from out-of-plane motions (316 nm) only (Vincent et al., 1982). Comparison of the b-values determined from the temperature dependence of the anisotropy and lifetime from these two exciting wavelengths allows us to isolate the b-values of the different membrane planes. We find that both in-plane and out-of-plane motions give the same **b** values and, thus, the **b** values in the planes parallel and perpendicular to the lipid chains are the same (Scarlata, 1989). The fact that the **b** values are identical in different locations in the membrane but the viscosity is not is because the thermal coefficient is reflecting the expansion of the membrane with temperature rather than the viscosity itself; **b** thus represents the thermal expansion coefficient.

Because membranes can be characterized by a unique value of the thermal coefficient of the viscosity, we can determine the **b** values of integral membrane proteins. The study mentioned in the previous section used anthroyloxy fatty acid probes in DOPC. Our initial protein studies focus on the peptide gramicidin in dimyristoylphosphatidylcholine (DMPC) bilayers in which the peptide has maximal solubility. DMPC undergoes a gel to liquid-crystal phase transition at $\sim 22\,^{\circ}C$. The **b** values of pure DMPC, as characterized by the anthroyloxy fatty acids, show single slopes in the gel, liquid-crystal and phase transition region of DMPC (Scarlata, 1989).

PERTURBATION OF SPECIFIC LIPID-PROTEIN CONTACTS BY VISCOSITY

Gramicidins (for review see Wallace, 1990) are small peptides having the sequence

$$HCO-l-Val-Gly-l-Ala-d-Leu-l-Ala-d-Val-l-Val-d-Val-l-Trp-d-Leu-l-X$$

$$-d-Leu-l-Trp-d-Leu-l-Trp-NHCH_2CH_2OH \qquad (5)$$

where $X =$ Trp, Phe, or Tyr, termed gramicidin A, B, or C, respectively. The function of gramicidin is believed to be the regulation of gene expression through binding to the α subunit of RNA polymerase or to superhelical DNA or both. In reconstituted membranes, gramicidin forms two major structures that allow for the specific passage of monovalent ions through the bilayer. The first of these, on which we focus here, is the formyl N to formyl N terminal dimer that corresponds to gramicidin channels. In this conformation, the aromatic residues are aligned closely to the bilayer surface. Alternatively, gramicidin can form an intertwined helical dimer that displays pore characteristics. In the latter conformation, the aromatic side chains are more evenly distributed along the peptide. The two forms can be readily distinguished by circular dichroism.

We followed the temperature response of the anisotropy and lifetime of the gramicidin tryptophans when the channel form of the peptide was reconstituted in DMPC bilayers at several different peptide:lipid ratios (Scarlata, 1988). The effects of temperature on the properties of the peptide-incorporated bilayers were determined by monitoring the rotational behavior of 11-anthroyloxyundecanoic acid (11 AU). The Y versus temperature plot is shown in Figure 21.2. Surprisingly,

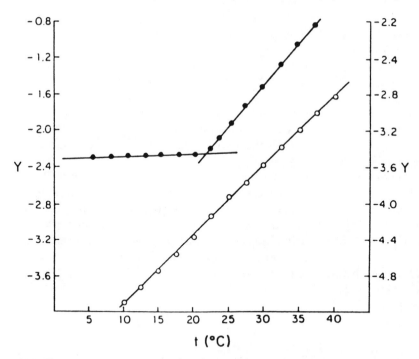

Figure 21.2. The Y versus temperature plot of gramicidin reconstituted into DMPC bilayers at 1:50 protein:lipid molar ratio (●) and 11 anthroyloxy-undecanoic acid (11 AU) incorporated into the gramicidin-DMPC bilayers (○). (Reprinted with permission from S. Scarlata, *Biophys. J.* **54**, 1152.)

below the phase transition temperature of the lipid, no change in rotational motion of the gramicidin tryptophans is observed, even through the viscosity of the membrane, as seen by the 11 AU probe, decreases substantially. However, after the phase transition, the rotational motion of the gramicidin tryptophans changes dramatically, giving a **b** value higher than that obtained for 11 AU, and, thus, higher than that for the bilayer solvent. This behavior is opposite to what is observed for aqueous soluble proteins.

What mechanism(s) can be holding the tryptophan residues firmly in place in the gel phase but in the fluid phase allow their rotational motion to increase more than the bilayer itself? One factor that could play a role is aromatic ring stacking between tryptophan-9 and tryptophan-15. Measurements of the change in tryptophan self-quenching indicate that part of this unusual temperature behavior is due to ring stacking interactions between tryptophan-15 and trypotphan-9 which are stable at low temperatures but break up as the temperature increases (Scarlata, 1988). However, aromatic ring stacking between two of the four tryptophan residues is only a partial explanation of the behavior seen in Figure 21.2.

Further insight into this problem was obtained by hydrostatic pressure studies. Before studying gramicidin, we needed to characterize the changes in bilayer properties induced by pressure. For these studies we took a slightly different approach (Scarlata, 1991a). The observed anisotropy can be related to the precession angle, θ, by

$$\langle \cos^2 \theta \rangle = [1 + 2A/A_0]/3 \tag{6}$$

The precession angle can be related to a volume element, and the viscosity can be related to the free volume (Batschinski, 1913). From this volume element, the compressibility can be calculated by its decrease with pressure. We determined the compressibility of a series of anthroyloxy fatty acid probes located at various membrane depths. We found that the local compressibility is similar for all of these probes and is thus constant with bilayer depth. However, values of the compressibility in the different membrane planes obtained from anisotropy data collected at excitation wavelengths of 381 nm and 316 nm indicate that the planes perpendicular to the hydrocarbon chains have compressibilities similar to organic solids but that the plane parallel to the hydrocarbon chains has a negative compressibility (that is, an expansion under pressure). This result is in accord with neutron diffraction studies (Braganza & Worcester, 1986) showing that even though the overall volume of bilayers decreases with pressure, the thickness of the bilayer increases due to the straightening of the hydrocarbon chains during pressurization.

With this in mind, we monitored the change in rotational behavior of gramicidin tryptophans as a function of pressure at three different temperatures (Scarlata, 1991b). At the outset we believed that the application of pressure should result in further limitation of tryptophan rotational motion due to increased lipid packing and to a stabilization of aromatic ring stacking. In Figure 21.3a we compare the change in polarization of the tryptophan of gramicidin under pressure with that seen for 11 AU embedded in the proteoliposomes (Figure 21.3b). We first note that 11 AU shows a systematic increase in polarization, corresponding to a decrease in rotational motion, with pressure at the three temperatures studied. The behavior for gramicidin is very different. At 40 °C, the rotational motion of the tryptophans as seen by the fluorescence polarization increased up to 1 kbar, at which the bilayers should undergo the fluid-to-gel phase transition, but decreased thereafter. At 28 °C, the polarization also decreased at higher pressure, indicating a pressure-induced increase in rotational motion. We note that the lifetime only changed by 50 ps in this 1–200 bar range. At 7.5 °C, at which the bilayers are in the gel phase throughout the pressure study, the polarization decreased continuously and significantly. Further studies showed that this increase in rotational motion with temperature was not due to elimination of the protein from the matrix or to a change in the position of the residues. Tryptophan self-quenching studies indicated a stabilization of ring stacking interactions with pressure.

What is the underlying reason for the increase in rotational motion of the gramicidin tryptophan with pressure? In examining the structure of the *N-N* dimer form of gramicidin, we find that all the tryptophans are close to the surface of the membrane (see Wallace, 1990). On the basis of functional studies (O'Connell et al., 1990), we then reasoned that the indole residues can hydrogen bond with the membrane surface. Increasing the pressure will destabilize these interactions

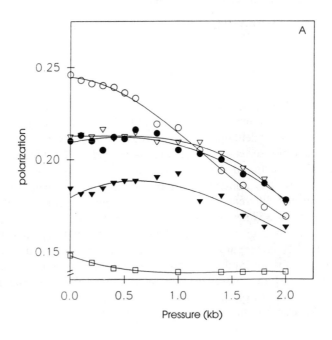

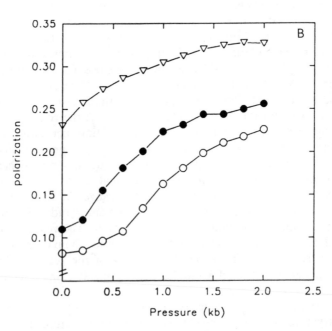

Figure 21.3. (**A**) The polarization of gramicidin in DMPC bilayers as a function of pressure at 7.5 °C (○), 28 °C (●), and (▽), 28 °C in the presence of 0.5 *M* KI, and 40 °C (▽). (□) are data for gramicidin incorporated into DOPC bilayers. (**B**) The polarization of 11-AU in gramicidin-DMPC bilayers at 7.5 °C (▽), 28 °C (●), and 40 °C (○). (Reprinted with permission from S. Scarlata, 1991, *Biochemistry* **30,** 9856. Copyright 1991 American Chemical Society.)

because pressure increases the thickness of the bilayer due to the straightening of the lipid chains, but the less compressible protein cannot increase its length to preserve these hydrogen bonds. Thus, we postulated that the increase in tryptophan rotational motion with pressure is due to the rupture of hydrogen bonds between the indole protons and the carbonyl of the lipid.

If the above hypothesis is correct, we would expect that fluorophores that could not form hydrogen bonds with the lipid would show a decrease in rotational motion with pressure. To explore this idea, we labeled the tyrosine at position 11 of gramicidin C with dansyl and followed the response of dansyl to changes in temperature and pressure (Teng et al., 1991). Unlike indoles, the dansyl group cannot donate a proton to hydrogen bond with the carbonyls of the lipid. We found that the rotational motion of the dansyl residue increased systematically with temperature and decreased systematically with pressure, showing behavior similar to fluorescent fatty acids (Figure 21.4). These data, along with subsequent NMR studies by others; support the notion that changes in indole–lipid hydrogen bonds are responsible for the unusual temperature and pressure behavior of gramicidin tryptophans.

Further support comes from fluorescence studies of a longer gramicidin in which alanine was attached to the N-terminus. This longer N-N terminal dimer shows a constant polarization with pressure and a smaller depolarization with temperature in the fluid phase than does native gramicidin, supporting the idea that rupture of hydrogen bonds in the native peptide is due to pressure-induced length mismatch of the lipid and the peptide.

We then considered whether indole–lipid hydrogen bonds were a general feature of integral membrane proteins and whether similar responses to increased lipid packing could be observed for other proteins. To test this idea, we conducted

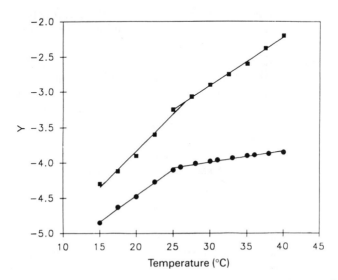

Figure 21.4. The Y versus temperature plot of dansyl gramicidin in DMPC bilayers (●) and 11 AU in DMPC bilayers (■). (Reprinted with permission from Q. Teng, R. Koeppe, and S. Scarlata, 1991, *Biochemistry* **30,** 7987. Copyright 1991 American Chemical Society.)

identical studies using the integral membrane protein bacteriorhodopsin (Scarlata, 1993). This protein contains eight tryptophan residues; five are close to the interface and two are close to the chromophore (see Hess et al., 1982, for review). We bleached the chromophore to prevent its quenching the intrinsic protein fluorescence and reconstituted the protein in DOPC bilayers. As with gel-phase gramicidin, the rotational motion of the tryptophan residues did not change with temperature. The anisotropy and lifetime remained constant from 1 to 2000 atm. Using a series of energy-transfer probes, we found no alteration in the location of these residues under pressure, although our data were consistent with the tryptophans becoming more surface orientated as the chains straighten. Thus, bacteriorhodopsin can accommodate the 1–3 Å increase in bilayer thickness caused by increased pressure. We note that since these studies a predominance of Trp residues located at interstitial positions has been observed. The advantage of having this less energetic interaction becomes clear when one considers that they offer more flexibility in accommodating changes in membrane thickness.

To answer our original question of how the membrane viscosity may control the function of integral membrane proteins by coupling and decoupling of local motions, we return to the temperature data collected for dansyl gramicidin (Figure 12.4) where we can view the changes in dansyl rotations without interference from specific interactions with the lipid head groups. We find that the **b** value of dansyl gramicidin is less than that of the free fatty acid probe in both the gel and liquid crystal phases, indicating that we have not achieved viscosities high enough to uncouple local motions. This result argues against control of protein function through this mechanism.

STABILIZATION OF PROTEIN CONFORMERS BY VISCOSITY

The studies discussed in the previous section show that increasing the lipid packing through pressure could destabilize specific interactions between the protein and the lipid head group. We then asked whether changes in lipid packing may stabilize or destabilize a particular conformational state of a protein through these surface interactions. We tested this idea with GLUT1, a facilitative transporter for glucose and other hexoses (for review see Baldwin, 1993), which has two binding sites for glucose, one on either site of the membrane, and a binding site for a non-competitive inhibitor, cytochalasin B, close to the inside binding site. GLUT1 can be purified from human erythrocytes as a protein suspension in endogenous lipids in which approximately 80% of the molecules are in an inside-out configuration. The secondary structure of GLUT1 has been proposed on the basis of both hydrophobicity analysis and biochemical studies (see Baldwin, 1993). In this structure, four of the six tryptophans and nine of the 13 tryosines are interfacial.

We chose a study GLUT1 since it appears to undergo a conformational change with ligand binding that is detectable by intrinsic fluorescence. Gorga and Leinhard (1982) noted a large decrease in intrinsic intensity of GLUT1 with glucose additions and with cytochalasin B additions. We also have observed this change (Figure 21.5). The extent of this decrease and its response to different ligands has led us, along with other groups, to assume that the change in fluorescence reflects two distinct conformational states of the protein.

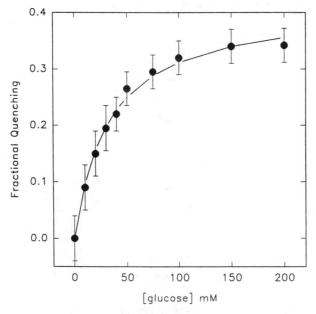

Figure 21.5. The increase in fractional quenching of the intrinsic fluorescence of GLUT1 as a function of glucose addition. (Reprinted with permission from S. Scarlata, H. McBath, and Haspel, 1995, *Biochemistry* **34,** 7705. Copyright 1995 American Chemical Society.)

We monitored the response of GLUT1 in its liganded and unliganded form to high pressure (Scarlata et al., 1995). In the absence of ligand, a 20% decrease in intensity occurs in the first 500 bar (Figure 21.6). However, in the presence of ligand, a much larger change in intensity occurs. Similar changes are seen when the protein is only half-saturated at atmospheric pressure and when cytochalasin B is used. The simplest interpretation of these data is that high pressure stabilizes the liganded form of the protein. Since the aqueous soluble quencher, I^-, has better access to the tryptophan residues in the liganded form than in the unliganded form, it appears that the tryptophan residues in the liganded form are closer to the surface. In this way, the increase in lipid length that accompanies pressure may serve to stabilize tryptophan and tyrosine lipid contacts and, ultimately, the liganded state of the protein.

MECHANISM THROUGH WHICH PRESSURE AFFECTS THE CATALYTIC RATE OF INTEGRAL MEMBRANE PROTEINS

We have presented examples showing that pressure can stabilize lipid-protein contacts and particular conformational states, but how are these effects manifested in the viscosity effects of integral membrane proteins? We approached this question by surveying the many examples of the dependence of the activity of integral membrane proteins with temperature (see Zakim et al., 1992, and references therein). We used temperature since very few pressure studies have been done.

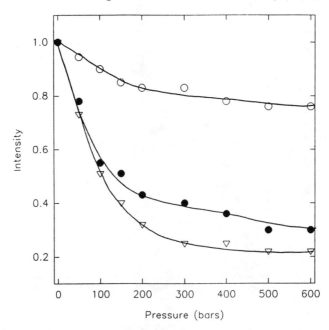

Figure 21.6. The decrease in intrinsic fluoresence of GLUT1: ($\bigcirc$) = without glucose; ($\bullet$) = half-saturating glucose (25 mM); and ($\bigtriangledown$) = saturating (200 mM) glucose. (Reprinted with permission from S. Scarlata, H. McBath, and Haspel, *Biochemistry* **34,** 7706. Copyright 1995 American Chemical Society.)

Returning to Kramers's relation, we find that if a protein is embedded in DMPC then its catalytic rate will show a discontinuity at 22 °C due to the 10-fold decrease in viscosity at the gel to liquid-crystal phase transition (Evans & Needham, 1987). In Figure 21.7 we show the plot for this behavior derived under the assumption that the slope, $-E_a$, remains constant. We note that for actual proteins, the slope includes terms for the work of expansion of the membrane and the activation energy for viscous flow, both of which are small compared to the activation energy.

We then surveyed integral membrane enzymes whose activity has been shown to respond to the gel to liquid-crystal phase transition, and found that all cases except two showed a discontinuity in the activation energy (which reflects the reaction pathway) rather than in the catalytic rate. The two exceptions had activity in the fluid phase but not in the gel phase. We conclude that Kramers's relation does not hold for integral membrane proteins.

CONCLUSIONS

Through these temperature and pressure studies, we find that the simple relationship between viscosity and reaction rate does not hold for membrane proteins, probably because these rates are governed by energy barriers in the protein that can become damped with increased viscosity. However, dampling by the lipid matrix does not appear to be extensive enough to uncouple the local coupled

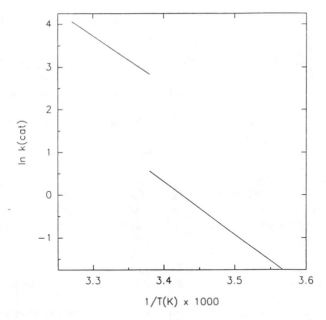

Figure 21.7. Simulated Arrhenius plot for a protein embedded in DMPC bilayers assuming a 10-fold decrease in viscosity at the phase transition temperature (22 °C). An activation energy of 22 kcal/mol was used. (Reprinted with permission from D. Zakim, J. Kavecansky, and S. Scarlata, 1992, *Biochemistry* **31,** 11590. Copyright 1992 American Chemical Society.

motions of the protein. Our studies indicate that the most likely mechanism through which the viscosity, in terms of lipid packing, could regulate protein function is by the alteration of specific interactions between the protein and lipid that may stabilize or destabilize a particular conformational state.

ACKNOWLEDGMENTS: I thank Dr. Gregorio Weber for initiating this study, and acknowledge all of the enlightening discussions with Drs. David Zakim, Olaf Andersen, Cathy Royer, and Walter Zurawsky. I thank Dr. Roger Koeppe for providing the interesting gramicidin analogues, Dr. Martin Teinze for the bacteriorhodopsin, and Dr. Howard Haspel for his help with the glucose transporter.

REFERENCES

Austin, R., Beeson, K., Eisenstein, L., Fraunfelder, H., & Gunsalas, I. C. (1975). Dynamics of ligand binding to myoglobin. *Biochemistry* **14**, 5355–5373.

Baldwin, S. (1993). Mammalian passive glucose transporter: members of an ubiquitous family of active and passive transport proteins. *Biochim. Biophys. Acta* **1154**, 17–49.

Batschinski, A. J. (1913). Untersuchungen unter die Ihnere Reibung der Flussigkeiten. *Z. Phys. Chem.* **84**, 643–655.

Beece, D., Eisenstein, L., Frauenfelder, H., Good, D., Marden, M., Reinisch, L., Reynolds, A., Sorensen, L., & Yue, K. (1980). Solvent viscosity and protein dynamics. *Biochemistry* **19**, 5147–5157.

Braganza, L., & Worcester, D. (1986). Structural changes in lipid bilayers and biological membranes caused by hydrostatic pressure. *Biochemistry* **25**, 2591–2596.

Cone, R. A. (1972). Rotational diffusion of rhodopsin in the visual receptor membrane. *Nature* **236**, 39–43.

Evans, E., & Needham, D. (1987). Elastic behavior of phosphatidyl choline membranes. *J. Phys. Chem.* **91**, 4219–4228.

Gavish, B. (1979). The role of geometry and elastic strains in the dynamic states of proteins. *Biophys. Struc. Mech.* **4**, 37–42.

Gavish, B., & Werber, M. (1979). Viscosity-dependent structural fluctuations in enzyme catalysis. *Biochemistry* **18**, 1269–1275.

Gorga, F., & Leinhard, G. (1982). Changes in the intrinsic fluorescence of the human erythrocyte monosaccharide transporter upon ligand binding. *Biochemistry* **17**, 1905–1908.

Hess, B., Kuschmitz, D., & Engelhard, M. (1982). In *Membranes and Transport*, ed. Martonosi. New York, Plenum, **3**: 309–318.

Kramers, H. A. (1940). Diffusion of small molecules in fluids as defined by the free volume. *Physica* (Amsterdam) **7**, 284–304.

O'Connell, A., Koeppe, R., II, & Andersen, O. (1990). Kinetics of gramicidin channel formation in lipid bilayers: transmembrane monomer association. *Science* **250**, 1256–1259.

Radda, G., & Smith, D. (1970). Retinol: a fluorescence probe for membrane lipids. *FEBS Lett.* **9**, 287–289.

Rholam, M., Scarlata, S., & Weber, G. (1984). Frictional resistance to local rotations of fluorophores in proteins. *Biochemistry* **23**, 6793–6796.

Rudy, B., & Gitler, C. (1972). Microviscosity of the cell membrane. *Biochim. Biophys. Acta* **288**, 231–236.

Scarlata, S. (1988). The effects of viscosity of the resistance to fluorphore rotation in model membranes. *Biophys. J.* **54**, 1149–1157.

Scarlata, S. (1989). Evaluation of the thermal coefficient of the resistance to fluorphore rotation in model membranes. *Biophys. J.* **55**, 1215–1223.

Scarlata, S. (1991a). Compression of lipid membranes as observed at varying membrane positions. *Biophys. J.* **60**, 334–340.

Scarlata, S. (1991b). The effect of increased chain packing on gramicidin tryptophan-lipid interactions. *Biochemistry* **30**, 9855–9859.

Scarlata, S. (1993). Effect of solvent viscosity on the rotational motions of bacteriorhodopsin tryptophans. *Biophys.* (Life Sci. Adv.) **12**, 13–18.

Scarlata, S., Rholam, M., & Weber, G. (1984). Frictional resistance to local rotations of fluorophores in small peptides. *Biochemistry* **23**, 6789–6792.

Scarlata, S., McBath, H., Haspel, H. (1995). Effect of lipid packing on the conformational states of purified GLUT-1 hexose transporter. *Biochemistry* **34**, 7703–7711.

Shinitzky, M., & Barenholz, Y. (1978). Fluidity parameters of lipid regions as determined by fluorescence polarization. *Biochim. Biophys. Acta* **515**, 367–394.

Somogyi, B., Welch, G., & Damjanovich, D. (1984). The dynamic basis of energy transduction in enzymes. *Biochim. Biophys. Acta* **768**, 81–112.

Teng, Q., Koeppe, R., II, & Scarlata, S. (1991). Effects of salt and membrane fluidity on fluorphore motions of a gramicidin C derivative. *Biochemistry* **30**, 7984–7990.

Thulborn, K., & Sawyer, W. (1978). Properties and locations of a set fluorescent probes sensitive to the fluidity gradient of the lipid bilayer. *Biochim. Biophys. Acta* **511**, 125–140.

Tilley, K., Thulborn, K., & Sawyer, W. (1979). An assessment of the fluidity gradient of the lipid bilayer as determined by a set of n-(9-anthroyloxy) fatty acids ($n = 2, 6, 9, 12, 16$). *J. Biol. Chem.* **254**, 2592–2594.

Vincent, M., deForesta, J., Galley, J., & Alfsen, A. (1982). Nanosecond fluoresence anisotropy decays of n-(9-anthroyloxy) fatty acids in dipalmitolylphosphatidyl-choline vesicles with regards to isotropic solvents. *Biochemistry* **21**, 708–716.

Wallace, B. (1990). Structure of gramicidin A. *Annu. Rev. Biophys. Biophys. Chem.* **19**, 127–157.

Weber, G., Scarlata, S., & Rholam, M. (1984). Frictional resistance to local rotation of fluorophores in solvents. *Biochemistry* **23**, 6785–6789.

Zakim, D., Kavecansky, J., & Scarlata, S. (1992). Are membrane enzymes regulated by the viscosity of the membrane environment? *Biochemistry* **31**, 11589–11594.

22

Pressure- and Temperature-Induced Inactivation of Microorganisms

HORST LUDWIG, WILHELM SCIGALLA, and BERND SOJKA

Bacteria are unstable when the temperature or pressure is sufficiently high. Their inactivation by pressure is the result of a complicated interplay of both temperature and pressure effects. The p-T stability diagram of bacteria is similar to that of proteins. But inactivation kinetics of bacteria indicate that the lethal event cannot be the denaturation of the most sensitive proteins in the cell. For in that case one would expect a lag time followed by a sudden inactivation when the last copy of those sensitive proteins was destroyed. On the contrary, it appears as if the kinetics is caused by a single damage mechanism. In addition, some evidence suggests that the membrane is involved. Therefore, it seems that membrane-associated proteins play a major role in the activation of bacteria. The inactivation of bacterial spores shows an even more complex T-p interrelationship. The reason is that two different processes are combined in spore inactivation: the germination of dormant spores at comparatively low pressures and the inactivation of the germinated specimens at high pressures. Thus, special procedures are needed for effective spore inactivation.

Microorganisms are killed when the surrounding hydrostatic pressure is sufficiently high. This finding provides the basis for developing a physical sterilization method for drugs and food. Initial experiments in this area were carried out nearly 100 years ago (Hite, 1899), but technical shortcomings and incomplete scientific knowledge impeded their utilization at that time. Recently, the application of high pressure in food preservation and processing has garnered new interest (Hayashi 1989; Balny et al., 1992).

METHODOLOGY

To collect precise kinetic data for pressure-induced degermination, we constructed a device which consisted of 10 pressure vessels that could be thermostated in two

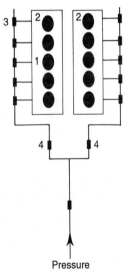

Pressure

Figure 22.1. High-pressure device for doing kinetic work: (1) pressure vessel; (2) temperature bath; (3) valve; (4) pressure-gauge.

groups of five each (Figure 22.1). Each vessel had an inner diameter of 1.2 cm and an inner length of 12 cm. The samples were separated from the pressure medium, water, by polyethylene tubes or bags. The maximum pressure was 7 kbar.

The vegetative bacteria were always freshly cultured from one single organism before each experimental run. The preparations were allowed to grow to the exponential phase and were used in experiments just before the stationary phase had been reached. Since they descended from one cell, a largely uniform bacterial population was realized. In these investigations, the bacteria were suspended in nutrient broth. (When the experiments were duplicated in physiological NaCl solution, the result was in every case the same as in broth.) The bacteria were grown from the following strains: *Pseudomonas aeruginosa*, ATCC 25102; *Escherichia coli*, ATCC 11229; *Serratia marcescens*, unnamed strain used in the Federal Research Center for Nutrition, Karlsruhe; and *Staphylococcus aureus*, NCIMB 50080.

Spores of *Bacillus subtilis* were produced from the strain ATCC 9372, purified by washing with 0.9% NaCl solution, centrifuged at 4000 g, and stored in the NaCl solution. By using this procedure, spores free from vegetative cells could be obtained. For the experiments, the spores were suspended in a salt solution containing amounts of D-glucose and L-alanine (Sojka & Ludwig, 1994).

The number of surviving microorganisms was evaluated by counting colonies on agar plates. If no colonies were present, the samples were tested for sterility according to USP XXII. The experiments were run three times in most cases. However, results from only a single run are given in the figures.

RESULTS

Inactivation of Vegetative Bacteria

Pseudomonas aeruginosa

Figure 22.2 shows the kinetics of inactivation for two bacteria populations with different initial concentrations of 10^9 and 10^6 cells per ml. The number of surviving bacteria per ml, shown as colony-forming units, is plotted logarithmically on the ordinate; the time on the abscissa. The runs are perfectly linear and parallel to each other. Clearly, the inactivation is a first-order reaction. At 20 °C, the inactivation rate accelerates as the applied pressure increases until a limit to the effectiveness of the pressure is reached at 3 kbar (Figure 22.3). Increases to 4 or 5 kbar had no additional effect.

The degree of inactivation depends only on the time under pressure. It does not depend on the rates of compression or decompression. We proved this for all the bacteria described here.

The dependence of inactivation on temperature at 2 kbar (Figure 22.4) looks complicated. Minimal inactivation occurs at room temperature. This feature changes at higher pressure levels (Figure 22.5), where a normal temperature dependence is found. The kinetics at 2 kbar at different temperatures (Figure 22.6) in the first minutes parallels the results shown in Figure 22.4, in that low and high temperatures give better inactivation than do medium temperatures, but after the

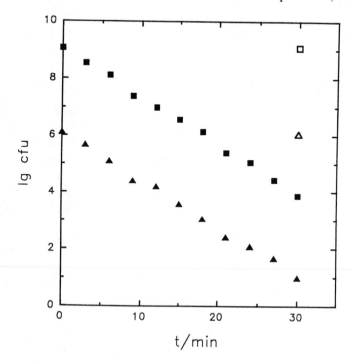

Figure 22.2. Inactivation of *Pseudomonas aeruginosa* at 2 kbar and 25 °C: ■ and ▲ are different initial concentrations; open symbols are controls.

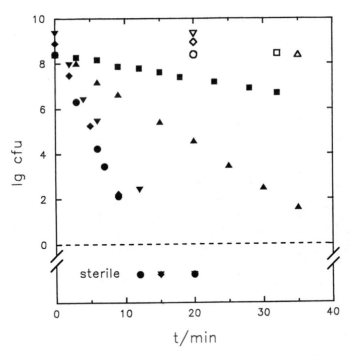

Figure 22.3. Inactivation of *Pseudomonas aeruginosa* at 20 °C and at different pressures: 1.7 kbar (■), 2 kbar (▲), 3 kbar (▼), 4 kbar (●), and 5 kbar (◆); open symbols are controls.

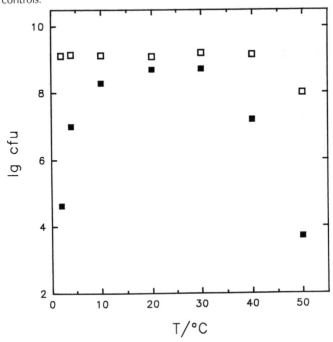

Figure 22.4. Temperature dependence of *Pseudomonas aeruginosa* inactivation, 10 minutes at 2 kbar; □ control.

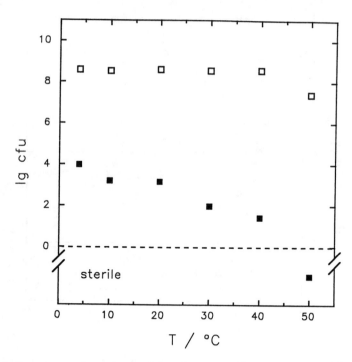

Figure 22.5. Temperature dependence of *Pseudomonas aeruginosa* inactivation, 4 minutes at 3 kbar; □ control.

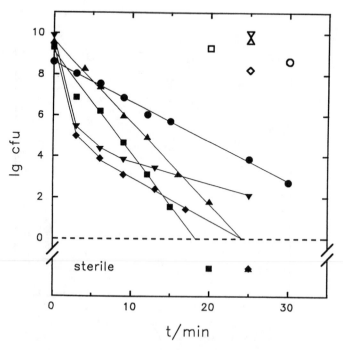

Figure 22.6. Inactivation of *Pseudomonas aeruginosa* at 2 kbar and different temperatures: 1 °C (■), 10 °C (▲), 30 °C (●), 40 °C (▼), and 50 °C (◆); open symbols are controls.

initial minutes one finds a new phenomenon at temperatures above 30 °C. Here the formerly linear curve is bent and appears to be composed of two linear segments, as if the formerly uniform population now consisted of two parts, each with a different sensitivity to pressure. Remarkably, the inactivation rate for the less sensitive part is slower at 40 °C than the rate at 30 °C; not before 50 °C is the faster rate once again attained.

Figure 22.7 demonstrates how the addition of ethanol to the solution causes much faster killing of the bacteria. The rate is doubled for each additional 10% of ethanol. Figure 22.3 suggested that a pressure limit may exist, since pressure values above this limit did not further increase the inactivation. This limit seems to hold also in 10% ethanolic solution, even though the limit can be exceeded with an ethanol concentration of 20% (Figure 22.8). In contrast to ethanol, glycerol stabilizes the bacteria, with the degree of stabilization being dependent on the glycerol concentration (Figure 22.9).

Escherichia coli

Figure 22.10 shows the death of *E. coli* at 2.5 kbar at two different temperatures. Similar to *Ps. aeruginosa*, the characteristics of the kinetic curve change near 30 °C, but the other way around. Here all temperatures below 30 °C give biphasic inactivation curves; higher temperatures show a simple first-order reaction. The curved lines are shifted in parallel in the case of different initial concentrations (Figure 22.11). This indicates first-order reactions in both phases of the biphasic runs. The progeny of one of the surviving pressure-resistant cells show the same

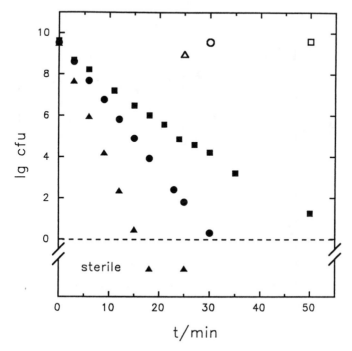

Figure 22.7. Inactivation of *Pseudomonas aeruginosa* at 2 kbar and 25 °C as a function of ethanol concentration: 0% (■), 10% (●), and 20% (▲); open symbols are controls.

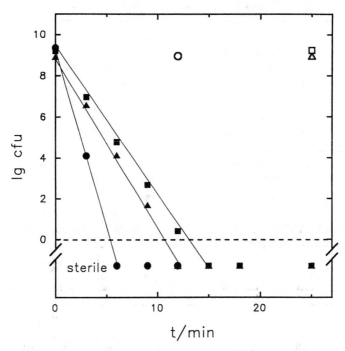

Figure 22.8. Inactivation of *Pseudomonas aeruginosa* at 25 °C in ethanol-water mixtures: 10% ethanol at 3 kbar (■), 10% ethanol at 4 kbar (▲), and 20% ethanol at 3 kbar (●); open symbols are controls.

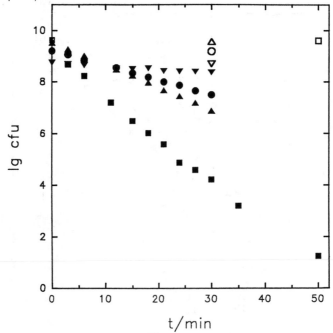

Figure 22.9. Inactivation of *Pseudomonas aeruginosa* at 2 kbar and 25 °C as a function of glycerol concentration: 0% (■), 20% (▲), 50% (●), and 100% (▼); open symbols are controls.

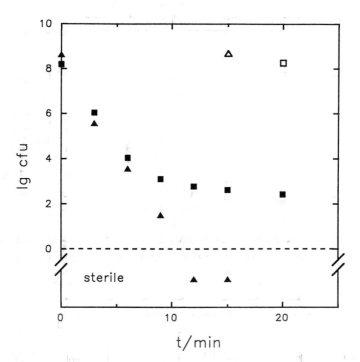

Figure 22.10. Inactivation of *Escherichia coli* at 2.5 kbar and at 30 °C (■) and 50 °C (▲); open symbols are controls.

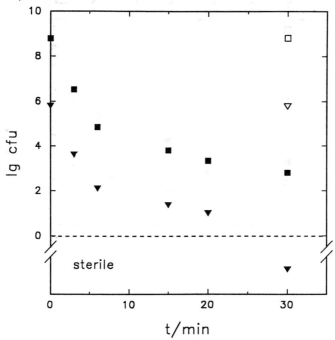

Figure 22.11. Inactivation of *Escherichia coli* at 2.5 kbar and 25 °C: ■ and ▼ are different initial concentrations; open symbols are controls.

inactivation curve as their antecedent population. Thus, each individual cell of the population retains the potential to reproduce the whole inactivation curve.

The temperature dependence of *E. coli* inactivation is similar to that shown for *Pseudomonas* in Figures 22.4 and 22.5, with a minimal rate at room temperature at a pressure of 2 kbar (Ludwig et al., 1992). The pressure dependence of *E. coli* inactivation is also similar to that of *Pseudomonas*. Figures 22.12 and 22.13 give the results at 25 °C in the form of the *D* values for the two parts of the biphasic curves. The *D* value is the time in minutes needed to kill 90% of the bacteria. *D1*, representing the fast inactivation, reaches a minimal end value of about half a minute at 3 kbar, while *D2*, which represents the slow part, diminishes exponentially with increasing pressure. Simultaneously, the ratio of sensitive to insensitive specimens increases considerably with pressure (Figure 22.14).

Figure 22.15 gives the time needed for the sterilization of solutions containing 10^{10} *E. coli* bacteria at 2.5 kbar at different temperatures. Again the break is visible at 30 °C, at which the change from single- to biphasic behavior occurs. Below 30 °C the overall sterilization time is nearly constant because the time is determined by the inactivation of the insensitive fraction which is relatively indifferent to temperature.

As in the case of *Pseudomonas*, the addition of ethanol weakens the *E. coli* bacteria, and the addition of glycerol stabilizes them against inactivation. An example of the effect of ethanol is seen in Figure 22.16. Inactivation is increased mainly in the slow phase.

From the kinetic data a pressure-temperature stability diagram for *E. coli*

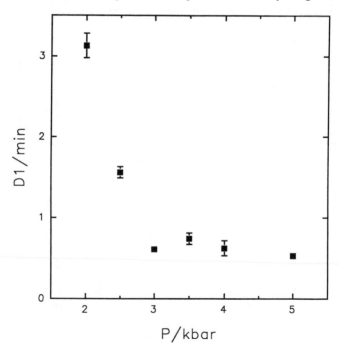

Figure 22.12. Decimal reduction time of *Escherichia coli* at 25 °C and different pressures; *D1* describes the fast part of the biphasic inactivation curve.

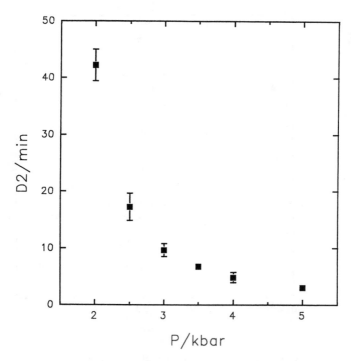

Figure 22.13. Decimal reduction time of *Escherichia coli* at 25 °C and different pressures; *D2* describes the slow part of the biphasic inactivation curve.

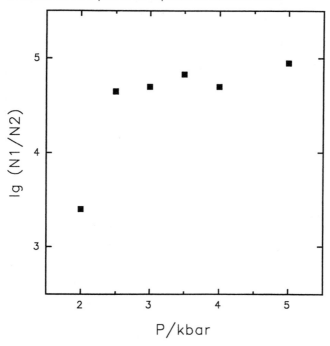

Figure 22.14. Ratio of the amplitudes, *N1/N2*, of the biphasic *Escherichia coli* inactivation at 25 °C and different pressures.

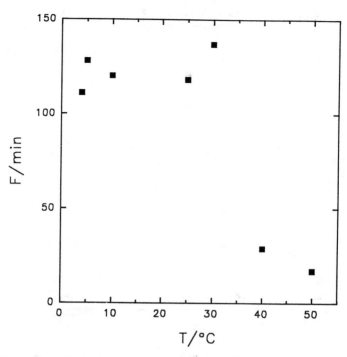

Figure 22.15. F values of *Escherichia coli* inactivation at 2.5 kbar as a function of temperature. (The F value is the time in minutes required to inactivate a population of 10^{10} bacteria at a given pressure and temperature.)

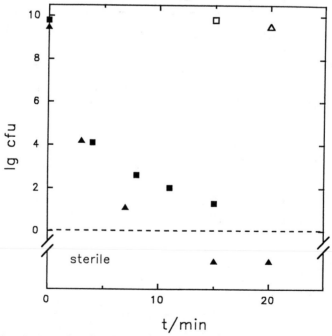

Figure 22.16. Inactivation of *Escherichia coli* at 4 kbar and 25 °C in ethanol-water mixtures: 0% (■), and 20% ethanol (▲); open symbols are controls.

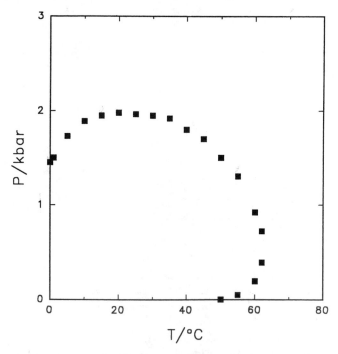

Figure 22.17. Stability diagram of *Escherichia coli*, showing the decrease of bacterial count by two orders of magnitude within 5 minutes.

can be derived. This phase diagram, shown in Figure 22.17, resembles that of a protein (Suzuki, 1960; Hawley, 1971).

Serratia marcescens

The inactivation of this organism is similar to that of *E. coli*. A biphasic course is found below 30 °C, and a single straight line is found above this temperature (Figure 22.18). The same resemblance is seen in the temperature dependence in Figure 22.19.

Staphylococcus aureus

These bacteria show biphasic inactivation behavior, as do *E. coli* and *S. marcescens* (Figure 22.20), but the pressure levels needed are much higher.

Inactivation of Bacterial Spores

Figure 22.21 shows the inactivation of a *Bacillus subtilis* preparation by 2 kbar and 4 °C. Within a few minutes, 99.99% of the specimens are killed, this percentage represents the vegetative bacilli. The remaining 0.01% seem stable under the given conditions; this percentage represents spores. The inactivation of pure spores at 3 kbar and 40 °C is shown in the upper curve of Figure 22.22. Even after 1 hour, inactivation reaches only 99%. Much faster inactivation is obtained when a period

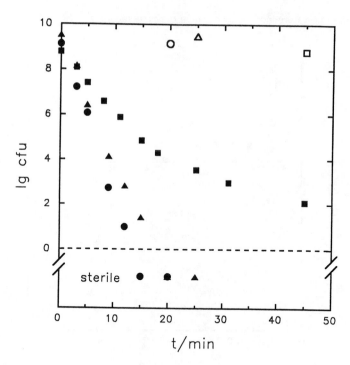

Figure 22.18. Inactivation of *Serratia marcescens* at 2.5 kbar and different temperatures: 25 °C (■), 40 °C (▲), and 50 °C (●); open symbols are controls.

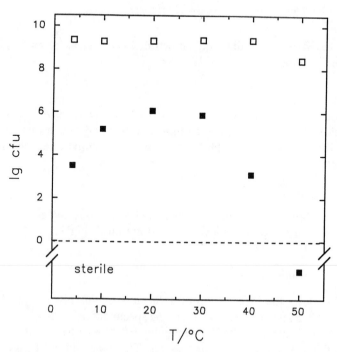

Figure 22.19. Temperature dependence of *Serratia marcescens* inactivation after 13 minutes at 2 kbar; □ control.

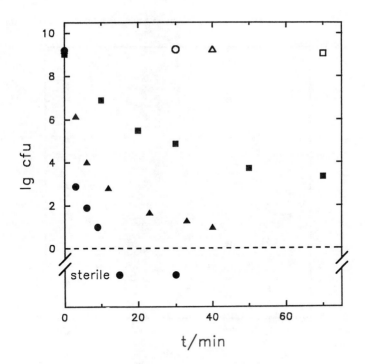

Figure 22.20. Inactivation of *Staphylococcus aureus* at different pressures and temperatures: 4 kbar and 25 °C (■), 5 kbar and 25 °C (▲), and 5 kbar and 40 °C (●); open symbols are controls.

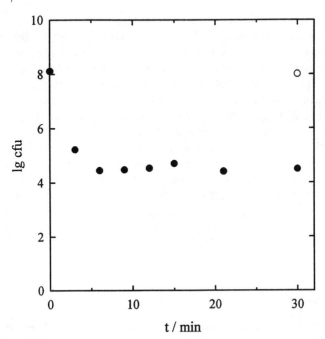

Figure 22.21. Pressure inactivation of vegetative germs of *Bacillus subtilis* at 2 kbar and 4 °C; ○ control.

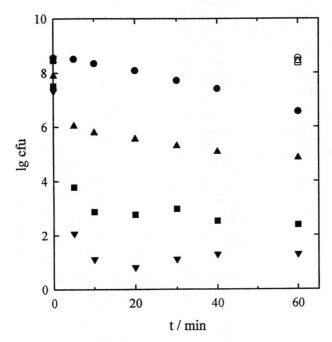

Figure 22.22. Inactivation of *Bacillus subtilis* spores at 3 kbar and 40 °C after pretreatment with 600 bar at 40 °C for different times: 0 h (●), 0.5 h (▲), 2 h (■), and 20 h (▼); open symbols are controls.

with a comparatively low pressure of 600 bar precedes the high-pressure treatment. The inactivation effected by 600 bar alone is small: one order of magnitude within 20 hours at most. This drop in activity can be located at zero on the ordinate of Figure 22.22. Spore suspensions pretreated in such a way are then very sensitive to an inactivating high pressure. The sensitive specimens, whose fraction depends on the duration of pretreatment, are killed in a few minutes by 3 kbar. Figure 22.23 shows that a very long pretreatment with 600 bar and only 10 minutes of 5 kbar gives the same result as four alternating pressure cycles after a total time of 4 hours at 40 °C. When the temperature is higher, the pressure cycles are more effective than a single pretreatment followed by high pressure. This can be seen in Figure 22.24 for 50 °C. Nevertheless, it would be economically preferable to use long periods of low pressure and short periods of high pressure in the treatment cycles.

DISCUSSION

Vegetative Bacteria

The stability characteristics of bacteria are similar to those of proteins. That is most clearly shown in Figure 22.17. It could then be stated that the cell dies when the last copy of an important protein is denatured and cannot be rebuilt. If this were true, the kinetics of inactivation would show a lag time, as long as $(n-1)$

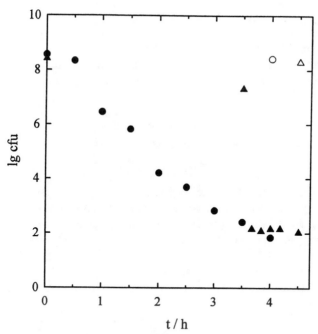

Figure 22.23. Inactivation of *Bacillus subtilis* spores by alternating pressures of 0.6 and 5 kbar at 40 °C with different intervals; 30 min (●) and 210 min (▲); open symbols are controls.

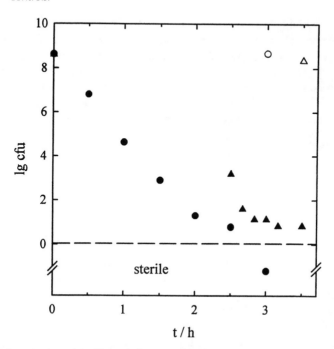

Figure 22.24. Inactivation of *Bacillus subtilis* spores by alternating pressures of 0.6 and 5 kbar at 50 °C at different intervals: 30 min (●) and 150 min (▲); open symbols are controls.

of *n* copies of the protein are destroyed. Actually, we found such a behavior recently in the case of vegetative *Clostridium stiglandii* (v. Almsick & Ludwig, unpublished experiments). But this seems to be an exception. In the majority of cases, the inactivation of bacteria starts as soon as the inactivating agent is present. Moreover, first-order kinetics is usually found. Even flexed curves prove not to be inconsistent because they are composed of two or more parts which are themselves first-order reactions. We proved this for all the bacteria described here. An example is given in Figure 22.11. First-order inactivation means that in a uniform population the single cells are killed according to a given probability by random events in one hit. Then one has to look for those possible targets which need only one hit to destroy the whole cell. Such targets could be the DNA or the membrane. Of these two, the nucleic acid is relatively resistant to pressure (Suzuki & Taniguchi, 1972). Therefore, the most probable point of attack is the membrane and, most likely, a protein or protein complex in the membrane. This would explain the temperature dependence of inactivation (Figures 22.4 and 22.19) and the similarity of the stability diagram in Figure 22.17 with that of proteins. At high pressures the killing rate reaches a limit (Figures 22.3 and 22.12). That means an activation volume of zero, which is consistent with the denaturation of a protein (Miyagawa & Suzuki, 1962). There is some evidence to support the idea of membrane involvement. First, there is the sharp bend in the inactivation curve that appears above (Figure 22.6) or, in most cases, below about 30 °C (Figures 22.10, 22.11, 22.18, and 22.20). Biological membranes sometimes show phase transitions in this temperature region (Eze, 1990). We therefore assume a small fraction of the cells to have an altered membrane composition with the consequent phase transition and corresponding break in sensitivity. Second, the promotional effect of ethanol and the inhibiting effect of glycerol could be explained by interactions of these agents with the membrane and membrane-bound proteins. It is known that the virucidal effect of ethanol is best in the case of lipid-coated viruses (Klein & Deforest, 1963). Third, Smelt and Rijke (1993) prepared cells of *Lactobacillus plantarum* which differed in the fluidity of the membrane. These authors were able to show that cells with a more fluid membrane are more resistant to pressure.

Spores

Dormant bacterial spores are very resistant to high pressure. Only when germination is induced can the spores be killed by pressure (Sojka & Ludwig, 1994; Butz et al., 1990). It was shown that pressures of medium intensity induce the germination process (Gould et al., 1970a, b; Ludwig et al., 1992). How pressure induces germination and its mechanism is not yet known and is under study. From Figure 22.22 it is clear that germination is a lengthy process for a small fraction of spores; in contrast, the inactivation of the germinated spores is as fast as that of vegetative cells. The results shown in Figures 22.23 and 22.24 prove that pressure cycles are useful for the sterilization of spores. Again, the underlying mechanism needs clarification.

CONCLUSION

Our data have given us a clue to the solution of the problem, but our conjectures about the mechanisms of pressure-induced inactivation remain incomplete and the precise elucidation must still be worked out.

ACKNOWLEDGMENTS: This work was supported by the European Community (AIR 1-CT 92-0296) and by the Max-Buchner-Forschungsstiftung.

REFERENCES

Balny, C., Hayashi, R., Heremans, K., & Masson, P., eds. (1992). *High Pressure and Biotechnology*. London, Colloque INSERM-John Libby.

Butz, P., Ries, J., Traugott, U., Weber, H., & Ludwig, H. (1990). Hochdruckinaktivierung von Bakterien und Bakteriensporen. *Pharm. Ind.* **52**, 487–491.

Eze, M. O. (1990). Consequences of the lipid bilayer to membrane-associated reactions. *J. Chem. Educ.* **67**, 17–20.

Gould, G. W., Sale, A. J. H., & Hamilton, W. A. (1970a). Inactivation of bacterial spores by hydrostatic pressure. *J. Gen. Microbiol.* **60**, 323–334.

Gould, G. W., & Sale, A. J. H. (1970b). Initiation of germination of bacterial spores by hydrostatic pressure. *J. Gen. Microbiol.* **60**, 335–346.

Hawley, S. A. (1971). Reversible pressure-temperature denaturation of Chymotrypsinogen. *Biochemistry* **10**, 2436–2442.

Hayashi, R. (1989). Application of high pressure to food processing and preservation: Philosophy and development. *Engineering and Food*, Vol. 2, ed. W. E. L. Spiess & H. Schubert, Oxford, Elsevier, pp. 815–826.

Hite, B. H. (1899). The effect of pressure in the preservation of milk. *Bull. West VA. Univ. Agr. Exp. Sta.*, no. 58.

Klein, M., & Deforest, A. (1963). Antiviral action of germicides. *Soap Chem. Spec.* **39**, 7, 70–95.

Ludwig, H., Bieler, C., Hallbauer, K., & Scigalla, W. (1992). Inactivation of microorganisms by hydrostatic pressure. In *High Pressure and Biotechnology*, ed. C. Balny et al. London, John Libby, pp. 25–32.

Miyagawa, K., & Suzuki, K. (1962). Pressure inactivation of enzyme: some kinetic aspects of pressure inactivation of Chymotrypsin. *Rev. Phys. Chem. Jpn.* **32**, 51–56.

Smelt, J. P. P. M., & Rijke, A. G. F. (1993). Influence of ultra high pressure on the behaviour of *Lactobacillus plantarum*. Abstracts of the 31st Annual Meeting of the European High Pressure Research Group. 30th August—3rd September 1993. Belfast.

Sojka, B., & Ludwig, H. (1994). Pressure-induced germination and inactivation of *Bacillus subtilis* spores. *Pharm. Ind.* **56**, 660–663.

Suzuki, K. (1960). Studies on the kinetics of protein denaturation under high pressure. *Rev. Phys. Chem.* (Japan) **29**, 91–98.

Suzuki, K., & Tanigushi, Y. (1972). Effect of pressure on biopolymers and model systems. In *The Effects of Pressure on Organisms*, (ed. M. A. Sleigh & A. G. MacDonald). Cambridge, Cambridge University Press, pp. 103–124.

23

Osmotic and Hydrostatic Pressure as Tools to Study Molecular Recognition

STEPHEN G. SLIGAR, CLIFFORD R. ROBINSON, and
MARK A. McLEAN

The question of molecular recognition is a central paradigm of molecular biology, playing central roles in most, if not all, cellular processes. Failed recognition events have been implicated in numerous disease states, ranging from flawed control of gene regulation and cellular proliferation to defects in specific metabolic activities. Historically, questions of molecular recognition have been approached through organic synthesis and through actual structural studies of biomolecular complexes. Fundamental insight into the mechanisms of molecular recognition can be realized through the use of broad interdisciplinary tools and techniques. In particular, the use of recombinant DNA technology in concert with hydrostatic and osmotic pressure methodologies have proven to be ideal for understanding the fundamental mechanisms of recognition. In our presentation, we will focus on recent results from our laboratory that examine three major classes of recognition events in biological systems:

1. Protein–protein recognition: here we seek to define the role of specific surface interactions; electrostatic, hydrogen bonding, and hydrophobic free energies provided through surface complimentarity, which define the specificity and affinity in the formation of complexes between the metalloproteins involved in electron transfer events in cytochrome P-450-dependent oxygenase catalysis and in the assembly of tetrameric hemoglobin.
2. Protein–small molecule recognition: here we seek to ascertain how the same fundamental forces of electrostatics, hydrogen bonding, and the hand-glove fit of a substrate into the active site of an enzyme can give rise to the observed high degree of control of regio- and stereo-specificity in catalysis and in the interfacial interactions of proteins at electrode interfaces.
3. Protein–nucleic acid recognition: here again the same fundamental forces control recognition processes, but in this case we will focus on our exciting,

recent discovery of a role for solvent water in mediating recognition between protein and nucleic acid components. Representative systems in the binding/catalytic class of restriction endonucleases and recombinases will be discussed.

In all cases, the use of pressure as a variable has provided unique understanding for the molecular details of these processes.

Pressure, both hydrostatic and osmotic, has proven to be an enabling experimental technique in understanding the mechanistic origins of molecular recognition events. In particular in this article, we will present three very different examples of how pressure can be used to understand molecular function and the basics of molecular recognition events that occur between proteins and either small molecules, nucleic acids, or heterologous protein systems. First, we will examine two fundamentally different types of pressure—hydrostatic, which is also the subject technique in other chapters in this volume, and osmotic. The latter type of experimental pressure variable is critical to understanding how solvent water can partition into the interstitial spaces between proteins and nucleic acid complexes. In particular, a major focus will be on the class of restriction enzymes that recognize and cleave DNA in a sequence-specific manner. Second, we will examine a case of subunit assembly as a molecular recognition question, using as an example the microbial expression of tetrameric human hemoglobin. The mechanisms of cooperativity, after all, are really nothing except a case of protein-protein recognition and subunit communication.

PRESSURE EFFECTS ON PROTEIN–NUCLEIC ACID RECOGNITION

The first system we examine is that of restriction enzyme recognition of cognate and noncognate sequences, and how a combination of hydrostatic and osmotic pressure can be used to elucidate the role of bound water in mediating molecular recognition. Since precise three-dimensional structure of various protein–nucleic acid complexes has been determined over the past decade, a central question has emerged regarding the potential role of bound solvent in mediating recognition between peptide and nucleotide base or backbone. We explore this nontraditional role of water by first using osmotic pressure to alter the partitioning of solvent into sequestered spaces within the macromolecular complex, thus proving that osmotic pressure is the key colligative property responsible for the effects on recognition. Then we directly implicate solvation by using a hydrostatic pressure head to reverse the effects of the osmotic gradient.

Our experiments utilize several type-II restriction endonucleases in which x-ray structural information is available. For example, consider the *Eco*RI endonuclease from *E. coli* which recognizes the hexameric GAATTC sequence. The enzyme cleaves between the G and A residues on both strands, leaving a five base pair sticky end overhang, and was one of the first widely available reagents to initiate recombinant DNA technology. Due to the early cloning of this enzyme, relative ease in acquiring large quantities allowed the solution of the three-dimensional structure of the protein-oligonucleotide complex (Kim et al., 1993;

Rosenberg, 1991). Competitiveness in the marketplace among biotechnology companies producing restriction endonucleases drove the detailed documentation of precise experimental conditions of buffer, temperature, cosolvent composition, and metal ion concentrations required for maintaining maximal activity and fidelity in recognition of the correct cognate DNA sequence. It was noted early on that many restriction enzymes displayed a so-called star activity in which additional sequences are cleaved. For *Eco*RI, this activity is depicted in Figure 23.1. Numerous catalogs from the suppliers of restriction endonucleases document the solvent conditions that give rise to this unwanted specificity of DNA cleavage. In a general discussion at a Parisian cafe of cosolvent effects on enzyme activity (Di Primo et al., 1992), Pierre Douzou, Gaston Hui Bon Hoa, Carmelo Di Primo, and one of us (SGS) realized that a feature common to many of the reagents that would elicit this star activity in restriction endonucleases, cosolvents themselves, were powerful osmolytes. This suggested that perhaps solvent, in the form of bound water in the interstitial spaces between protein and nucleic acid components, could be responsible for the mediation of hydrogen bonds involved in the recognition of the restriction endonuclease for its cognate base sequence. Indeed, an intense debate had taken place over the identification of ordered water in cocrystals of trp repressor-operator complex (Otwinowski et al., 1988; Carey et al., 1991). In the case of the *Eco*RI structure, x-rays revealed that the bound water structure was stabilized by hydrogen bonds between the G base and a diarginine pair in the protein. Considering the numerous other bound waters present and the potential for differential hydration of the DNA backbone due to bending or other alterations in topology within the protein–nucleic acid complex, we decided to examine the role of bound water in mediating recognition in the type-II restriction endonucleases. Figure 23.2 illustrates the increased degree of star activity observed with an increase in the solution's osmotic pressure as generated by the addition of a

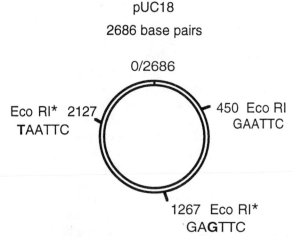

pUC18

2686 base pairs

0/2686

Eco RI* 2127
TAATTC

450 Eco RI
GAATTC

1267 Eco RI*
GAGTTC

Figure 23.1. The pUC vector system showing the single-cognate *Eco*RI site in the poly linker region, and the so-called star (*Eco*RI*) sites of cleavage under nonideal buffer conditions. Also illustrated is the fragmentation pattern following digestion with *Eco*RI.

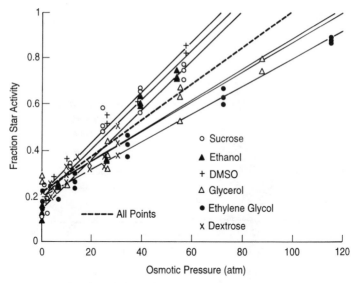

Figure 23.2. Increase in fraction of cleaved star sites in the pUC vector system as a function of osmotic pressure induced by the addition of cosolvents to the reaction buffer.

variety of cosolvents. Since each cosolvent added to the solution also alters other properties, such as viscosity, dielectric constant, and the actual mole fraction of water available, it is important to examine the fraction of relaxed specificity as a function of these other colligative properties. Interestingly, we found that the induction of star activity in the *Eco*RI restriction endonuclease was only correlated with the osmotic pressure of the solution (Robinson & Sligar, 1993). This strongly suggests that a water equilibrium is involved in the maintenance of faithful recognition of the enzyme for its cognate DNA sequence. If indeed it is a simple water partitioning and resultant different hydration of the nucleic acid and/or protein, then it should be possible to reverse the effect of osmotic pressure by applying a hydrostatic pressure head on the cosolvent-containing solution. Figure 23.3 unambiguously demonstrates that the *Eco*RI enzyme regains its selectivity for the cognate site with the application of hydrostatic pressure (Robinson & Sligar, 1994a–1994d). Hence, pressure as a thermodynamic variable can be a powerful tool to prove water equilibria in enzymatic systems. To further explore the role of hydration in maintaining the specificity of type-II restriction enzymes and to attempt to identify the particular step in the catalytic cycle of recognition and cleavage, we measured the overall observed maximal velocity and Michaelis constant for the cleavage of both cognate and noncognate sites carried on equal length oligonucleotides (Figure 23.4) (Robinson & Sligar, 1994c). The most exciting result is that osmotic pressure (and reversal by hydrostatic pressure) simultaneously decreases the reactivity at the cognate GAATTC site, while increasing the cleavage rates at the star sites.

Such effects of pressure are not unique to *Eco*RI. Figure 23.5 (Robinson & Sligar 1994b, 1994d) illustrates similar results for the restriction endonucleases *Pvu*II and *Bam*HI. Interestingly, a type-II restriction endonuclease with a very

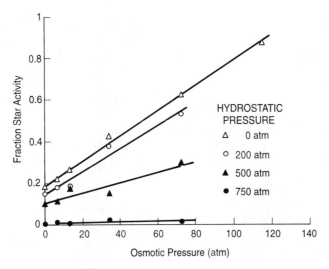

Figure 23.3. A hydrostatic pressure head will reverse the effects of a cosolvent-induced osmotic pressure gradient in the cleavage of the pUC vector by *Eco*RI.

different DNA recognition motif, *Eco*RV, is not sensitive to pressure, nor are those failed recognition events which are induced by extremes of pH, temperature, and metal ion concentration (Robinson & Sligar, 1994b, 1994d). Clearly, more than one mechanism can exist for induction of a failed recognition event.

These and further experiments using pressure as a variable offer unique opportunities to study fundamental recognition processes in restriction endonucleases. Are there other DNA-protein complexes which are also sensitive to the osmotic/hydrostatic pressure in the system? To answer this question, we began experiments with the *Hin* recombinase, an enzyme that recognizes specific regions

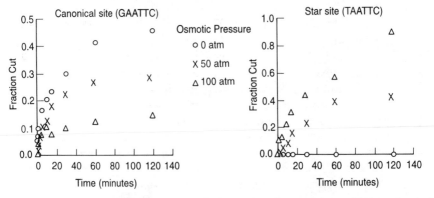

Figure 23.4. The rates of *Eco*RI cleavage of oligonucleotides containing (**A**) cognate or (**B**) noncognate sequences. In (A), a 351-bp sequence contains the GAATTC normal recognition sequence for *Eco*RI, while in (B) a 386-bp sequence introduces a single base pair mismatch as TAATTC.

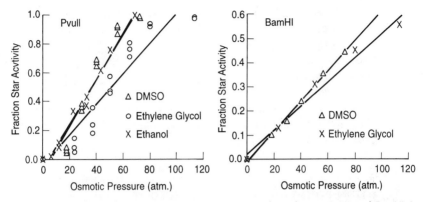

Figure 23.5. Star activity in two other type-II restriction endonucleases, *Pvu*II and *Bam*HI, is also induced by osmotic pressure and reversed by hydrostatic pressure.

of DNA and elicits strand recombination events (Hughes et al., 1992). Shown in Figure 23.6 is preliminary gel shift analysis that suggests that, indeed, osmotic pressure can bring about a failed molecular recognition event, and which implicates participation of water molecules in formation of the complex.

An immediate question arises as to whether one can specifically identify those solvent waters that are mediating the recognition between nucleic acid and protein. A major advantage of thermodynamics is also its disadvantage—equilibrium parameters are independent of specific mechanisms and structures, yet these values do not identify specific molecular properties of the interaction. For example, although several sequestered and immobilized water molecules can be visualized in a cocrystal structure of the *Eco*RI-oligonucleotide complex (Rosenberg, 1991),

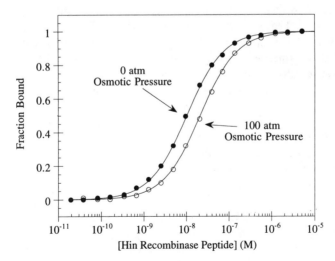

Figure 23.6. The 53 amino acid DNA recognition fragment of the *Hin* recombinase was synthesized by solid-state methods and used in a gel shift assay to determine its recognition of its correct base pair sequence. As the osmotic pressure of the buffer is increased, there is a corresponding loss in the fidelity of recognition.

recent high-resolution structures of *Eco*RI bound to both cognate and star sequences demonstrate large differential changes in the overall structure of the DNA. Hence, osmotic and hydrostatic pressure experiments explore the overall change in hydration in the protein–nucleic acid complexes, including important contributions from DNA backbone hydration and protein conformational equilibria. Nevertheless, the use of osmotic and hydrostatic pressure in concert with structure determination and site-directed alterations of protein and nucleotide sequences is a powerful tool to elucidate the molecular mechanisms of protein–nucleic acid recognition.

PROTEIN–PROTEIN RECOGNITION AND MULTISUBUNIT RECOGNITION

Weber, Chapter 1 (in this volume), has presented a comprehensive review and thought-provoking analysis of the role of pressure in multisubunit dissociation and reassociation equilibria. In our final example of pressure studies of molecular recognition, we examine human hemoglobin and the mechanisms of intersubunit communication that elicit the cooperative behavior of ligand binding.

More than 10 years ago we had the idea that it could be possible to express mammalian heme proteins in the bacteria *E. coli* by totally synthesizing the gene with proper attention to the nucleotide sequences controlling transcription and translation. Hemoglobin was a first target of these investigations and, following our successful expression of cytochrome $b5$ (von Bodman et al., 1986) and sperm whale myoglobin (Springer & Sligar, 1987), we were able to construct an operon of α- and β-chains which would express human hemoglobin in *E. coli* as a soluble holoprotein (Hernan et al., 1992). Unfortunately, it turned out that the observed cooperativity of oxygen binding, assayed either by standard Hill parameters or by the rate constant for carbon monoxide binding to the ferrous deoxy state, was a strong function of particular growth conditions. Yet all physical properties of the expressed tetramer were identical to the hemoglobin isolated from human blood. It occurred to us that this variability in functionality was in fact due to a misassembly of the individually folded α- and β-chains into the $\alpha_2\beta_2$ tetramer. Figure 23.7 displays a standard equilibrium binding and carbon monoxide (CO) association assay for native hemoglobin and that purified following expression in *E. coli*. The CO association assay is the most sensitive to small deviations in the overall cooperativity of ligand association. If the lack of full cooperativity for the recombinant protein is due to a misassembly of subunits, hydrostatic pressure could be used to study the dissociation and reassociation of the tetramer. Figure 23.8 presents an experiment completed by Tom Manning in our laboratory measuring the CO binding assay for the recombinant protein as a function of incubation pressure. Here the hemoglobin tetramer is placed in a sealed reaction volume and the indicated pressure applied for a period of 12 hours. At the end of the incubation time, the pressure is slowly released and the treated hemoglobin is isolated, concentrated, reduced, and assayed for the rate of CO combination using stopped flow. Clearly, an increase in pressure for a fixed time interval results in nearly a complete restoration of native function. While it is possible to rule out

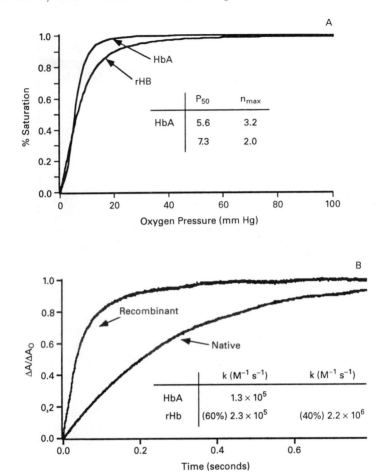

Figure 23.7. Altered functionality of native and recombinant human hemoglobin expressed from a synthetic operon in *E. coli*: (**A**) Standard saturation curves of oxygen binding and (**B**) the kinetics of carbon monoxide (CO) binding to the stripped ferrous deoxy state.

a contribution of tertiary structure differences, the relatively low pressures utilized suggest that the major origin of the low cooperativity of recombinant human hemoglobin expressed in *E. coli* is due to misassembly of subunits. Pressure is an ideal variable to elicit the control of multimeric protein assembly as discussed in detail by Weber, Chapter 1 (in this volume).

To summarize, in this chapter we have presented three cases in which the use of pressure as a variable, either hydrostatic or osmotic, can be used to provide critical information regarding the mechanisms of macromolecular recognition and dynamics. As the technology for utilizing high pressure in concert with well-understood spectroscopies, structural tools, rapid kinetic methods, and chromatography becomes widely available, a new generation of scientists will emerge and

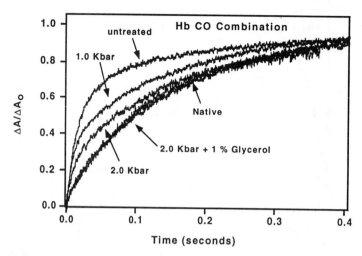

Figure 23.8. Restoration of native cooperativity of recombinant human hemoglobin following the application of hydrostatic pressure to control dissociation and reassociation of the tetramer.

bring with them a more complete understanding of important biological and chemical processes.

ACKNOWLEDGMENTS: We thank the National Institutes of Health for generous support.

REFERENCES

Carey, J., Lewis, D. E. A., Lavoie, T. A., & Yang, J. (1991). How *does trp* repressor bind to its operator? *J. Biol. Chem.* **266**, 24509–24513.

Di Primo, C., Sligar, S. G., Hui Bon Hoa, G., & Douzou, P. (1992). A critical role of protein-bound water in the catalytic cycle of cytochrome P-450$_{cam}$. *FEBS Lett.* **312**, 252–254.

Feng, J. A., Johnson, R. C., & Dickerson, R. E. (1994). *Hin* recombinase bound to DNA—the origin of specificity in major and minor groove interactions. *Science* **263**, 348–355.

Hernan, R. A., Hui, H. L., Andracki, M. E., Noble, R. W., Sligar, S. G., Walder, J. A., & Walder, R. Y. (1992). Human hemoglobin expression in *E. coli*: importance of optimal codon usage. *Biochemistry* **31**, 8619–8628.

Hughes, K. T., Gaines, P. C. W., Karlinsey, J. E., Vinayak, R., & Simon, M. I. (1992). Sequence-specific interaction of the *Salmonella Hin* recombinase in both major and minor grooves of DNA. *EMBO J.* **11**, 2695–2705.

Kim, Y., Grable, J. C., Love, R., Greene, P. J., & Rosenberg, J. M. (1993). Refinement of *Eco*RI endonuclease crystal structure: a revised protein chain tracing. *Science* **249**, 1307–1309.

Otwinowski, Z., Schevitz, R. W., Zhang, R.-G., Lawson, C. L., Joachimiak, A., Marmorstein, R. Q., Luisi, B. F., & Sigler, P. B. (1988). Crystal structure of *trp* repressor/operator complex at atomic resolution. *Nature* **335**, 321–329.

Robinson, C. R., & Sligar, S. G. (1993). Molecular recognition mediated by bound water: a mechanism for star activity of the restriction endonuclease EcoRI. *J. Mol. Biol.* **234**, 302–306.

Robinson, C. R., & Sligar, S. G. (1996). Change in solvation upon binding and cleavage by EcoRI endonuclease at canonical and alternate sequences in DNA. *J. Biol. Chem.* (submitted).

Robinson, C. R., & Sligar, S. G. (1994b). Heterogeneity in molecular recognition by restriction enzymes: osmotic and hydrostatic pressure effects on PvuII, BamHI, and EcoRV specificity. *Proc. Natl. Acad. Sci. USA* **92**, 3444–3448.

Robinson, C. R., & Sligar, S. G. (1995). Hydrostatic and osmotic pressure as tools to study macromolecular recognition. *Meth. Enzymol.* **259**, 395–427.

Robinson, C. R., & Sligar, S. G. (1994d). Hydrostatic pressure reverses osmotic pressure effects on the specificity of EcoRI-DNA interactions. *Biochemistry* **33**, 3787–3793.

Rosenberg, J. (1991). Structure and function of restriction endonucleases. *Struct. Biol.* **1**, 104–113.

Springer, B. A., & Sligar, S. G. (1987). High-level expression of sperm whale myoglobin in Escherichia coli. *Proc. Natl. Acad. Sci. USA* **84**, 8961–8965.

von Bodman, S. B., Schuler, M. A., Jollie, D. R., & Sligar, S. G. (1986). Synthesis, bacterial expression, and mutagenesis of the gene coding for mammalian cytochrome b5. *Proc. Natl. Acad. Sci. USA* **83**, 9443–9447.

Index